Advanced Cutting Tool Technology and Machine Processes

Today's industry heavily relies on advanced cutting tool technology and manufacturing techniques to enhance quality and efficiency in production. This undergraduate-level textbook is dedicated to exploring the latest developments that enhance tool performance, lower costs, and boost productivity. The mission of this textbook is to provide readers with a thorough understanding of cutting-edge technologies and methods used in cutting tools and industrial processes.

Advanced Cutting Tool Technology and Machine Processes steps into the world of modern cutting tools used in the metallurgical industry. Each tool is meticulously dissected, showcasing its distinctive features and methods of operation. The text explores advanced metal-cutting and processing techniques, including laser, water, and plasma cutting. By covering these cutting-edge methods, students and professionals can remain at the forefront of industry advancements. In addition to detailed tool descriptions, this textbook offers practical guidance on utilizing tools effectively and safely, as well as tips on tool maintenance to ensure longevity and peak performance. To enhance comprehension, this textbook incorporates exercise problems, case studies, and practical examples that illustrate how theoretical knowledge is applied in real-world scenarios. This hands-on approach aids in the development of problem-solving skills and the practical application of concepts. Lastly, this textbook provides comprehensive information on the properties of various metals and how to handle them effectively. This knowledge is crucial in selecting the appropriate tools for each type of metal, guaranteeing precision and efficiency in cutting processes.

This textbook is ideal for undergraduate students in production, materials, industrial, mechatronics, marine, mechanical, and manufacturing engineering programs, and is also useful for graduate programs related to higher-level machining topics, as well as professional engineers and technicians.

Figure slides and a solutions manual are available for qualified textbook adoptions.

Advanced Cutting Tool Technology and Machine Processes

Riyadh A. Al-Samarai and Yarub Al-Douri

CRC Press
Taylor & Francis Group
Boca Raton London New York

CRC Press is an imprint of the
Taylor & Francis Group, an **informa** business

Designed cover image: Shutterstock

First edition published 2026
by CRC Press
2385 NW Executive Center Drive, Suite 320, Boca Raton FL 33431

and by CRC Press
4 Park Square, Milton Park, Abingdon, Oxon, OX14 4RN

CRC Press is an imprint of Taylor & Francis Group, LLC

ISBN: 978-1-032-99299-0 (hbk)
ISBN: 978-1-032-99526-7 (pbk)
ISBN: 978-1-003-60464-8 (ebk)

DOI: 10.1201/9781003604648

Typeset in Times
by codeMantra

Access the Instructor and Student Resources: www.Routledge.com/9781032992990

Contents

Preface

A key method for reducing wear and friction is ensuring several factors are met, including the characteristics of the material being processed, the availability of appropriate tool materials, and the speed and feed capabilities of the machine tools. When determining "high-speed machining" (HSM), four possibilities can be distinguished: cutting operation, work material, process stability, and cutting force. The relative temperature drop for ferrous and non-ferrous metals occurs at different rates. For aluminum and other non-ferrous metals like magnesium, the reduction in cutting temperature accompanied by a corresponding increase in cutting speed is more noticeable. It is minimal for steel and cast iron, which is in sharp contrast.

HSM technology significantly enhances machining processes by reducing cutting forces, which minimizes tool wear and prolongs tool life. Compared to traditional machining, less heat is transferred to the workpiece and cutting tool from the cutting zone. Thus, thin-walled products can be processed with extreme precision. More sophisticated cutting tool materials are needed to compensate for reduced tool life. Due to their poor chemical and thermal stability, the carbides are unsuitable for most rapid machining processes. Although ceramics can be cut at speeds up to 1000 m/min, their use in high-speed steel is severely limited due to their high brittleness and low thermal and mechanical shock resistance. HSM pairs well with specific cutting tools like CBN (cubic boron nitride) for ferrous metals and PCD (polycrystalline diamond) for non-ferrous metals, thanks to their unique properties. To minimize the use of metalworking fluid, dry and near-dry machining techniques, along with Minimum Quantity Lubrication (MQL), have been developed as alternatives to internal high-pressure coolant and flooding. Although very little lubrication is supplied to the cutting zone during near-dry machining, compressed air is delivered to the cutting zone during dry machining. Both of these strategies aim to reduce the use of traditional lubricants and coolants. The main tasks of a metalworking fluid are cooling and lubrication, which reduce the force of contact that processes occur at boundaries between the cutting tool, chips, and workpiece. Changes in contact temperature, normal and shear stresses, and their distribution along interfaces significantly affect the type and mechanism of tool wear, the integrity of the machined surface, and residual stresses in the workpiece during machining. As a result, the metalworking fluid plays a crucial role in influencing the tribological conditions at these boundaries. In addition, chip shape and frequency segmentation are strongly influenced by the use of coolant and lubrication. It is assumed that in some cases, the metalworking fluid can additionally perform secondary service functions, such as washing workpieces or transporting chips. There are several basic approaches to mitigating the environmental, financial, and health impacts of cutting fluids: careful selection of plumbing fluids used, proper application, careful monitoring, and gradual reduction in the number of plumbing fluids used through the use of drier and MQL (or near-dry) plumbing fluids. Currently, numerous efforts are being made to develop sophisticated machining methods in which metalworking fluids are either not used at all or used in very small quantities. Nowadays, dry machining is becoming increasingly popular among metalworking industries as a material processing method for the future. Growing environmental consciousness, concern for human health, and stricter laws are gradually forcing industrial production to switch to advanced production technologies rather than traditional methods of cooling and lubricating machines. The environmentally preferred method for metal removal is dry machining, which eliminates the need for wet cutting fluids. However, the cutting tool is subject to increased mechanical and thermal stresses. As a result, there are three possible approaches to developing cutting tools for dry machining: using new materials, modifying existing geometry, or using unique coatings. The following are the main advantages of dry processing: (a) absence of significant air and water pollution risk to the environment and public health; (b) reduced costs associated with coolants and lubricants (purchase, adaptation, and disposal); (c) removal of residual coolant and lubricants from the machine and workpiece parts, which eliminates the need to clean the equipment and workpiece; and (d) removal of residual coolant and lubricants from the chips and reduced chip disposal costs. The three dry machining methods are cold air-dry machining, cryogenic machining, and standard compressed air-dry machining. This relatively new approach of using compressed cold air for cooling during machining operations has attracted scientific interest. To reduce the amount of metalworking fluid used, internal high-pressure coolant and priming have been replaced by MQL machining. This method, commonly called near-dry machining (NDM), lubricates the treatment area in very small quantities. The cooling medium used in MQL processing is an aerosol mixture of oil and air, commonly referred to as mist. A gaseous suspension (hanging) of solid or liquid particles in the air is called an aerosol. Aerosols are tiny droplets of oil dispersed in the air stream during MQL processing. For a practicing engineer or plant manager, making the right decision about MQL processing modes and required equipment is challenging, as there are no established MQL processing classifications in the literature or industry practice. Traditionally, the machine parts are processed with high-temper steel. Results with hardness values above 60 HRc (Rockwell C hardness scale) can be obtained using grinding procedures. This is often due to the material being work-hardened during the grinding process, which increases the surface hardness of the material. In addition, obtaining such high levels of hardness is significantly dependent on several other factors, including the type of grinding wheel used, the speed, and the characteristics of the material being processed. Leading industries such as automotive, roller bearings, hydraulics, die, and mold manufacturing are seeing an increase in their use. While high-speed milling is a common technology in the die and mold sector, gears, pinion shafts, bearings, and other gearbox parts are typically produced by turning. The hard-cutting procedure used by ceramic or CBN tools can often reduce production costs, reduce lead times to improve overall product quality, provide greater flexibility, and eliminate coolant to enable dry machining. The first two advantages lead to a significant reduction in the length of the traditional process chain, including heat treatment and final grinding after roughing. Hard turning can generally provide a relatively high degree of accuracy for many hard parts; however, serious problems with surface integrity sometimes arise, especially due to unwanted residual stresses and changes.

Using a small depth of cut (0.05–0.3 mm) and feed rate (0.05–0.2 mm/rev) results in a tiny undeformed chip thickness and a tiny undeformed chip thickness to the cutting-edge radius ratio. This geometric connection creates very high pressure to remove material at the cutting edge, resulting in an effective rake angle of 60°–80°. A larger corner radius can affect the geometry of the cutting tool or grinding wheel, affecting how the material is captured and removed. It is established when hardness is more than 45 HRc and cutting forces increase significantly. One of the main features of creating sawtooth chips is that hardened steels are processed by cutting tools with precise geometric characteristics. When sawtooth chips are used, catastrophic failure within the primary shear zone is usually attributed to either the occurrence of thermoplastic instability, cyclic initiation, or propagation of failure. Many industries, including automotive, aerospace, bearings, hydraulics, stamping, and molding, have discovered the high efficiency of hard machining. It can be implemented in a variety of machining processes using cermet, ceramic, CBN, and PCD tools including turning, milling, drilling, broaching, reaming, and tapping. This book covers practical applications of cutting tools, as well as extensive educational content that combines theory and practice. Engaging exercises enhance real-world understanding, and pictures and diagrams explain ideas. This book includes appendices with important technical terms to help students, industry representatives, and researchers apply their knowledge to practical situations. It covers a variety of topics such as different types of modern cutting tools. This book on advanced manufacturing technologies aims to provide researchers, business professionals, and students with the information and skills needed to succeed in today's rapidly changing industrial environment. To address common challenges in manufacturing processes, each chapter provides comprehensive problem-solving advice and explains critical methods for improving productivity and product quality.

The content will be updated frequently to reflect the latest developments and industry best practices as cutting tool technologies and manufacturing methods continue to change. Expert opinions from industry experts will ensure that the information presented is credible and applicable. The eBook will feature interactive sections to enhance the learning experience; these will include short comprehension quizzes and opportunities to network with peers and educators. This book is designed to be an active resource and will guide readers through each chapter, creating a stimulating learning environment. To encourage innovation and meet the needs of today's industrial sector, this book provides students with access to online resources (www.routledge.com/9781032992990) and encourages team collaboration. The content of this book consists of eight chapters covering the following topics: Using a preferred selection index approach, Chapter 1 presents the cutting tool geometry, nomenclature, and applications. Chapter 2 describes modern cutting tool technologies specifically designed for machining difficult-to-cut materials. This includes innovative methods that have been completely replaced by traditional grinding procedures. The manufacturing business has evolved at a faster pace than previous tools, and as a result, the demands on modern machining have become higher, as presented in Chapter 3. An overview of a unique machining method that uses mixed ceramic or CBN tools to machine workpieces hardened to 45–70 HRC or higher is presented in Chapter 4. To be used for a long period, a machine must have significant advantages. Here are a few benefits that these devices offer. The high rotation speed of a multi-blade cutter ensures a high metal removal rate. Consequently, these devices provide an excellent surface quality, as presented in Chapter 5.

This chapter covers cermet, sintered titanium carbonitride carbides, and titanium nitride. A unique class of cermet, known as TiN and TiCN compounds, offers an excellent combination of high-temperature resistance and hardness when exposed to aggressive, oxidizing, or abrasive conditions. TiN and TiCN coatings have many applications, but their mechanical and tribological properties depend on the substrate they are applied in Chapter 6. HSM is a new development in machining technology. This is a metal-cutting method to prioritize high feed rates and speeds for increasing productivity and improving surface quality as described in Chapter 7. This is followed by the processing of complex surfaces, whether ruled or sculpted, typically applied to injection molds, forging and stamping dies, and components, as discussed in Chapter 8.

Riyadh A. Al-Samarai
University of Samarra, Iraq

Yarub Al-Douri
Al-Bayan University, Iraq
University of Malaya, Malaysia
Piri Reis University, Turkey

About the Authors

Prof. Dr. Riyadh A. Al-Samarai was born in Salah Al-Din in 1970. Al-Samarai is a graduate of the University of Technology in Baghdad, holding B.Sc. and M.Sc. in mechanical engineering. He obtained his Ph.D. in mechanical engineering-applied mechanics from the Universiti Malaysia Perlis, Malaysia. He has held several positions: Director of Missions and Social Relations, Assistant Dean for Scientific Affairs, and Head of the Department of Electromechanical Engineering at Samarra University. Al-Samarai is the winner of the Iraqi Scientific Excellence Award 2024. He has launched an M.Sc. program in electromechanical engineering at Samarra University. He has received many medals and awards, including the Creativity Award for Scientists in Malaysia, many medals in international scientific forums and conferences, Iraqi Forum for Intellectuals and Academics (IFIA), and the Best Research Paper Award at the Asian Universities Conference in Malaysia, with other awards. Al-Samarai is the editor-in-chief of the *Samarra University Journal of Engineering and Science.* Al-Samarai has 32 years of research experience, 23 years of university teaching experience, 16 years of administrative experience, and more than 60 publications including citations >350, an h-index of 7, and research grants of US$1 million.

Prof. Dr. Yarub Al-Douri is a Fellow of the European Academy of Sciences. Al-Douri is the winner of the prestigious Khalifa International Award for Date Palm and Agricultural Innovation 2024. He is one of the Middle East, North Africa, and Southeast Asia's most renowned scientists, known for his contributions in renewable energy and nanotechnology. Al-Douri is a scientist, researcher, administrator, and an excellent lecturer who has students all over the world. Al-Douri has Doctorat D'etat in materials science (2000), M.Sc. in physics (1995), and B.Sc. in physics (1991). He has been appointed full-professor, visiting professor, adjunct professor, consultant expert, associate professor, assistant professor, research fellow (a), scientific collaborator, and post-doc in UAE, KSA, Iraq, Malaysia, Turkey, Pakistan, Algeria, Yemen, Singapore, Germany, and France, respectively. Al-Douri initiated an M.Sc. Program in nanotechnology engineering and Nano Computing Laboratory, in addition to founding Applied Materials Laboratory. Al-Douri was Head of the Department of Physics, Head of the Department of Nanomaterials and the Department of International Networking and Collaboration, and Secretary of the Department of Physics. He has received 78 national and international prizes and awards including IAAM Scientist Award from the International Association of Advanced Materials, Sweden 2022. He is also the winner of World's Top 2% Scientist Career-Long Citation Impact by Stanford University, USA 2020; World's Top 2% Scientists by Stanford University, USA 2024, 2023, 2022, 2021, and 2020; OeAD Award, Austria 2020; Japan Society for the Promotion of Science (JSPS) Award 2019; Asian Universities Alliance (AUA) Award 2019; Iraqi Forum for Intellectuals and Academics Award 2019; Best Researcher Award at Cihan University Sulaimaniya 2021 and 2019; Best Paper at Global Conference on Energy and Sustainable Development, Coventry, UK 2015;

Distinguished Researcher Award at University Malaysia Perlis 2011–2015; Gold Award at ITEX Kuala Lumpur 2013; TWAS-UNESCO Associateship 2009–2012 and 2012–2015; and others. He has more than 31 years' experience in research, teaching, administrative and editorial board management, organizing events, and research grants and consultations, in addition to more than 900 publications with renowned international publishers and US$5.1M research grants. Al-Douri has notable citations > 14.2K and h-index > 65. He has supervised more than 25 students. Al-Douri is the associate editor of *Nano-Micro Letters* (Q1, IF = 31.6), editor-in-chief of *Experimental and Theoretical Nanotechnology* (Scopus-indexed), editor-in-chief of *World Journal of Nano Science and Engineering*, editor and peer-reviewer of 34 international journals, member of 13 international scientific bodies, organizer of 33 international conferences, and chairperson of 11 international conferences. He appointed 38 internal and external examiners for post-graduate students, presented 41 training courses and public lectures, and has been invited 66 times as keynote, invited, and guest speakers to many reputed universities in Europe, Asia, and Africa that ended up in generating more than nine MOUs and MOAs. His research field focuses on renewable energy, nanotechnology, nanoelectronics, nanomaterials, modeling and simulation, semiconductors, and optical studies.. Finally, Al-Douri is a public figure among international media in the UK, Singapore, Malaysia, Qatar, and UAE.

1 Cutting Tool Geometry, Nomenclature, and Designation

1.1 CUTTING TOOL

Single-edge cutting tools, or tools with only one edge in contact with the workpiece, are used in virtually all turning procedures. Cutting tool geometry refers to the precise shapes and angles of the cutting tools that determine its performance in manufacturing processes. Important elements include the following:

- The portion of the cutting edge that contacts the material: its shape affects cutting efficiency.
- Bend angle: Positive angles reduce cutting forces; this is an angle formed by the cutting edge and surface of workpiece.
- Relief angle: The angle reduces wear by preventing the tool from rubbing against the workpiece.
- Design: Chips are removed, and coolant flows through the flutes of drills and end mills.

Manufacturing involves several processes:

- Material selection: Common materials include carbides, ceramics, and high-speed steel.
- Forming: The desired geometry is created using methods such as electrical discharge machining (EDM), milling, and grinding.
- Coating: Materials such as titanium nitride can be applied to tools to improve durability and performance. Surface smoothness, tool life, and cutting efficiency of mechanical parts are improved by these geometries and manufacturing technologies.

The angle of inclination, from side view or angle of seat or pocket of the holder plate, measured from front to back, is called rake. This trend can be neutral, negative, or favorable. The angle formed by the cutting edge and actual cut is known as the rake angle of the cutting tool. It can be neutral, negative, or good. The rake angle of the tool holder plus the rake angle built into the insert equals the effective rake angle [1]. The angle formed by the feed direction of the cutting tool and cutting edge is known as the angle of inclination or the angle of entry. The angle formed by the tip of the

DOI: 10.1201/9781003604648-1

tool is called the nose radius of the tool. This radius can be sharp for precise turning or large for force. Since a sharp edge is brittle and breaks easily, the cutting edge of attachment is enhanced by preparing a special die. These shapes can be burr, chamfer, tapered radius, or any combination of the three.

The viewer will gain knowledge and understanding of the shapes, angles, and other geometric features of single- and multi-point cutting tools after watching the program and studying the printed materials. The geometry of multi-point cutters is explained; the geometry of a turning tool is explained; the mechanism of chip formation based on the geometry of the tool is demonstrated; and the geometry of end mills is studied. Metal cutting tools come in a variety of shapes; many can be distinguished by their angles or geometry. Each of these tool shapes performs a specific function in metalworking. The most efficient separation of chips from the workpiece is the primary goal of machining. Therefore, choosing the appropriate cutting tool shape is of utmost importance. Other factors that can affect chip formation include the workpiece material, cutting tool material, machine force and speed, and various process variables including vibration and heat [2].

Single-edged cutting tools, or tools with one edge in contact with workpiece, are used in almost all turning procedures. Although most turning operations use replaceable coated carbide inserts, high-speed steel, carbide, ceramic, cubic boron nitride, or polycrystalline diamond can also be used as tool materials. Some basic tool geometry is used in 75% of turning operations. When inserts are used for turning, most of geometry is in tool holder itself rather than insert itself. However, first, let us focus on inserts. Insert dimensions include the type of insert, its offset or clearance angle, its basic shape, its nose radius, engraved "IC" circle or dimension, and chip cutter design. The balance of force and machinability determines the choice of insert shape during turning. For example, a higher angle is stronger such as round inserts for forming and square inserts for roughing and finishing. Smaller angles, 35° and 55°, are most useful for complex applications. Rotary inserts can be milled or given the desired shape for use. Castings are widely used and more cost-effective. To achieve the highest level of accuracy and create clear, clean contours, ground plates are required [3] (Basic Machining Processes for Cutting Tool Engineering, DV03PUB13-2). When inserting the cutting tool edge into a rotating workpiece, angles are important. These angles consist of the following:

- rake angle that works
- angle of entry or ascent
- tool nose radius

The angle of insert seat or socket in tool holder, measured from front to back, is the angle of inclination from front and side views. This trend can be neutral, negative, or favorable. The angle formed by the cutting edge and actual cut is known as the rake angle of the cutting tool. It can be neutral, negative, or good. The rake angle of the tool holder plus the rake angle built into the insert equals the effective rake angle, which is the angle between the direction of entry that approaches the cutting edge and feed of the cutting tool [4].

The angle that the tip of the tool makes is called the tool nose radius. For strength, its radius can be large. When turning a thin radius, it can be sharp. To strengthen the cutting edge of insert, special shapes are made, since sharp edges are fragile and break easily. These shapes can be a protrusion, chamfer, pointed radius, or any combination of the three. The largest circle that can be inscribed within the perimeter of insert, called the inscribed circle, indicates the size of insert. Insert size and tool holder size are closely related. The components of an insert holder for turning include an insert pocket, head, fixture, and shank. Tool holders can be left-handed, right-handed, or unbiased. The turning operation, feed direction, cut size, machine design, accessibility requirements, and workpiece shape all affect the size and type of tool holder [5]. The chip breaker is necessary for efficient processing of work and high-quality finishing. The balance between tool shape and depth of cut results in correct chip breaking. Chip-breaker grooves are molded into many inserts. Turning produces four main types of chips:

- Spiral or spiral chips
- Long, fibrous chips
- Chips with corrugations

The ideal chip is represented by the first type, shaped like numbers “6” or “9”. Other types indicate the need to change feed and speed or select an alternative chip breaker design (DV03PUB13: Tutorial on the basics of manufacturing processes). Cutting tools with two or more chip-forming edges on one body are called multi-point tools. As shown in Figure 1.1, the structure of turning tool consists of rake face, angle, small side surface, large and small cutting edges, and shank. The angle of turning tool serves as the basis for design, manufacturing, grinding, and measurement of tool [6].

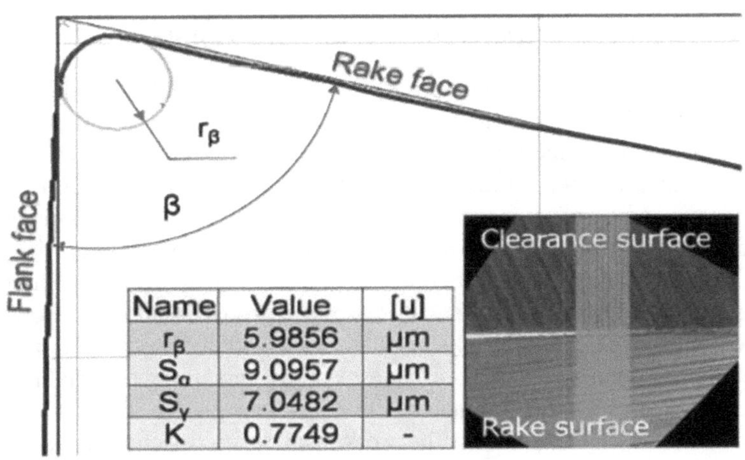

FIGURE 1.1 Measuring the cutting edge with Alicona microscope. ([6] [Open access].)

Short, continuous chips are produced during surface milling, as each insert enters and exits the workpiece in sequence. The majority of additive milling is performed in climbing milling mode. This shows that the cutting edge of insert cuts the material by first creating the thickest part of the chip and then gradually thinning it toward the exit point. The conventional milling procedure is the opposite. Design of Experiments (DOE) for milling tools typically involves analyzing the effects of various parameters on the performance and quality of milling process after machining. With the edge radius of the SiO_2 abrasive particles as a consideration, the draw time in each set is fixed at 10 minutes. The milling tool is cleaned by ultrasonic cleaning [7].

Intensive surface milling is not the only use of face mills. With pocket milling, you can out-cut, down-cut, and surface-cut using smaller diameter end mills. The basic elements of face mill body design that influence tool selection are cutter diameter, left- or right-side kerf, cutter geometry (comb and front angles), insert pocket design, cutter tilt, and machine assembly technique are some of variables to consider [1,6–8]. The insert pockets in most face mills click into place. Some insert chucks are "modular", meaning that they can be inserted at different angles and support different insert designs. The operating range of cutter body grows as a result. The number of chucks or chuck is determined by their holders and diameter. The greater the chuck (fewer inserts), the larger the chip chuck, which carries the chips as cutter passes through the workpiece [6–8]. Cutter pitch options include extra fine, fine, and coarse. There are many brands and shapes of milling inserts. Careful and precise insert placement is required to achieve acceptable finish quality, minimize tool life, and maintain tight tolerances. In addition, the milling tool must be mounted on the machine [7]. In the first step, the influence of process factors on the grinding of an end mill is investigated. Therefore, cylindrical ground blanks made of carbide type HP920 from KG with a diameter of $D = 12$ mm and a cobalt concentration of 9% are used. The grinding experiments are carried out using 5-axis sharpening machine HELITRONIC VISION 400L from Walter. The tool shape remaining after grinding is shown in Figure 1.2 [7].

1.2 CUTTING TOOL ANGLES AT A SINGLE POINT

The curvature can be seen when looking at the auxiliary tool from a side, as shown in Figure 1.3. The directions to which the cutting angle tool belongs are called single-point wide and wide-range line data and have opposite purposes [9]. For the experimental measurement of the cutting angle, Gildemeister CTX 420 linear spindle milling center is used (Figure 1.3a). Heat-treatable steel AISI 1045 is used as a test specimen. The holes are filled with pre-fabricated brass pins (Figure 1.3b and d). The workpiece is clamped in a three-jaw chuck of the machine and supported by the machine head (see Figure 1.3a). For each test run, a groove is cut in the workpiece to ensure a free perpendicular cut. The result is a disk with a diameter, 3 mm, which is produced during the test run. After each test run, the remaining material is removed from the disk and a new disk is made by creating a new groove. The backshift angle is useful for removing chips from the workpiece during machining. It makes the machining process even easier. A negative slope angle occurs when the surface is inclined downward toward the nose. This weakens the cutting edge of tool. However, this further enhances the advantage [8].

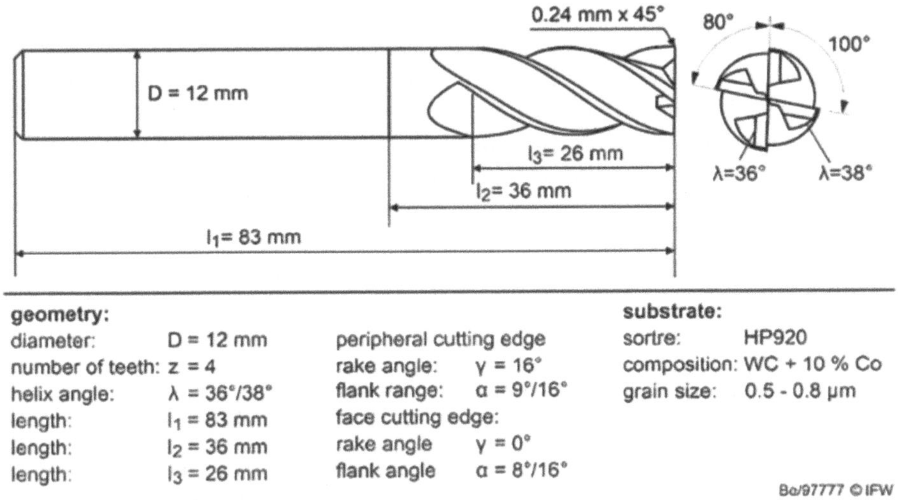

FIGURE 1.2 Tool geometry as a result of grinding. ([7] [Open access].)

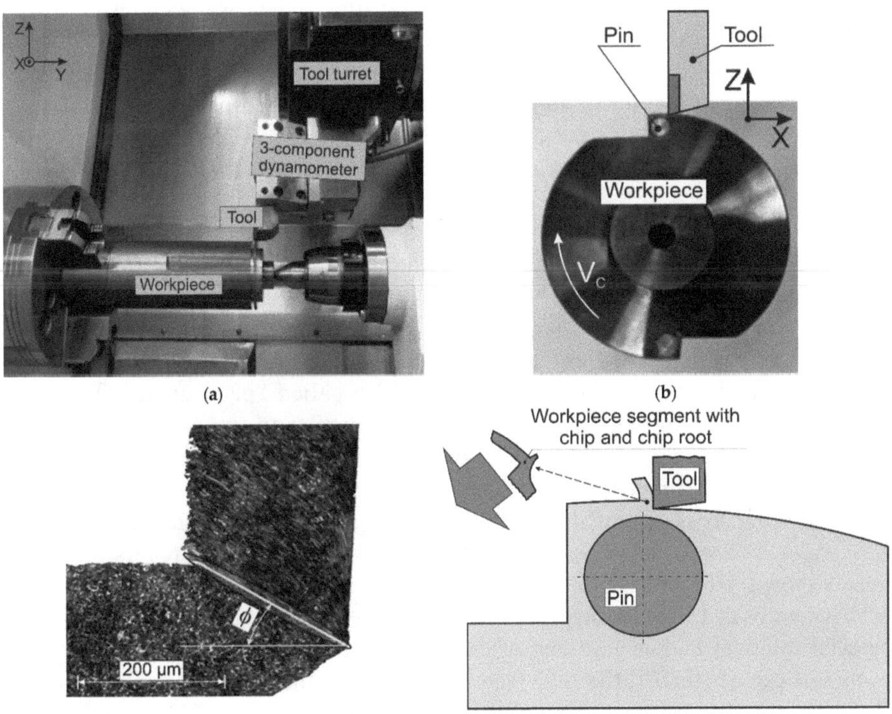

FIGURE 1.3 Experimental configuration for measuring cutting forces: (a) experimental setup, (b) general view of the workpiece using the chip root extraction tool, (c) chip root, and (d) extraction method. ([8] [Open access].)

In addition to increasing the required cutting forces and power, this will also increase the friction between the tool and the workpiece, but can help improve the surface quality of the product. "The term 'positive back rake' refers to the inclination of the cutting edge upward, which helps improve chip flow and reduce cutting resistance." In conventional bevel cutting, the tool moves continuously along the direction of cutting speed it generates, while the workpiece is fixed. To avoid friction, space is provided between the side surface and the workpiece.

When cutting and forming metals, the shape of the cutting tool plays a critical role in productivity and efficiency. Key angles, such as the clearance angle, the front and side cutting angle, and the bend angles, are essential for reducing friction and facilitating chip removal. These angles affect tool wear, surface finish, and cutting forces during machining. Understanding the function of each angle helps optimize forming processes and extend tool life.

- Relief Angle: The relief angle refers to the amount of clearance provided behind the cutting edge of the tool to prevent it from rubbing against the workpiece. It ensures smooth cutting and reduces friction. This angle is also commonly referred to as the relief angle. Face cut angle: The angle that makes a right angle with a flank when the minor flank or minor cutting edge is located concerning a line perpendicular to the base. As a result, this prevents lateral friction between the work piece and the tool.
- Side cut angle: The angle that forms a right angle with the side between the main side surface or main cutting surface and the perpendicular plane to base is known as side cut angle. In addition, this prevents the work piece from rubbing against the longitudinal angle equal to the final clearance angle.
- Side rake: This is a line perpendicular to shank and has an angled edge. The back rake angle helps clear chips during machining just like the side rake angle [6].

The innovative single-sided cutting-edge angle is formed between the line parallel to shank and side cutting edge, which is also called "plant angle". The angle formed by the end-cutting edge and line perpendicular to the shank is known as end-cutting-edge angle. Figure 1.4 illustrates how the cutting speed direction is parallel to the tool's travel at any point along cutting edge [1,6–8].

The specific cross-sectional shape of the milling tool edge is indicated by presenting K value as a significant parameter after the edge preparation process of milling tool, as shown in Figure 1.5. The tool edge shows a symmetrical profile with $S\gamma$ equal to $S\beta$ when K is 1, which suggests that the tool surface has undergone equivalent material removal both before and after the edge preparation procedure. However, when K is greater than 1, the rake face removes more than the side surface, and the edge becomes asymmetric. As a result, the rake angle of the tool decreases, which has a great influence on cutting forces and integrity of machining surface and vice versa. Therefore, the difference in material removal between the tool surfaces before and after edge preparation can be analyzed using K values, which shows that it is successful [6].

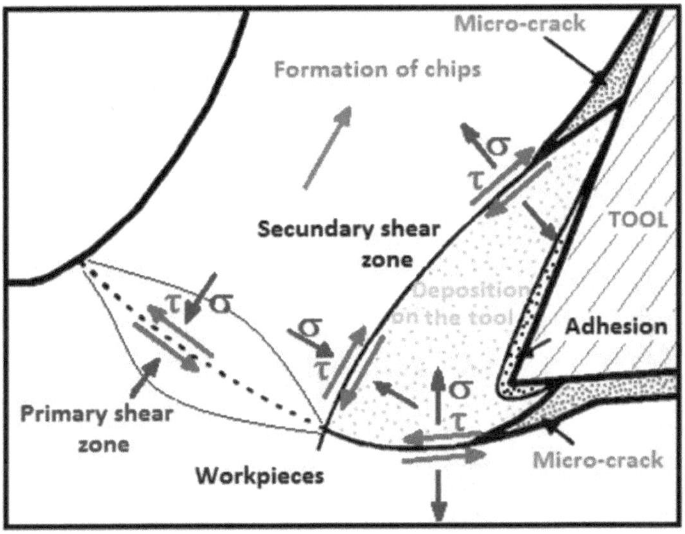

FIGURE 1.4 Diagram showing the directions of speed during oblique cutting. ([6] [Open access].)

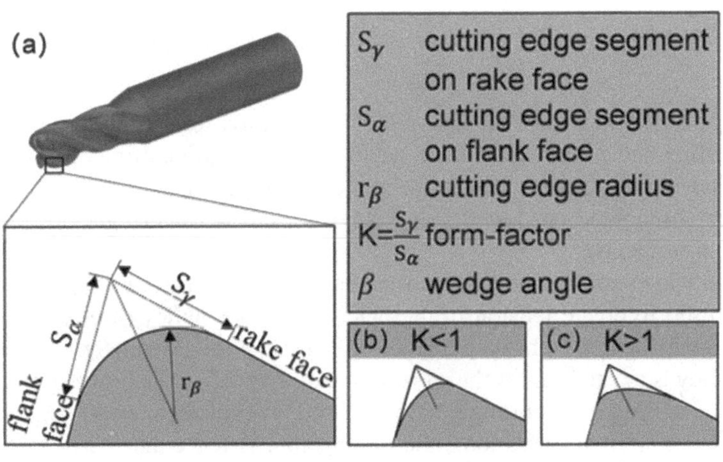

FIGURE 1.5 Edge geometry definition parameter and form factor approach: (a) tool edge parameters are characterized, (b) waterfall sharpening ($K < 1$), and (c) tube honing ($K > 1$). ([6] [Open access].)

1.3 WHAT IS THE BASIC OPERATING PRINCIPLE OF A LATHE?

The most common and important tool in metalworking is the lathe. "Horizontal longitudinal access lathe" is abbreviated to lathe. Unlike milling machines, lathes are designed to rotate the workpiece relative to the cutting edge of the tool. The entire form states that it operates on the principle of a stationary cutting tool and

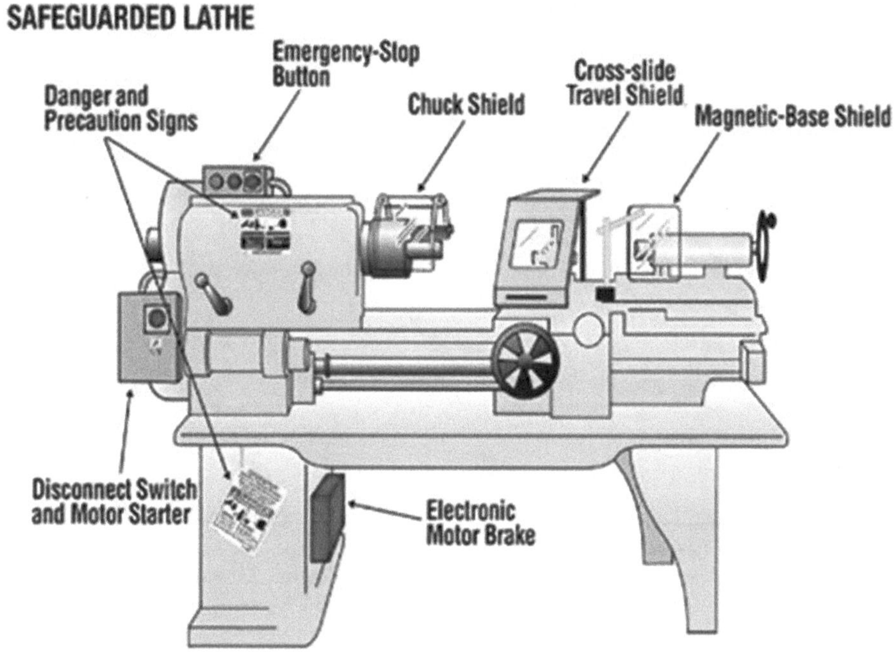

FIGURE 1.6 Universal lathe with several additional devices. ([1] [Open access].)

rotating part [10]. A 1.1 kW electric motor transmitted rotary motion to the vortex thread-cutting device via SPA V-belt, performing the main cutting motion of vortex device. Four cutting speeds and frequency converter are used for machining. Each workpiece is mounted on mandrel and clamped with a Böehringer three-jaw lathe chuck (Figure 1.6) [1].

The electric motor is fixed to the vortex device housing through egg-shaped holes, which is used to create tension in the belt. The V-belt pulley of electric motor on the thread-cutting device has a diameter of 94 mm. With a diameter of 140 mm, the driven pulley is located across the cutting ring of the tool on the inside of radial roller bearings. A cutting tool is brought to the rotating workpiece to give it an appropriate shape. During normal operations, the cutting tool is fed either perpendicularly or parallel to the working axis.

To understand the cutting tool, its corresponding angles and its operation must be determined. Figure 1.7 shows the response surfaces and measurement locations used to obtain each sensor's vibration measurement data [11]. The resulting surfaces and 3D plots illustrate how the vibration amplitude levels depend on the processing parameters. The R_a value for surface roughness is calculated similarly. Figure 1.8 shows the response surface. The physical units of the following plots are omitted for simplicity, but they can still be identified. The shape of the current $z2$ model is significantly different from the response surface of Sensor 1. Additional measurements are planned to obtain a more accurate data for the actual model. The roughness measurement data are used to create a response surface model using the Wolfram

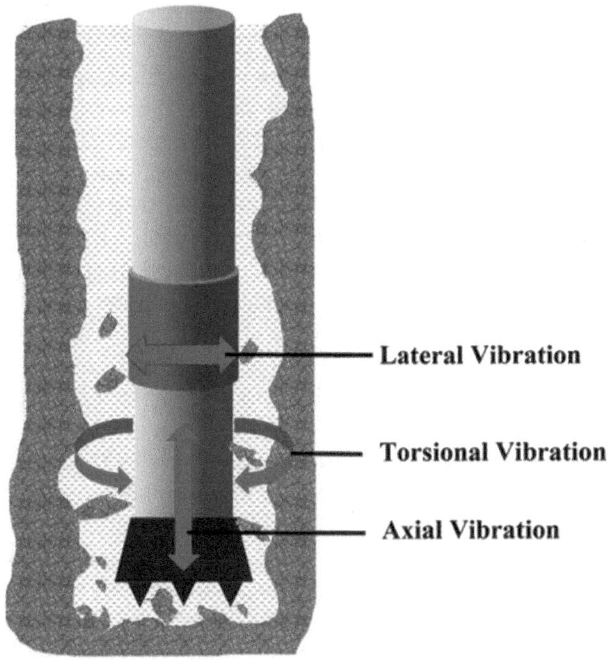

FIGURE 1.7 Measurement points and response surfaces used to extract vibration data. Data sensors 1 and 2 represent response surfaces, respectively. ([1] [Open access].)

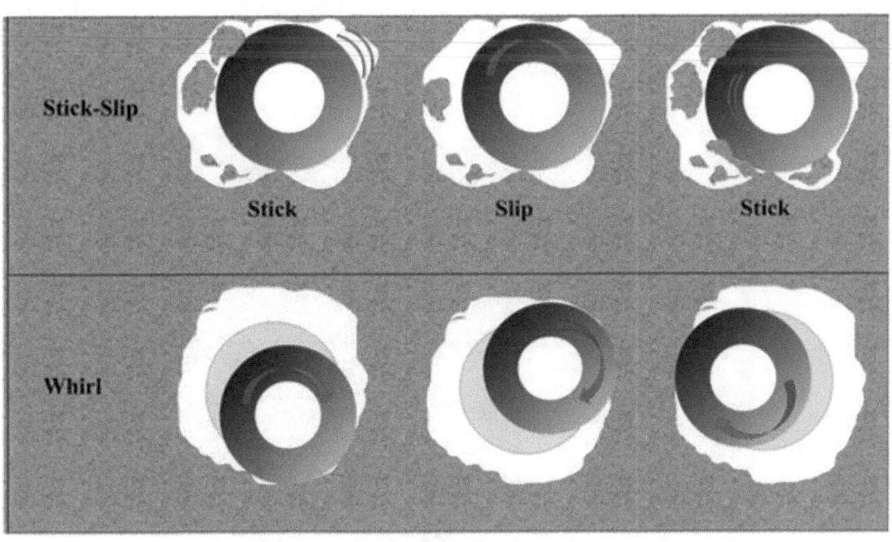

FIGURE 1.8 Surface roughness data is obtained from the response surface data and measurement points. ([1] [Open access].)

Mathematica 10.2 function. It demonstrates how the processing parameters affect the surface roughness, as illustrated in Figure 1.8 [1].

The workpiece is rotated, and the turret is moved along the workpiece during turning to allow the insert to perform the cutting operation. The workpiece is 300 mm long and 90 mm in diameter, which was made of Al6061, a material that is often used in mechanical engineering. The inserts with different conditions are placed on the tool holders for turning. Tungsten carbide is used to make the insert. For this study, inserts in different situations are collected from several metal component manufacturers. The machining in this study is performed on computer numerical control (CNC) lathe, as shown in Figure 1.9. The tailstock provided stability, and the 3-jaw chuck clamped the workpiece. The CNC controller is used to control the movement of the turret that held the tool holder with the insert [2]. The main function of a lathe is to rotate the workpiece, while applying the cutting tool to it to shape it. Here is a brief explanation:

- Rotation: The workpiece is rotated around its axis by the lathe. The speed and accuracy of rotation may or may not be controlled.
- Cutting tool: The lathe has a cutting tool with several adjustment options. The cutting tool is fed into the material as the workpiece rotates, removing material to create the desired shape.
- Motion: The lathe allows the cutting tool to move in two separate directions: longitudinally, along the axis of rotation, and transversely, perpendicular to the axis. This movement makes it easier to create different sections and shapes.
- Auxiliary device: The lathe head, which houses the spindle and motor and lathe tail, which provides additional support and alignment, supports the workpiece. This ensures accuracy and stability during cutting.

A lathe shapes materials such as metal, wood, or plastic using a rotary motion and fixed cutting tool. Different setups and fixtures may be required depending on the type of lathe and complexity of job.

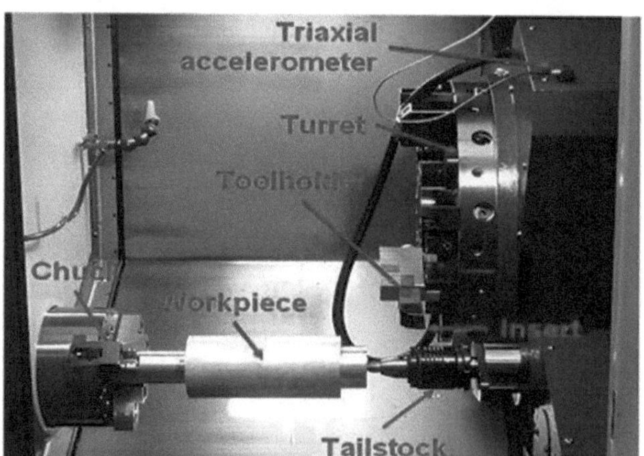

FIGURE 1.9 Equipment for the architecture of the turning plant [2].

FIGURE 1.10 Four insertion circumstances: built-up edge (a), flank wear (b), normal wear (c), and crack (d). ([2] [Open access].)

The four states of the insert, namely built-up edge, flank wear, normal wear, and fracture, are shown in Figure 1.10. Built-up edge is defined as the phenomenon of material accumulation on the edge of the insert. High temperatures usually occur at the interface between the insert and the workpiece during turning. This can lead to the formation of a built-up edge when the insert breaks due to the separation of material on its edge from the insert [1,2,7,8].

1.4 WHAT COMPONENTS MAKE UP THE MAIN BODY OF A LATHE?

The lathe bed, lathe head, lathe shank, tool holder, bracket, carriage, quick-change reducer, and other components are important parts of lathe. We have studied the conditions of various angles and single-point cutting tools. We use this tool on the lathe.

To rotate the workpieces at higher speeds, steam engines and water wheels are installed after evolution [12]. A new era in the history of industrial revolution began with the advent of electric motors to rotate the chuck. In addition, the working time and machining accuracy have improved over time.

1.4.1 End Milling Tools

The machining technique known as end milling uses an end mill, which is a rotating cutter, to cut or shape materials. It is used for contouring, slotting, and profile cutting in most manufacturing industries. To remove materials from the workpiece, the cylindrical end mill cutter is equipped with cutting blades on both side and bottom. The part moves along several axes and rotates [13]. End mills have a variety of tip and flute designs that are tailored to different applications and hole/slot periphery shapes [3]. The drawing of experimental solid ceramic end mills, which are being developed and tested in the field, is shown in Figure 1.11a, where the general appearance is demonstrated in Figure 1.11b. The authors have adopted these experimental end mill samples as a standard after comparing them with ceramic tools of similar design and geometry, which are currently commercially produced from SiAlON-based ceramics including Yb-containing stabilizing phases (commercial ceramic cutters) [3].

The tool is tested using a program written in the GeMMa-3D CAD/computer-aided manufacturing (CAM) system (GeMMa Scientific and Technical Center, Zhukovsky, Moscow, Russia) using a strategy for machining the end face of workpiece using an end mill moving in spiral. Figure 1.12b shows the machining plan, and Figure 1.12a shows the general view of the ceramic end mill arrangement relative to NiCr20TiAl nickel alloy workpiece during the tests. The cutting modes during the tests are cutting speed $V_c = 376.8$ m/min (rotation frequency $n = 12{,}000$ rpm), feed $S = 1500$ mm/min, and feed per tooth $S_t = 0.031$ mm/tooth. The dry machining strategy is used without using cutting fluids.

The experimental data are presented in Figure 1.13, which illustrates how different powder compositions (80% (90α10β) + 20% TiN, 90% (90α10β) + 10% TiN, and

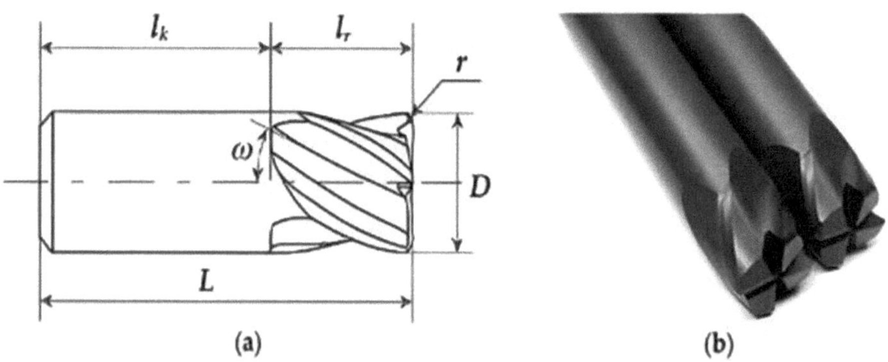

FIGURE 1.11 Illustration (a) and overview (b) of authors' experimental solid ceramic end mill samples. ([3] [Open access].)

FIGURE 1.12 Tool life testing method for solid ceramic end mills: (a) general perspective of end mill position relative to NiCr20TiAl nickel alloy workpiece during the test and (b) visualization of machining process. ([3] [Open access].)

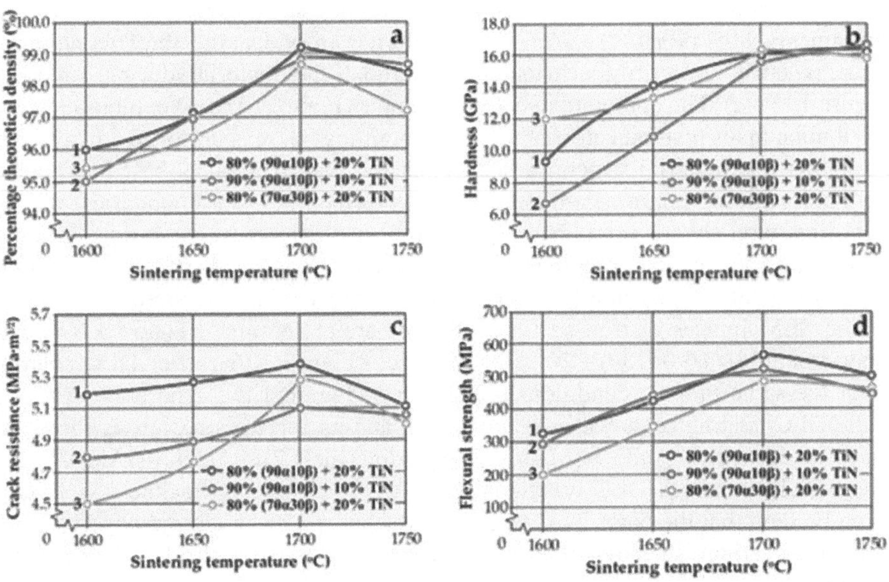

FIGURE 1.13 Dependences mechanical and physical properties on spark plasma sintering temperature: (a) theoretical density in percent, (b) hardness, (c) crack resistance, and (d) flexural strength of sintered ceramic samples made from different powder compositions: 80% $(90\alpha10\beta)+20\%$ TiN (1), 90% $(90\alpha10\beta)+10\%$ TiN (2), and 80% $(70\alpha30\beta)+20\%$ TiN (3). ([3] [Open access].)

(3) 80% ($70\alpha30\beta$) + 20% TiN) affect the main mechanical and physical properties of sintered ceramic specimens [1–3,7,8]. The spark plasma temperatures are found to be in the range of 1600°C–1750°C. The properties include percent theoretical density (Figure 1.13a), hardness (Figure 1.13b), fracture toughness (Figure 1.13c), and flexural strength (Figure 1.13d).

The results of a microstructural study of fracture surfaces of two different types of SiAlON-based ceramics are presented in Figure 1.14, where the experimental sample 80% ($90\alpha10\beta$) + 20% TiN (Figure 1.14a, c, and e) and commercial ceramic samples (Figure 1.14b, d, and f) differ significantly from each other. This makes sense since the structure of sintered ceramics is mainly determined by the type and number of additives and sintering components used, as well as technical characteristics of the process. When comparing the microstructures of fracture surfaces of ceramic samples using scanning electron microscopy (SEM), it becomes obvious that both samples include a significant amount of α-modification of SiAlON as a dominant phase [3].

For the commercial ceramic sample, this phase shows a high degree of homogeneity and dispersion. The grain sizes in experimental ceramics are typically 3–5 μm, and they can be either equiaxed or non-equiaxed, in contrast to the grain size of 1.5–3.5 μm in conventional ceramics. In addition, both samples contain significant amounts of β-modification of SiAlON. Due to elongated (oblong) particles in this phase, commercial ceramics exhibit more characteristically. Using energy-dispersive X-ray (EDX) analysis, element distribution maps are generated along the fracture surface of samples, which allow for a qualitative assessment of major elements and Ti- and Yb-containing phase homogeneity in ceramic samples (Figure 1.15). In the experimental 80% ($90\alpha10\beta$) + 20% TiN sample, it is an evident that the Ti-containing phase is fairly uniformly distributed throughout the material during sintering (Figure 1.15a–c). The commercial ceramic sample (Figure 1.15d–f) exhibits element distribution maps that indicate that the Yb-containing phase is uniformly distributed throughout the material structure. It is concluded that in SiAlON-based ceramics, the Yb-containing phase (stabilizing phase) is formed at grain boundaries, which, according to reliable research data [2,4,5], prevents their expansion during sintering and ensures the creation of more homogeneous and fine-grained structure compared to experimental samples of 80% ($90\alpha10\beta$) + 20% TiN ceramics (Figure 1.15) [3].

The SEM images of thin sections of commercial ceramics (Figure 1.16b) and experimental 80%($90\alpha10\beta$) + 20%TiN ceramics (Figure 1.16a) after DLC coating under the same process conditions are shown in Figure 1.16d. The bulk structure of the DLC coating is shown in Figure 1.16d, and its surface structure is shown in Figure 1.16c, which is the SEM image. Including the 1.7 μm thick (CrAlSi)N sublayer and 1.9 μm thick DLC functional layer, the total coating thickness is 3.6 μm. It can be seen that the outer DLC layer is characterized by an amorphous structure, which is indistinguishable by microscopic methods, while the nitride sublayer has a columnar structure. The thickness of the DLC layer does not show any discernible grain boundaries, which are usually created during the nitride coating process, while the surface of the DLC layer has a clear relief consisting of intergrown spherical segments, with sizes ranging from 0.2 to 1.5 μm. It should be noted that the state of the surface layer of the ceramic base before coating does not have any effect on the creation of clear reliefs. Selected area diffraction (SAED) of complex coating

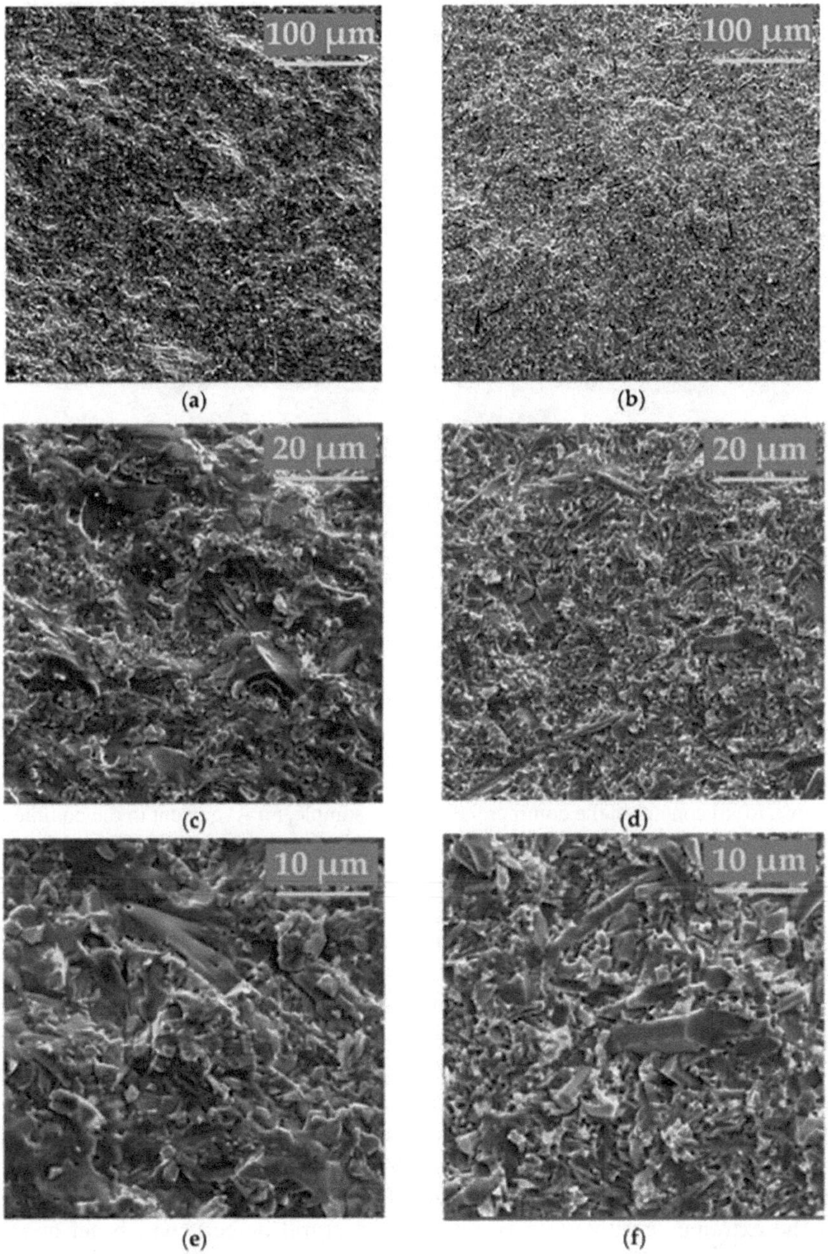

FIGURE 1.14 SEM images of SiAlON-based fracture surface microstructures of sintered ceramic specimens. The following specimens are used in experiment: (a) 80% $(90\alpha10\beta)+20\%$ TiN, ×1000; (b) commercial ceramic specimen, ×1000; (c) experimental 80% $(90\alpha10\beta)+20\%$ TiN specimen, ×5000; (d) commercial ceramic specimen, ×5000; (e) experimental 80% $(90\alpha10\beta)+20\%$ TiN specimen, ×10,000; and (f) commercial ceramic specimen, ×10,000. ([3] [Open access].)

FIGURE 1.15 Based on SiAlON samples obtained by EDX analysis, the distribution of chemical components on the fracture surface of sintered ceramic material is shown: Si and Al contents of experimental samples are as follows: (a) 80% ($90\alpha10\beta$) + 20% TiN in the Si sample, (b) 80% ($90\alpha10\beta$) + 20% TiN in the Al sample, (c) 80% ($90\alpha10\beta$) + 20% TiN in the Ti sample, (d) Si content in the commercial ceramic sample, (e) Al content in the commercial ceramic sample, and (f) Yb content in the commercial ceramic sample. ([3] [Open access].)

(Figure 1.16e and f) shows an amorphous structure of the DLC layer together with a complex phase of a (CrAlSi)N sublayer [3].

The surface topographies of 80%($90\alpha10\beta$)+20%TiN ceramic samples are shown in Figure 1.17, both before and after DLC coating (Figure 1.17a and b, respectively). As can be seen in Figure 1.17a, the surface of the ceramic sample has a noticeable network of abrasive scratches, which are typical for diamond-sharpened ceramic tools [3].

In addition, the ceramic surface shown in many earlier works has numerous craters from grains chipped during grinding [14–16]. The observed features are essentially abrasive machining defects that are a part of physics of the process. The risk arises from the extreme sensitivity of ceramic to structural defects, which act as stress concentrators and often cause chipping and spalling on the tool's working surface.

1.5 END MILLING: WHAT IS IT?

A machining technique called end milling is used to shape and remove materials in the metal and woodworking industries. This uses materials to be cut or shaped using a rotating end mill. Numerous tasks are performed using this method, including

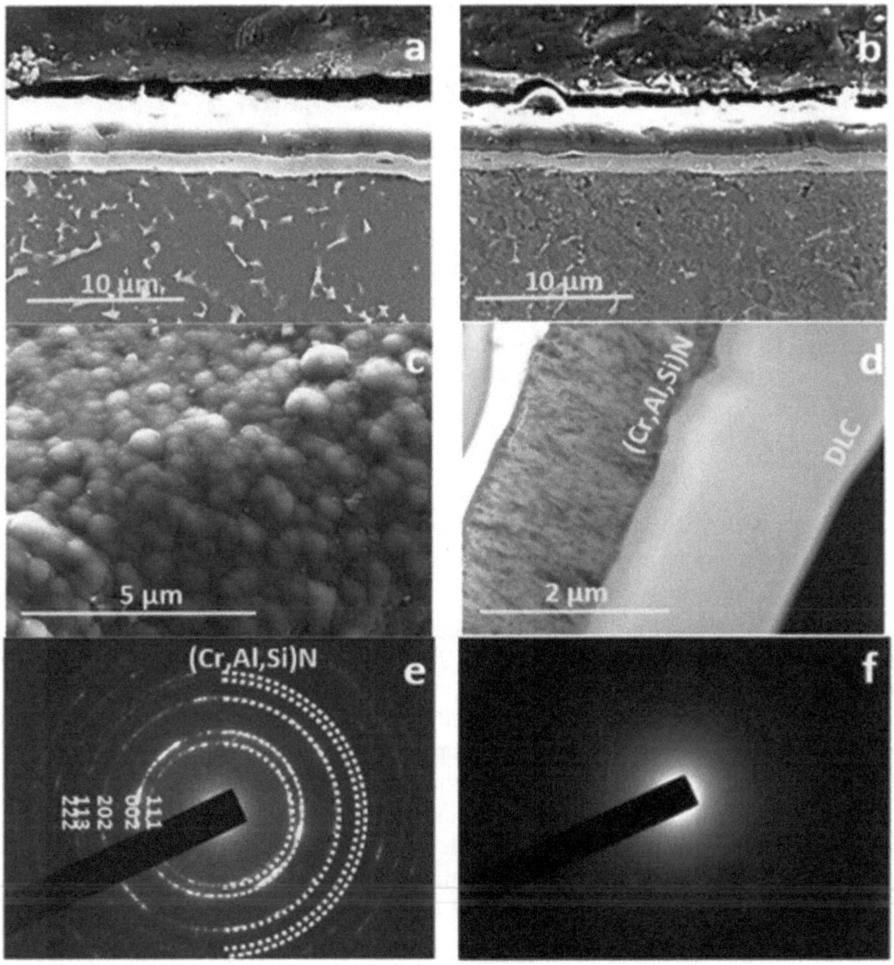

FIGURE 1.16 DLC-coated ceramic specimens and their microstructure. The images show DLC coating on thin section of experimental specimen consisting of 80% (90 ± 10β) + 20% TiN ceramics. They show commercial ceramic specimen with thin section of DLC coating, SAED of (CrAlSi)N sublayer, TEM image of two-layer structure of DLC coating, and SEM image of surface structure of DLC coating. ([3] [Open access].)

grooving, contouring, trimming, and profiling of workpieces. A workpiece has a wide variety of end milling features, including holes, pockets, slots, complex curves, and mixed shapes. Precision cutting and shaping of metals, polymers, wood, stone, and composite materials are a fundamental manufacturing process used in all industries [17]. The material being processed and specific machining needs to determine the type, size, and cutting parameters of the end mill used. End milling is used to precisely cut and shape materials to create specific surfaces, profiles, and features. It effectively removes materials from workpieces by cutting off pieces

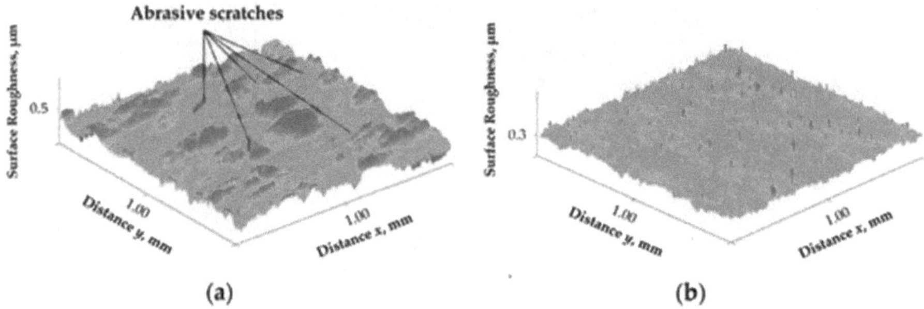

FIGURE 1.17 Surface topography of 80% (90α10β)+20% TiN ceramic samples both before and after DLC coating. ([3] [Open access].)

until the product is shaped to the desired size. In addition, end milling can produce parts with complex geometries, complex curves, and complex contours. Additionally, it can be used to make master holes, pockets, and slots in workpieces of varying depths and diameters [18].

Additionally, when used as a post-process after casting or other less complex cutting procedures, end milling can improve the surface finish of workpieces. Due to its precision and accuracy, it can be used for operations that require tight tolerances. To scale production from trial to moderate- and possibly large-volume production, end milling is an essential procedure in both prototyping and production. There are several milling machine types, sizes, and layouts available. However, they have several important common characteristics necessary for their functioning.

The cutting tool is usually held in a collet by a spindle, which is a rotating drive. It provides the rotating power needed to remove materials. The workpiece is mounted or secured on the work table. In its most basic configuration, it can move along the *X*, *Y*, and *Z* axes; in more complex machines, it can move up to 12 degrees of freedom. The work table and other machine parts move along guides, which are linear bearings or rails. Even when the cutter exerts force, it guarantees precise and smooth movement. The CNC machine control panel contains programming functions and machine control system. It allows the user to observe the machining process and enter G-code and machining settings. Manual control is used in manual machines, which typically incorporate a three-axis position measuring system. The end mill is held firmly on the spindle by a tool holder or chuck. The coolant system is typically used in milling machines and distributes cutting fluid or coolant to the workpiece and cutting tool. It improves surface quality, extends tool life and dissipates heat. Chip removal features such as chip conveyors or chip augers are often added to machine tools to rid the cutting area of chips and chips generated during machining. Electrical shut-off enclosures protect operators from flying debris and improve workplace safety. The number of parts required, type of material, and accuracy are determined [4]. Figure 1.18 details the resulting cutting-edge geometry. The cutting-edge angles measured here are determined by the ISO nomenclature. The angle created by the cutting speed and cutting edge is known as the rake angle [4].

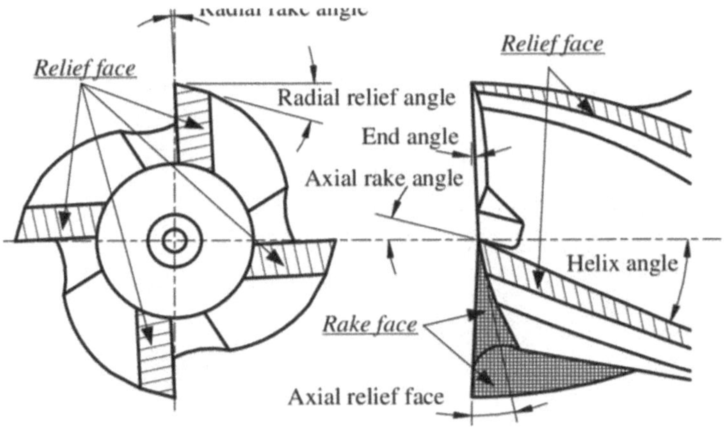

FIGURE 1.18 Geometry of an end mill. ([4] [Open access].)

The tool changer, measured in plane and defined by CNC machining centers, is used to automatically change the tools mounted on the column and pre-calibrated, ready for the subsequent machining steps. A variety of gripping tools, including clamps and vices, are used to hold the workpiece securely in the workspace. These devices can be more versatile, quick-release, or customizable discrete element AB of the cutting edge and cutting speed vector V_c [19]. The local helix angle, denoted as i, is measured in the P_p plane, which is defined by the Z axis and vector V_c [15]. The maximizing tool path and cutter guidance for 5-axis freeform machining on the CNC machine (see Figure 1.19) are the same [4].

1.5.1 The Rake and Clearance Angles

The purpose of traditional machining or metal cutting is to gradually remove excess materials from a workpiece to give it the desired shape, size and finish. Removal of materials is done using a tiny tool known as a cutter or cutting tool. The workpiece and cutter are firmly fixed to the machine during the cutting process. To consistently cut the material, a relative speed between the cutter and the workpiece is necessary. This is achieved through the use of cutting speed, feed, and depth of cut. In reality, the cutter compresses a thin layer of the material, which is then gradually removed as chips. In addition, the correct material and correct shape with a sharp cutting edge are required for continuous and efficient material removal. Cutter geometry indicates the direction or inclination of various surfaces of the tooltip. The rake surface, main flank surface, and minor flank surface are the three tooltip surfaces that make up the cutting tool. The inclination of such surfaces in different directions is indicated by different angles [20]. A HAAS VF2 CNC machining center (HAAS Automation, Oxnard, CA, USA) with a maximum rotation speed of 10,000 rpm and a maximum power of 20 kW is used to design the milling tests. With deep grooves and relatively thin walls, some of the difficult-to-machine shapes are targeted for reproduction.

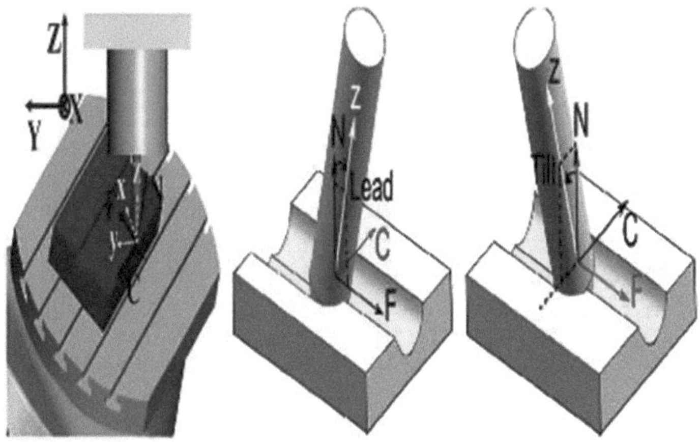

FIGURE 1.19 Optimization of cutter orientation and tool path for 5-axis CNC freeform machining. ([4] [Open access].)

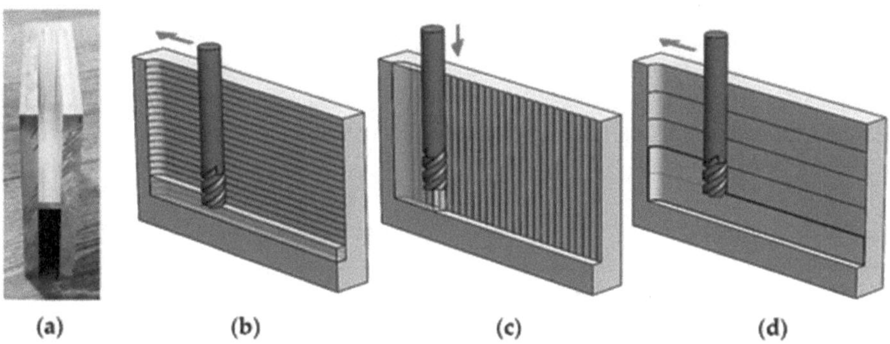

FIGURE 1.20 Front view of the machined part (a) after finishing and plunge milling operations. The schematic of the conventional strategy is shown in (b); plunge milling strategy is shown in (c); schematic of the finishing strategy for both machining strategies is shown in (d); and $V_c = 160$ m/min, $f_n = 0.07$ mm/rev, $a_p = 14.8$ mm, represents the productivity of machined part after adjusting by 3.7 mm. The red arrows indicate the movement. ([5] [Open access].)

The aim is to simulate some of typical machining methods. An ER32 DIN6499 collet, an ISO 7388-2 tie rod, and ISO 40 DIN69871 collet holder are used to attach the milling tools to spindle [5]. The chosen geometry, shown in Figure 1.20, is dictated by a part that is frequently machined by a Portuguese company in the aeronautical sector [5].

The rake angle, by definition, is an orientation angle of the rake face of the cutter, measured in another plane and compared to the reference plane. Its value can be positive, negative or even zero, but is usually in the range from +15° to −15°. Tooltip strength, cutting force, power consumption, shear strain and machinability are determined by this critical angle. The cutter rake angles can be positive, negative, or zero; each has unique significance and benefits for the machining efficiency. For example, a positive rake angle provides a strong cutting edge, allowing you to cut easily with little effort. On the other hand, a negative rake angle has a stronger tool nose and can withstand a higher cutting force. However, the cutter clearance angle cannot be negative or even zero, since in these cases, the side surface of the tool will rub against the final surface of the workpiece [21]. The rake angle is typically in the range from +15° to −15°, while the clearance angle is typically in the range from +3° to +15°.

Conversely, the clearance angle is the angle at which the side surface of the cutter deviates from the speed vector and is measured in a different plane. Its value must be positive and often ranges from +3° to +15°. Three separate areas of the milled surface are examined, with the direction of each point indicated on the point map in Figure 1.21. The probe consists of seven segments with a total length of 5.6 mm and a threshold value of 0.8 mm. To allow for acceleration and deceleration of the probe, the first and last parts are ignored. Based on ISO 4288:1996 [1–5,22], surface roughness values are calculated for each working condition. In an attempt to depict the start and end points of the machining process in both upper and lower cases, Figure 1.21 shows the locations that are analyzed along with any changes due to vibration induced by the tool length. Each position is measured six times, and the average values are obtained [5].

Contribution to accuracy, product quality, and chip deflection: Chip shear deformation and chip flow direction are directly affected by the rake angle. The thickness of AnChip will increase as a result of increase in shear strain caused by the negative tilt. With a negative rake, the chip reduction rate (CRC) will also be greater. However, this does not have a significant impact on the quality of the machined part

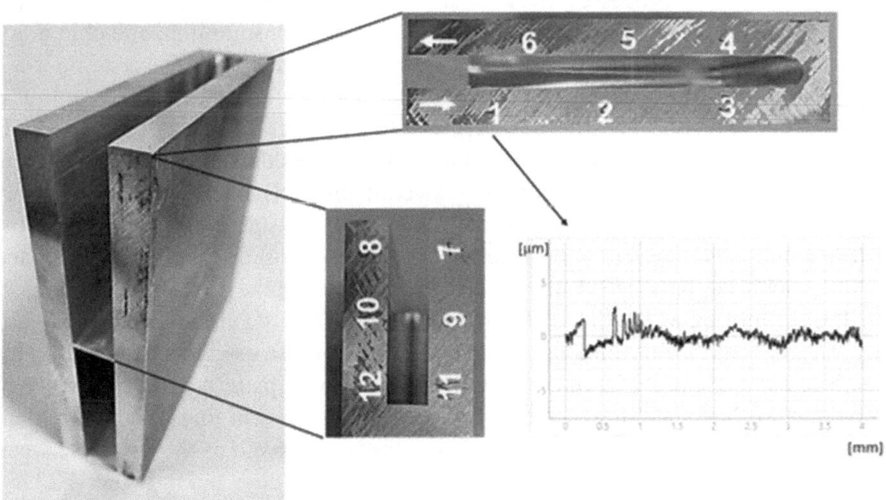

FIGURE 1.21 Position for measuring the roughness on the surface of the specimen: the lower and upper positions are used to measure the starting position of finishing (1) and ending position of finishing (6), respectively. The width of specimen is shown by points 7–12, which correspond to beginning and end of ending position of finishing, respectively (12). ([5] [Open access].)

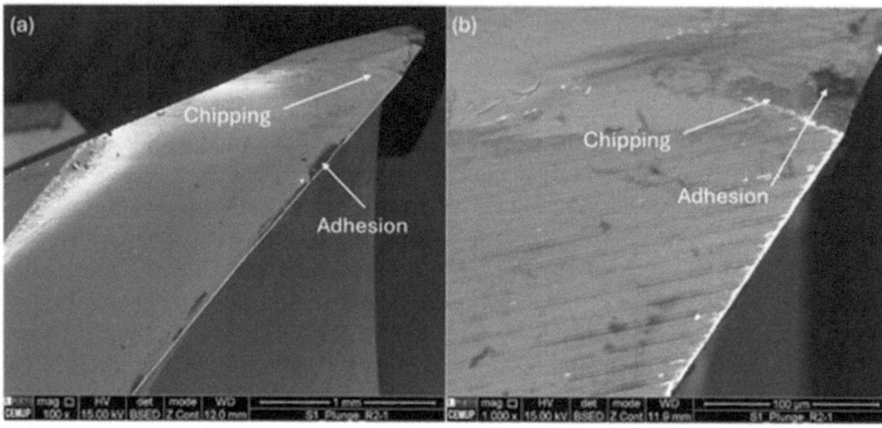

FIGURE 1.22 SEM micrograph examination of cutting edge of tool used in plunge milling (PM): (a) general perspective of the side region of the cutting edge and (b) in-depth view of the tip of the cutting edge. ([5] [Open access].)

or dimensional accuracy. These parameters play a key role in the clearance angle. The first analysis concerns the equipment used in the plunge milling (PM) approach. Figure 1.22 shows two images of the side cutting edge, showing the tools used in this strategy; one illustrates the general appearance of the cutting edge, while the other highlights the detailed appearance of some of failures. The side cutting edge is the most representative of all observations [3–5].

In manufacturing processes, the rake angle has a significant impact on some variables such as shear stress, cutting force, torque, energy consumption, and chip properties like type and color. In manufacturing mechanics, this angle is very important because it affects the behavior of the material and performance of the tool. The analysis highlights the built-up edge effect (BUE), which allows ductile materials to adhere to the cutting edge of the tool at best locations during cutting. Improving manufacturing processes and increasing overall productivity and product quality require understanding the effects of the rake angle. As stated earlier, because the different methods correlate with different parameters, primarily in terms of ap and ae, the machining parameters are matched as closely as possible to the PM process [5]. The authors are aware that different tool forces affect the vibration levels of each tool. The tools of the conventional strategy, as well as those used for the PM strategy, are observed using SEM in an attempt to find a behavior pattern that could serve as an indicator of the state in which the tools are left after machining. Particular attention is paid to any wear mechanisms that are observed, as well as the degree of wear. The general pattern found near the cutting edge of tool is depicted in Figure 1.23 [5].

1.5.2 Several Applications for Milling Tools

Milling cutters are indispensable equipment in the world of machining. They come in many different types, each designed for use with specific materials and applications. The SEM image of the micro milling cutter at different wear stages is shown

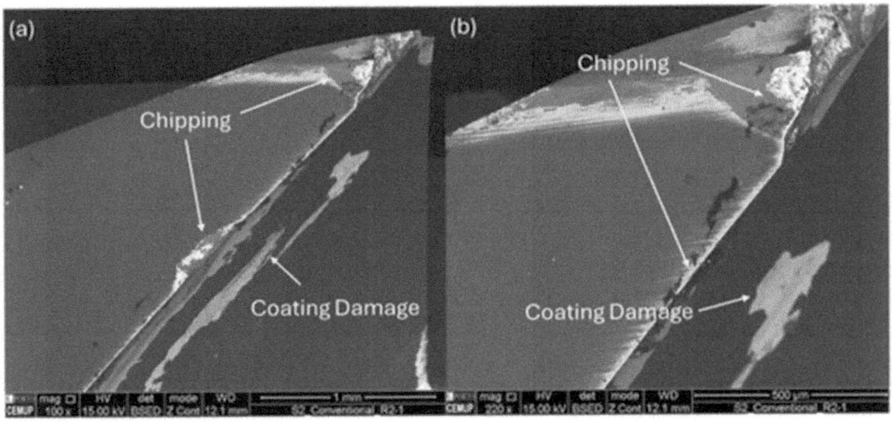

FIGURE 1.23 Side view of SEM of cutting edge of tool used in conventional milling (TM) technology: (a) general view of the cutting edge and (b) in-depth view of the cutting edge. ([5] [Open access].)

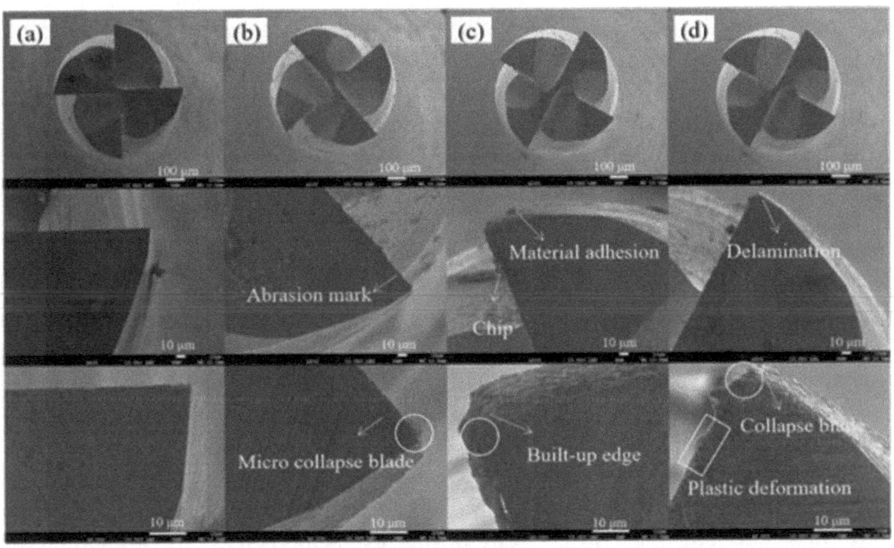

FIGURE 1.24 SEM of micro milling cutter under different wear conditions. The images are as follows: (a) initial 100/500/2000×, (b) wear I 100/500/2000×, (c) wear II 100/500/2000×, and (d) wear III 100/500/2000×. ([22] [Open access].)

in Figure 1.24. The initial wear/first wear, second wear, and third wear are presented in order of (a), (b), (c) and (d), respectively. The magnification factor represented by each row is 100/500/2000 times per cycle. There are two categories of wear modes: abnormal wear and gradual natural wear. BUE, physical adhesion, blade micro-cracks, and wear marks are main components of gradual natural wear. The most common abnormal wear modes include catastrophic cracks, cutting-edge peeling, broken blades, and severe plastic deformation. Since there is no wear on a new blade, the surface is very smooth. The first wear is characterized by an initial blade micro-crack and wear marks as it progresses [22].

The purpose of this study is to classify and compare these many types of cutters, clarify their unique applications, and offer helpful tips for their ideal application in a range of manufacturing processes . Professionals can improve the accuracy and efficiency of their work by better understanding these tools [22]. The milling options are offered by the safety tool manufacturer. These options include ball-end cutters, tools with 90° inserts for slotting and shouldering, and milling wheels with variable insert angles for trimming. Specialized tools called cutters are used in milling machines to perform a variety of tasks such as drilling, profiling and slotting [15]. There are many different types of these incisors; each is designed for a specific use and can be identified by its design, use, and grooves or surfaces it creates.

1.6 MOLD AND HARD MACHINING

High temperatures occur at the cutting edge during high-speed hard machining. If coolants and oil are supposed to be used for cooling, then a sudden temperature change will cause stress on the device, leading to minor destruction in the installed carbide body of the device. Therefore, the gadget will explode due to negligence. As a result, when performing fast hard machining, dry hard machining should be used instead of coolant and oil. However, when the hardness of the workpiece is not too high, a small amount of lubrication can be taken into account in fast hard machining because the cutting temperature in hard machining increases with the increasing workpiece hardness and cutting speed. When the workpiece's hardness is at its minimum, it is incredibly high and this small amount of oil creates impact zones for carbide tools. Therefore, to prevent thermal shock loads in this type of high-speed hard machining, compressed cold air must be used to remove chips and tool life can often be increased by 20%–30%. In addition, chip trapping during machining is prevented by blowing with compressed air, especially in six-edge machining [20].

Currently, when blasting chips with cold compressed air, the compressed air pressure is significantly higher than 0.6 MPa. For example, for fast hard machining, a bucket with a hardness of 61 HRC is designed to compress air with a voltage of 0.6 MPa and cool it to –30°C. Compressed cold air is not only useful for blowing away chips, but in the most difficult circumstances, it can increase the strength of the device by 30%. It should be noted that the packaged viral air used in this case uses an unusual vortex tube mechanism to accelerate the compressed air to the speed of sound. During the acceleration process, it is divided into hot and cold sections and each is output independently. The –30°C virus is directed through a plastic hose with a spout so that it aligns with the sharp edge of processing mold. This sequence made sense for two reasons:

- One-fifth of the cost of high-speed machines per hour is spent on roughing machines, which have minimal precision requirements. In addition, the low hardness of workpiece material resulted in less tool wear.

- Roughing is not possible because the most common high-speed spindles available at time cannot provide sufficient torque below 1500 rpm.

In some cases, the strong thermodynamic coupling effect between the cutting tool and workpiece during the hard-cutting process results in a hardening effect on finished surfaces. However, the phase transition mechanism of material during heat treatment during the cutting process is different from a traditional method, which leads to a modification of the mechanism of creating a damaged layer on the surface of the workpiece. In this work, the mechanism of formation of a damaged layer on the machined surface during hard cutting of PCBN-hardened Cr12MoV steel is investigated. Using finite element modeling, rules governing temperature changes on machined surfaces and subsurfaces are obtained. Fast conversion tools are used in phase transformation temperature studies, and the effect of hardening under cutting conditions on the creation of damaged layers is revealed. This is used to determine temperature phase transition points under cutting conditions. The influence of tool wear and cutting speed on the thickness of white layer in damaged layer is demonstrated experimentally. By constructing a numerical prediction model, it is possible to obtain the temperature distribution law of the third deformation zone and also take into account the effect of tool wear when predicting the thickness of the white layer inside the damaged layer. The actual results demonstrate the prediction accuracy of the model, and establishment of the prediction model serves as a guide for the reasonable selection of parameters in the precise rigid cutting procedure. Using a predictive model, the formation of a damaged layer on a machined surface can be accurately controlled in machining processes that place great importance on surface integrity [23].

1.6.1 Sculpting Surfaces with a Ball Cutter

Machining complex surfaces is a common practice in many industries. The most important technological advancement in the field of precision shaping of dies, punches, dies, or blades is manufacturing. The mission is to create precise freeform shapes with tight tolerances and good economics from difficult-to-machine materials. This point is a critical due to low-wage countries, which are now competitive in the mold and die industry. Thus, several factors must be considered, including the use of three- or five-axis machines, powerful computer-aided manufacturing (CAM) systems, sophisticated tooling, and availability of experienced programmers and machine operators. There are two types of complex surfaces: ruled surfaces, which are found in compressor disks and blades, and sculptured surfaces, also known as freeform surfaces, which are commonly found in dies and dies. Regarding industrial use, four categories are defined and presents their main characteristics:

- Forging dies: the component is steel, carefully processed and prepared for hot working, with a hardness of 30–60 HRC. Both the final roughness and tolerance are quite high. Deep and narrow places are rare.
- Stamping dies: The components are ductile iron castings. However, the use of advanced high-strength steel (AHSS) for automobiles has increased over the past 5 years.

- The body is now more modular. As a result, it is necessary to machine a higher percentage of hardened iron, 60 HRC hardness input blocks and stronger forging tool surfaces. In addition, the final processing step may take several hours. For this reason, many users choose weaker but more reliable tooling materials (such as carbide) rather than stronger but more sensitive materials (such as PCBN); in the latter case, an unspecified allowance can lead to unplanned tool failure, which can cause catastrophic consequences for the die.
- Injection molds made of plastic or aluminum alloy. The mold is made of hardened steel with a hardness, 50–55 HRC. The casting process requires good finishing and precision of form. Because plastic components are so complex, mold designs must contain very deep and small areas, some of which are on the verge of EDM cutting and machining.
- Base surfaces: This type of surface is found in some aircraft or power engine components such as Inconel 718 aluminum alloy (or similar superalloy) and Ti_6Al_4V workpiece material. These applications require high precision of interface regions. A special type of finishing process called side milling [6] aims to take advantage of controlled geometries by following tool paths where part positions touch the tool being created, which can be a cylinder or a cone. Otherwise, the same techniques used for freeform surfaces are used to produce this surface.

To find out how the best-performing samples are lubricated, SEM-EDX analysis is performed. Acetone and heptane are used to clean the balls after tribological testing. The worn and unworn ball surfaces are then photographed using electron microscopy and EDS analysis is performed to obtain a combined morphological and chemical overview of the samples. As can be seen in Figures 1.25 and 1.26, the worn areas of the blend-lubricated samples contain trace amounts of components that are not detected in the wear tracks of samples lubricated with base oils alone. However, the peak strengths of these additional features (weight percent) are not sufficiently visible in EDS analysis to suggest that a chemical tribofilm has formed after tribological testing [24].

A typical ball milling tool used for polishing hard steels. Combined with an advanced TiAlN coating (hardness greater than 3700 HV), the four-flute geometry with a full cutting edge to the ball core provides the necessary cutting efficiency, as well as high heat and wear resistance. Due to the nature of shape, this procedure requires milling with an allowance, 0.3 mm, which is typically performed using ball cutters less than 20 mm in diameter. Ball mill or ball mill is a general name for ball mill tool used to polish hard steel. By rotating a mill that holds the balls and material, hard steel or ceramic balls are used in the ball milling process to grind or polish the material. Here is a general flow of how it works: A cylindrical container known as a ball mill cylinder is used to hold both the material to be polished and steel or ceramic balls. Depending on the desired finish and hardness of steel to be polished, they can be made of hard steel, ceramic, or other materials. The balls move and interact with the material as a result of the rotation of the cylinder. The purpose of these studies is to determine whether a tribofilm has formed during tribological testing (Figure 1.26).

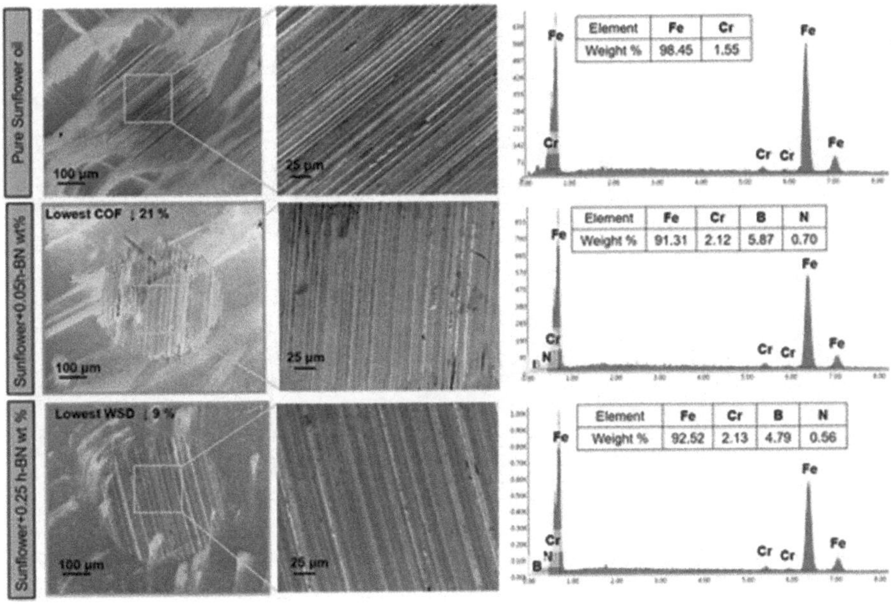

FIGURE 1.25 Wear pattern of the best-performing sunflower oil samples analyzed using SEM and EDS. ([24] [Open access].)

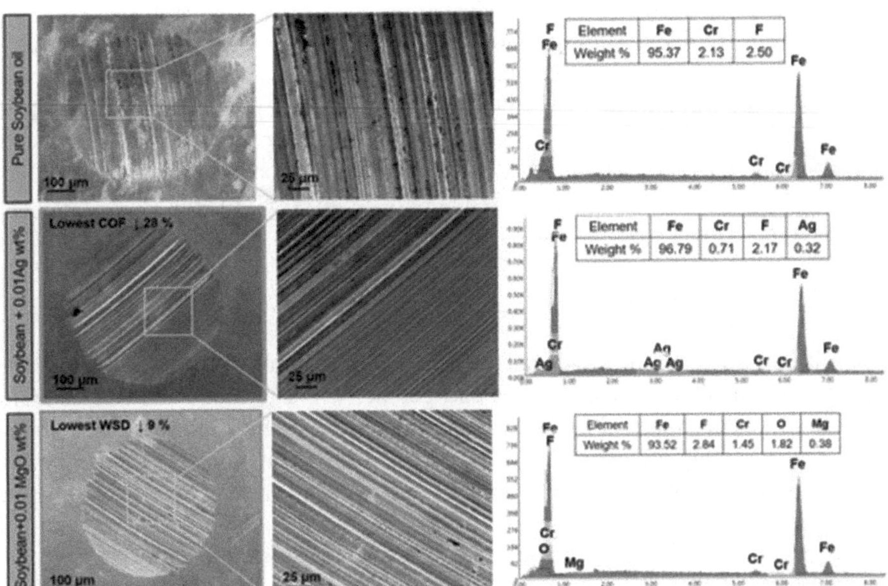

FIGURE 1.26 Wear pattern of the best-performing soybean oil samples analyzed by SEM and EDS. ([24] [Open access].)

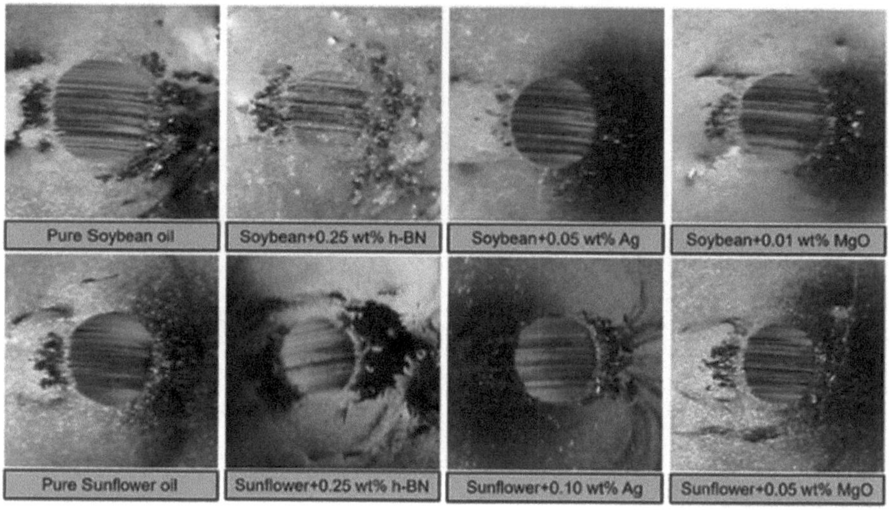

FIGURE 1.27 Microscopic images of wear marks were taken before cleaning the samples. ([24] [Open access].)

Depending on the hardness, the nanostructures can function as third bodies that are abraded when added. As can be seen in Figure 1.27, the wear marks did not show any noticeable deep grooves, indicating that h-BN is not a significant feature of abrasion. This indicates that the shear stress is reduced by h-BN nanosheets that formed on the surface as soft third-body particles.

The debris from the sliding materials is not as stable as nanostructures are obtained. From the microscopic images of wear marks taken before cleaning the samples (Figure 1.27), it can be seen that after evaluations, the debris did not form a transfer film or weld to the sliding surfaces. However, after the tribological process, a black residue is observed on most of samples. Oxidation or carbonization of oil has led to the formation of black sludge. When they are exposed to atmospheric oxygen or high temperatures, vegetable oils can oxidize [24].

1.6.2 Five-Axis Ball-End Milling

Five-axis ball milling is an advanced manufacturing technique that uses ball milling to precisely machine complex shapes and geometries. The tool's ability to handle complex shapes is enhanced by its five-axis capabilities, which allow for simultaneous movement along multiple axes. The spherical tip of this tool makes it ideal for shaping curved surfaces and fine details in a range of materials, such as composites, metals and polymers.

The machine allows the tool to move in five directions, including three primary linear movements along the *X*, *Y*, and *Z* axes, which represent motion in three-dimensional space: left-right (*X*-axis), forward-backward (*Y*-axis), and up-down (*Z*-axis). *A* and *B* axes are rotary movements that enable complex angles and orientations by tilting the tool and workpiece.

- Complex geometries can machine complex shapes such as molds, dies, and complex surfaces that traditional three-axis machines would find difficult or impossible to machine.
- Improved surface finish: Continuous five-axis motion reduces tool marks and improves surface quality.
- Reduced setup time: By operating multiple features in a single operation, efficiency increases and fewer setups are required.
- Increased accuracy: Tighter tolerances and more precise measurements are enabled by synchronous motion.

1.6.2.1 Applications

- Aerospace: Manufacturing complex sections of turbine blades and other aerodynamic components.
- Medical devices: Manufacturing complex parts with precise features, such as implants and surgical instruments.
- Automotive: Manufacturing custom parts, housings and intricately designed engine components.
- Prototyping: The process of quickly creating complex-shaped prototypes for verification and testing.

Five-axis ball milling is a powerful manufacturing method that improves manufacturing capabilities. It enables the creation of complex shapes with exceptional surface finish and accuracy. Due to its adaptability, it can be used in a variety of industries for a wide range of purposes. Milling can be done in five axes without cutting the tooltip. Another example provided by machine tool manufacturer StarragHeckert [1] is the Sturz machining method known as P-milling, which uses round-head tools to mill freeform surfaces, cutting production time in half thirds. Additionally, five-axis milling helps minimize tool overhang, which is always a concern when cutting deep cavities. As a result, the tool becomes more rigid, which improves machining accuracy and reduces the likelihood of tool breakage [3].

1.7 TOOL CLAMPING SYSTEMS AND HOLDER

If the best machine and cutting tool performance is to be achieved, selecting the appropriate CNC tool clamping system is an important part of tool assembly. This CNC tool clamping section has everything from the simplest HSK, ISO, BT, or SK collets to the highest precision tool holders such as hydro or shrink chucks. Starting with the basics leads to precision. For example, the ball-bearing collet nut supplied with standard tool holders performs a vital function in ensuring that the tool and collet are securely secured within the tool holder. A rigid clamping base (with additional tool measurement) and a professional tool tightening wrench (torque wrench) are the first steps in effectively setting up the tool to ensure the correct torque is transmitted. When securing the tool in the holder, a collet nut must be used. Since the combined tool tolerances associated with the collet and collet nut are eliminated, tool holders with the highest clamping tolerances (shrink or hydrodynamic) provide the longest tool life for certain tool types such as polycrystalline diamond tools or indexable carbide shank tools [1,2,6–8]. Our range of tool clamping systems is complemented by precision collets, saw blade adapters, universal adapters for drilling and boring, tool settings and measuring instruments. Our products meet the highest German-Designed Precision (GDP) quality standards and are manufactured in Germany by Deutsches Institut für Normung (DIN) and International Organization for Standardization (ISO). Our tools are installed on all renowned CNC machining centers including Homag, HolzHer, Weeke, IMA, Onsrud, SCM, Biesse, Busellato, Laguna, Masterwood, and others that typically use HSD. Tool holders generally need to be able to do the following:

- The system must be easy to install and disassemble.
- It should provide automatic tool change (ATC) as directed by CNC.
- It should have the maximum coaxial accuracy of tool-holder-spindle assembly.
- It should have the maximum rigidity throughout.
- It should transmit the maximum amount of torque from the spindle to the tool.

1.7.1 Tool holders for Turning Operations

You have looked at the different types of machines used in manufacturing, as well as the machining process. This section will focus on the holder-cutting tool interface where the machine tools and the workpiece meet. In today's industrial environment, there are many advanced machine tools available, such as CNC machines with specialized fixtures to support high-speed machining and manual control, and CNC machines. These new machines can be fast and accurate, but many of them are also very expensive. However, no matter how advanced the machine is, poor product quality will still be the result of a thoughtless choice of a cutting tool and holder. This module will teach you some basic features and other parts of turbomachinery that are common to most tool holders and what information is needed to select the right tool holder. A wide range of machine tools are used in manufacturing environments. All of these machines are unique in their way [6–8].

In addition to drilling, boring, milling, and turning, many other operations are possible. A schematic representation of machining operations is shown in Figure 1.28. On some machines, the cutting tool is fixed in place, while the workpiece moves. In other cases, the work element remains stationary, while the tool is moved. Some machines, such as various types of grinders, have moving parts, including the cutting tool and the workpiece. Besides metal cutting or "chip making", the procedures indicated that tasks such as tube bending, stamping, welding, flame cutting, punching, and even inspection require specialized equipment. Since their first use over a century ago, machine tools have become increasingly complex as processing techniques and knowledge of materials has increased. As far back as the 1600s, lathes were used to work metal and wood. As knowledge of materials, especially steels, has increased, precision machines have been developed to perform cutting and forming operations on a wide range of materials. During and after the First and Second World Wars, machines that could cut metal quickly and accurately were in demand. Since the introduction of numerical control, both the DIN 69880 tool holder and the Capto system have been developed, as shown in Figure 1.26. Notably, the DIN 69880 system introduces a VDI serrated holder interface designed for use with interchangeable tool turrets [1–3,8].

The dual interaction between the outer and inner surfaces of the tool and holder forms the basis of the Capto clamping mechanism. While the internal clamping device simplifies tool clamping and unclamping, the external surfaces, which combine polygonal and radial design, transmit torque. The system is built on a segmented expandable sleeve. The cutting block and the clamping block are pressed together by their edges at the outer periphery of the segment, locking into the inner groove. The front ends of the segmented bushing advance toward the centerline of coupling when the tongue is in the forward position or not clamped. The outer edges and diameter are smaller. The inner groove of the cutting block and edge of bushings is separated [12]. The cutting unit is extended using a drawbar. The drawbar is in retracted position when it is clamped. The drawbar arm pushes the front ends of the segmented sleeve out and away from the centerline of the sleeve. When the cutting unit is moved into a working position, the protrusions on the outer edge of bushings fit into the inner groove of the device. Capto, now used in multitasking applications, is a solution not only for turning tools but also for milling tools (see section 1.6.2 of this chapter).

1.7.2 Tool Holders for Milling Operations

The accuracy and efficiency of the CNC machines used today can be improved with reliable tool holders. These popular tool holder shapes can be used for certain manufacturing processes or as a replacement for conventional end mill holders. We can learn more about servicing or replacing tool holders using these faulty holder indicators. Although tool holders have different designs, they use the same basic components to perform the same function. Each type of tool holder holds a tool specifically for using in CNC machining [2–4]. While some tool holders allow tools to float to avoid damage and premature wear, others have a tight grip for precise measurements. The most common types of CNC router tool holders that you can purchase for your machine are as follows:

- Chuck collette
- Shrink-fit holder
- Faucet holder
- Special tool holder for end mills

Collet chucks hold the end mill in place using a slotted collar. This allows them to easily change tools using the same chuck. Based on the required size in millimeters, the collet size is selected. A stronger, more durable connection is provided by shrink-fit holders. These holders provide better-clamping force through heating and cooling. This will not only save you the trouble of replacing tools and reusing the tool holder but will also increase the life of the tool [23]. The tool can float and adapt to minor changes in material or other circumstances, thanks to tap holders.

These faucet covers are the easiest way to replace your fixtures. End mill holders come in small diameters and varying lengths and secure the cutting tool with a set screw. End mill holders can be manufactured to suit a variety of tools and applications, just like custom tool holders. As a result, systems with dual contact surfaces—lateral and perpendicular to the tool axis—replace traditional milling holders that rely on a single lateral contact surface of a tapered shank.

1.7.2.1 HSK Tool Holders

Like other tool holders, Hohlschaftkegel, or Hollow Shank Taper (HSK), is a tool used in metal and wood milling. Most tool holders' cone (also known as the male component) has a conical shape that matches the internal (female) structure of the machine spindle receiver. The part of the machine that rotates at different speeds is attached to the gearbox that is called a spindle. The tools are secured in the part of tool holder that protrudes from the spindle. Tool holders are available in a variety of sizes and specifications, with a variety of shank and flange options. HSK is one of the most used standards for tool holders. This is a relatively new design. There is no holding handle (pull) on these tools (pin), and hollow shank makes them special. High-speed machining (HSM) is taken into account when creating this design [9–11,24,25]. The segments are placed in a hollow cup-shaped machined holder. Because of this, as spindle speed increases, the segments expand due to centrifugal force, which increases the clamping force. Compared to systems based on ISO holders, this feature provides a greater rigidity and accuracy to allow for working in harsher cutting conditions [14,17,18].

In CNC machining, HSK tool holders (Hohlschaftkegel or Hollow Shank Taper) are a type of tool-holding method. They have a special design that allows objects to be manufactured with exceptional stability and precision.

- Important design elements: HSK tool holders fit securely into the machine axis, thanks to their hollow and conical shape. This design ensures stable tool holding with minimal deflection.
- Automatic tool change: In automatic tool change systems, HSK tool holders are often used to change tools quickly and efficiently without the need for human intervention.
- High precision: Thanks to their high-precision design, HSK tool holders are ideal for applications that require precise surface finishes and tight tolerances.
- Versatility: Milling cutters, drills, and tapers are just some of tools that HSK tool holders handle, making them adaptable to a variety of manufacturing processes.

Types of HSK Tool Holders

- HSK-A: The most common type, suitable for a variety of machining tasks.
- HSK-E: Designed to provide improved balance at higher speeds and in heavy-duty machining.
- HSK-F: Used in unique situations where a flat surface is required.

Benefits

- Reduced vibration: By reducing vibration during cutting, the design extends tool life and improves workpiece quality.
- Extended tool life: Reliable clamping helps reduce tool wear and extend overall tool life.
- Quick setup: In production environments, downtime is reduced due to easy tool changes.

Applications

- High accuracy is essential in areas such as aerospace, automotive, and medical applications where precision machining is used.
- High-speed machining is suitable for applications that require little downtime and quick tool changes.

HSK tool holders are reliable and efficient modern CNC machining system that improves accuracy, reduces setup times and supports a range of cutting tools. They are a popular choice in a variety of high-production industrial environments due to their design and adaptability [9–13,25].

1.7.2.2 Other Tool-Holder Systems

The main component connecting the machine and tooling is the tool holder. Depending on the interface, each installation technology is unique. They can be mounted using legacy R8 types, VDI mounts, or HSK tool holders. The cone, flange, and collet pocket are the three separate components that make up any type of tool holder. There are two types of non-powered tools: static non-powered tools and powered (powered) non-powered tools.

- Taper: The taper of a tool holder is a conical-shaped component. When changing a tool, this component is attached to the spindle.
- Flange: The spindle and tool changer are moved by a flange which is connected to and attached to the automatic tool changer.
- Collet pocket: The pocket is fixed with a collet. This is the area where the collet inserts attach to the various collet nuts.

The cutting tool (machine) is surrounded by a variety of tool holders that hold it in place, while a variety of other cutting tools provide maximum clearance for various sizes [14,15,17–19]. There are many varieties of tool holders, which are indexed in industrial machine tool databases:

- Machine mandrels: These are driven by motors and are responsible for the rotation mechanism of machine tools.
- Lateral cutter holders: The function of these tool holders is to hold the cutting tools.

- Saw blade holders: These hold the saw blades securely in place.
- Boring heads: Used to hold various types of tools and boring bars in holders.
- Threading chucks: These hold threading tools in place and facilitate smooth threading.
- Adapter plugs: These can be adapted for different applications depending on the processing processes used.
- End mill holders: These holders essentially hold the milling tools in place throughout the entire milling process.
- The outer diameter (OD) and inner diameter (ID) of these universal holders can be used with a variety of cutting tools .
- Collet chucks: Various collets are used to hold various industrial tools.
- Milling and drilling chucks: Specially designed to accommodate tools used in drilling and millions of operations.

1.7.3 CNC Tool Clamping Systems

Depending on the precision of machining that needs to be done, numerical controlled machines whether CNC milling machines, machining centers, or lathes use different tools. Since different types of tools are designed for specific purposes, precise machining can be done in minimal time [19]. Additionally, different tools have different connections and therefore different fittings and connections to machine spindle. This study discusses various tool couplings and tightening methods. Therefore, to ensure accurate machining, at least three components are required, a competent CNC machine, high-quality diamond cutting tool, and an accurate, fast, and reliable clamping mechanism. Since this is true for all industries, CNC machines used to process stone, glass, wood, metal, or plastic also fall into this category. The following are four of the most popular families of tool fixtures, along with some of their outstanding features:

- Straight Shank Tool ER Push-in Connector
- Threaded connection for M10, M12, and M14 tools or ½″ gas tools
- Use a flange connection to secure wheels and flanged tools or extensions
- Wheel connection profiling

1.7.3.1 Tool Clamping Systems of Collet Chuck

The most popular method involves inserting the tool into a segmented collet located inside the holder. ER collet chucks are widely used tool holders designed by specific standards, particularly DIN ISO 15488-B. They are primarily intended for clamping tools with cylindrical shanks using collets, ensuring precise gripping and reliable performance. The wide and adaptable clamping range of collets and shanks with different tolerances can be clamped. The delivery set includes a clamping nut. The gap between the tool flange and front of machine spindle is closed by ER collet chuck in Dual Contact version with a simultaneous cone and a flat working surface at the junction of flat working surface. According to DIN ISO 15488-B, the ER collet chuck is suitable for gripping tools with cylindrical shanks in collets. Thanks to the wide and adaptable clamping range of collets, shanks with different tolerances can be clamped.

The nut used for clamping is included in the package. The gap between the tool flange and front of machine spindle is closed by ER assembly chuck in the Dual Contact version with a simultaneous cone and a flat working surface at the junction of flat working surface [16,26–28].

1.7.3.2 Tool Clamping Systems of Heat Shrink

Heat-shrink tool holders provide affordable prices, high accuracy, and minimal runout. Unlike hydraulic systems, there are no internal systems that provide the pressure to hold the tool. Figure 1.30 shows the cross-sectional structure of coil and diagram of electromagnetic induction heating. The electromagnetic induction heating of a shrouded tool holder works based on the internal magnetic lines creating a loop through the clamping part of shrouded tool holder when the high-frequency alternating current creates an electromagnetic field through the induction coil. Then, the tool holder is rapidly heated due to the generation of induced eddy current on the inner wall and surface of the clamping part. The thermal energy has been completely converted from electrical energy. Four effects are involved in the whole energy conversion process: ring effect, proximity effect, skin effect, and eddy current effect [15,18,19,29]. The hole is slightly smaller than the tool diameter at ambient temperature. The hole of the tool body expands as cone is heated by an external heater. A heater can be as inexpensive as an air heater; however, more advanced induction heaters are used in industry. The instrument is inserted, and the holder is cooled to room temperature after the hole has been expanded. The tool is firmly fixed until the holder reaches its original dimensions. This technology provides excellent rigidity and minimal runout (values comparable to hydraulic holders). In addition, the tool is perfectly balanced because it does not require additional parts such as screws or nuts to secure it. However, every change in tool diameter requires a tool holder, which may result in additional costs [29].

PROBLEMS

1. Describe the engineering and manufacturing of cutting tools.
2. Mention several applications of milling tools.
3. Determine the carving surfaces using ball milling cutters in machining.
4. Describe the tool-holding systems in a lathe.
5. Explain the tool-holding systems in heat shrink.

REFERENCES

1. Botak, Z., Pisačić, K., Horvat, M., & Tomić, T. (2024). Determination of optimal machining parameters based on roughness and vibration measurements of pieces produced by whirling on a lathe machine. *Machines*, 12(5), 328.
2. Huang, Y. W. & Yeh, S. S. (2020). Development of insert condition classification system for CNC lathes using power spectral density distribution of accelerometer vibration signals. *Sensors*, 20(20), 5907.
3. Grigoriev, S. N., Volosova, M. A., Fedorov, S. V., Okunkova, A. A., Pivkin, P. M., Peretyagin, P. Y., & Ershov, A. (2021). Development of dlc-coated solid sialon/tin ceramic end mills for nickel alloy machining: Problems and prospects. *Coatings*, 11(5), 532.

4. ISO, I. (1989). 8688-2: 1989_Tool life testing in milling—Part 2: End milling. *International Organization for Standardization. ISO.* ISO 8688-1-1989: Tool life testing in milling.
5. Silva, F. J., Martinho, R. P., Magalhães, L. L., Fernandes, F., Sales-Contini, R. C., Durão, L. M., … & Sousa, V. F. (2024). A comparative study of different milling strategies on productivity, tool wear, surface roughness, and vibration. *J. Manuf. Mater. Process.*, 8(3), 115.
6. Zhou, L., Wang, Y., & Lv, D. (2024). Research on cutting edge form factor of milling tool after drag finishing preparation based on discrete element method. *Machines*, 12(4), 236.
7. Denkena, B., Krödel-Worbes, A., Beblein, S., & Hein, M. (2021). Influence of end mill manufacturing on cutting edge quality and wear behavior. *J. Manuf. Mater. Process.*, 5(3), 77.
8. Storchak, M., Stehle, T., & Möhring, H. C. (2022). Determination of the shear angle in the orthogonal cutting process. *J. Manuf. Mater. Process.*, 6(6), 132.
9. Wu, X., Li, C., Sun, S., Tong, R., & Li, Q. (2019). A study on the heating method and implementation of a shrink-fit tool holder. *Energies*, 12(18), 3416.
10. Fan, W., Zheng, L., Ji, W., Xu, X., Wang, L., Lu, Y., & Zhao, X. (2020). Function block-based closed-loop adaptive machining for assembly interfaces of large-scale aircraft components. *Robot. Comput. Integr. Manuf.*, 66, 101994.
11. Scott, C. T., & Daily, M. (1988). The development of a microcomputer CAD/CAM and decision support system for the manufacturing of aerospace vehicles. *Comput. Ind. Eng.*, 15, 14–23.
12. Zhang, P., Zhang, S., Zhang, J., Sun, Y., Zhou, H., & Yue, X. (2024). Influence of milling parameters on the microstructural evolution mechanism of 7075-T6 aluminum alloy. *Vacuum*, 222, 113100.
13. Ping, Z., Xiujie, Y., Penghao, W., & Xiao, Y. (2021). Surface integrity and tool wear mechanism of 7050-T7451 aluminum alloy under dry cutting. *Vacuum*, 184, 109886.
14. Lazoglu, I. (2003). Sculpture surface machining: A generalized model of ball-end milling force system. *Int. J. Mach. Tool Manuf.*, 43, 453–462.
15. Fernandes, G. H. N., Lopes, G. H. F., Barbosa, L. M. Q., Martins, P. S., & Machado, Á. R. (2021). Wear mechanisms of diamond-like carbon coated tools in tapping of AA6351 T6 aluminium alloy. *Procedia Manuf.*, 53, 293–298.
16. Geng, D., Sun, Z., Liu, Y., Liu, L., Ying, E., Cai, J., … & Zhang, D. (2024). Unravelling the influence of vibration on material removal and microstructure evolution in ultrasonic transversal vibration-assisted helical milling of Ti-6Al-4V holes. *J. Mater. Process. Technol.*, 326, 118320.
17. StarragHeckert. (2007). 4/5 axis machining for complex workpieces.
18. Bleicher, F., Ramsauer, C. M., Oswald, R., Leder, N., & Schoerghofer, P. (2020). Method for determining edge chipping in milling based on tool holder vibration measurements. *CIRP Ann.*, 69, 101–104.
19. Gao, S., Zheng, M., Zhu, J., & Chen, J. (2021). Mechanism of adhesion failure during interrupted cutting with cemented carbide tools: Experimental and ab-initio perspective. *Int. J. Refract. Met. Hard Mater.*, 98, 105549.
20. Abebe, R. & Gopal, M. (2023). Exploring the effects of vibration on surface roughness during CNC face milling on aluminum 6061-T6 using sound chatter. *Mater. Proc.*, 90, 43–49.
21. Montassar, F., Frédéric, M., Johanna, S., & Walter, R. (2021). Cutting parameters and tool geometry selection for plunge milling—Analysis of cutting forces at the bottom of deep titanium workpieces. *J. Manuf. Process.*, 62, 491–500.
22. Zheng, T., Song, Q., Du, Y., & Liu, Z. (2022). Development of tool wear standards and wear mechanism for micro milling Ti-6Al-4V alloy. *Metals*, 12(5), 726.

23. Swan, R., Penney, J., Corson, G., Nazario, J., & Schmitz, T. (2024). Surface location error in robotic milling: Effect of combined low frequency and high frequency vibration modes. *CIRP J. Manuf. Sci. Technol.*, 49, 203–215.
24. Granja, V., Jogesh, K., Taha-Tijerina, J., & Higgs III, C. F. (2024). Tribological properties of h-BN, Ag and MgO nanostructures as lubricant additives in vegetable oils. *Lubricants*, 12(3), 66.
25. Kechagias, J. (2024). Multiparameter signal-to-noise ratio optimization for end milling cutting conditions of aluminium alloy 5083. *Int. J. Adv. Manuf. Technol.*, 132, 4979–4988.
26. Martinho, R. P., Silva, F. J. G., Martins, C., & Lopes, H. (2019). Comparative study of PVD and CVD cutting tools performance in milling of duplex stainless steel. *Int. J. Adv. Manuf. Technol.*, 102, 2423–2439.
27. Silva, F. J. G., Sebbe, N. P. V., Costa, R. D. F. S., Pedroso, A. F. V., Sales-Contini, R. C. N., Barbosa, M. L. S., & Martinho, R. P. M. (2024). Investigations on the surface integrity and wear mechanisms of TiAlYN-coated tools in Inconel 718 milling operations. *Materials*, 17, 443.
28. Nogueira, F. R., Pedroso, A. F. V., Silva, F. J. G., Campilho, R. D. S. G., Sales-Contini, R. C. M., Sebbe, N. P. V., & Casais, R. C. B. (2024). A comparative study on the wear mechanisms of uncoated and TiAlTaN-coated tools used in machining AMPCO® alloy. *Coatings*, 14, 4.
29. Sebbe, N. P. V., Fernandes, F., Silva, F. J. G., Pedroso, A. F. V., Sales-Contini, R. C. S., Barbosa, M. L. S., … & Magalhães, L. L. (2024). Wear behavior of TiAlVN coated tools in milling operations of Inconel 718. *Coatings*, 14, 311.

2 Advanced Cutting Tools

2.1 MATERIALS FOR MANUFACTURING CUTTING TOOLS

Each material has advantages and disadvantages; the specific application and machining parameters determine the best option. Advances in materials science continue to improve the functionality of cutting tools, increasing their durability and efficiency. Evaluation of a specific application requires specific properties, and cost-effectiveness is important when selecting the appropriate material for a cutting tool. Improvements in materials technology continue to drive innovation and performance in cutting tools, increasing the versatility and efficiency of tools for a wide range of machining applications. Contact machining includes operations such as drilling, milling, and turning of components subject to severe stress, temperature, and/or corrosion; several cutting tool materials are used. Cutting tool life is limited, especially for single-point cutting tools, due to extremely high strain rates (several orders of magnitude higher than forming) caused by these harsh conditions [1]. The most commonly used cutting tool material is tungsten carbide, but unfortunately, the main components, W and Co, are very vulnerable to change in material supply and are classified as critical raw materials (CRM) for the European Union, for which sustainable use considerations must be ensured. The present study focuses on the trends in development and using CRM in cutting tool machining through an operational review. It should be emphasized that the purpose is to raise awareness and open the door to creative thinking, and not to serve as a comprehensive reference book on machining cutting tools.

In the friction test, a TiAlN-coated tool takes 102 seconds to reach a steady state. Meanwhile, the uncoated carbide tool takes 150 seconds to reach a steady state. Due to its rake surface roughness being higher than that of the carbide tool, the TiAlN-coated tool reaches a steady state faster. Figure 2.1 shows the surface roughness of two types of cutting tools and rake surface topography. On the rake surface of the uncoated carbide tool, asperities and grooves are formed for the first time. On the rake surface of the TiAlN-coated tool, a few particles and pits are appeared at the same time. Figure 2.1c and d illustrates the surface roughness of uncoated carbide tool and coated TiAlN tool, which are 0.432 and 0.361, respectively [1].

The properties, applications, and benefits of materials used in cutting tool manufacturing are as follows:

1. High-speed steel (HSS): Chemical composition carbon, tungsten, molybdenum, chromium, and iron make up the majority.
   - Toughness: Excellent toughness allows for deformation under stress without breaking.
   - Heat resistance: Maintains hardness up to 600°C (1112°F) or very high temperatures.

DOI: 10.1201/9781003604648-2

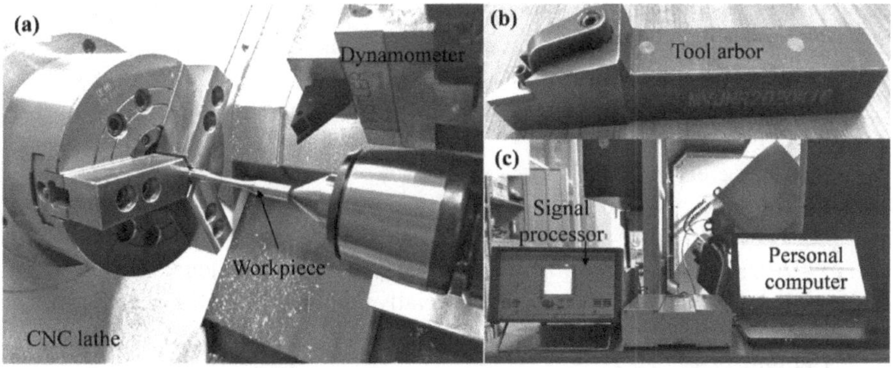

FIGURE 2.1 Topographies of tool rake surface: (a) surface topography of the carbide tool, (b) surface topography of the TiAlN-coated tool, (c) surface roughness of the carbide tool, and (d) surface roughness of the TiAlN-coated tool. ([1] [Open access].).

- Applications: Used for milling cutters, saw blades, and drill bits.
- Ideal for applications that require adaptability and versatility.

2. Carbide types:
   - The most popular type, tungsten carbide, is renowned for its hardness and wear resistance.
   - Ceramic carbide: Provides improved heat resistance.
   - Hardness: Extremely hard, two to three times harder than HSS.
   - More brittle than HSS, which can lead to chipping (CH) in severe conditions.
   - Applications: Often used for trimming inserts and end turning tools as well as milling cutters.
   - Preferred in production and high-speed machining applications.
3. Cermets composition: Combination of metals (often titanium or nickel) and ceramic elements.
   - Balance of qualities: Provides good wear resistance and hardness.
   - Heat resistance: Able to withstand high temperatures and cutting speeds.
   - Applications: Often used in milling and turning processes, especially where a fine surface finish is required.
   - Useful for machining hardened steels.
4. Ceramic types:
   - Aluminum oxide-based: Often used with cutting tools.
   - Silicon nitride: Renowned for its toughness.
   - Features: Exceptionally hard, meaning they resist abrasion.
   - Heat resistance: Performs well in high-temperature applications.
   - Application: Used for machining hard materials at high speeds.
   - Ideal for finishing applications where precision is critical.
5. Polycrystalline diamond (PCD):
   - Characteristics: hardness: One of hardest materials on the market, withstands abrasive machining of materials.

- Chemical inertness: Able to withstand wear from heat and abrasion.
- Application: Excellent for composites, non-ferrous metals, and other difficult-to-machine materials.
- Often used in plastics and woodworking cutting equipment.

6. Cubic boron nitride (CBN):
   - Characteristics of CBN.
   - Hardness: Second only to diamond in hardness.
   - Thermal stability: Continues to perform at elevated temperatures.
   - Application: Ideal for machining superalloys and hardened steel.
   - Used in grinding operations, especially those requiring a precise surface finish.
7. Titanium coatings:
   - Characteristics of titanium coatings.
   - Surface hardness: Increases wear resistance.
   - Friction reduction: Reduces the friction coefficient, extending the tool life.
   - Application: Often used as a coating for carbide or high-speed cutting tools.
   - Improves performance in a range of machining tasks, especially high-speed cutting.
8. Metal alloys:
   - Composition: Varies with different alloying elements (cobalt, nickel, etc.).

In addition to reducing the coefficient of friction between the chip and rake face of the cutting tool, TiAlN coating has reduced the tangential cutting forces by shortening the tool-chip contact length. Figure 2.2 shows that the tool-chip contact lengths are at different cutting speeds. On the rake face of the TiAlN-coated tool, the tool-chip contact length has decreased as cutting speed increases. On the other hand, as the cutting speed increases, on the rake face of the uncoated tool, the tool-chip contact length increases. As the cutting speed increases from 50 to 150 m/min, the tool-chip contact length for uncoated tools increases from 184 to 244 µm. However, on the other hand, for TiAlN-coated tools, the tool-chip contact length decreases from 136 to 108 µm [1].

Before getting into the key characteristics of each material, it is interesting to note the type of company that manufactures the tools. Consequently, in the modern tool market, there can be two types of companies; first, there are manufacturers of basic tool materials, which are usually large multinational corporations such as General Electric. These companies produce complete cutting tool systems that include tool holders, inserts and inline cutting tools. At the moment, these enterprises occupy 80% of the world market. Second, there are small and medium enterprises that start with calibrated rods from the material supplied by some previous companies, and then shape and size the cutting tools. This is where drilling tools, specialty tools and linear end mills, come into play. Natural markets for these businesses are highly specialized niches or unique, custom tools built with users in mind [2]. In this study, the Al_2O_3 oxide layer is formed on the rake face of the TiAlN-coated tool during the metal-cutting procedure. The fracture surface morphology of the TiAlN-coated tool

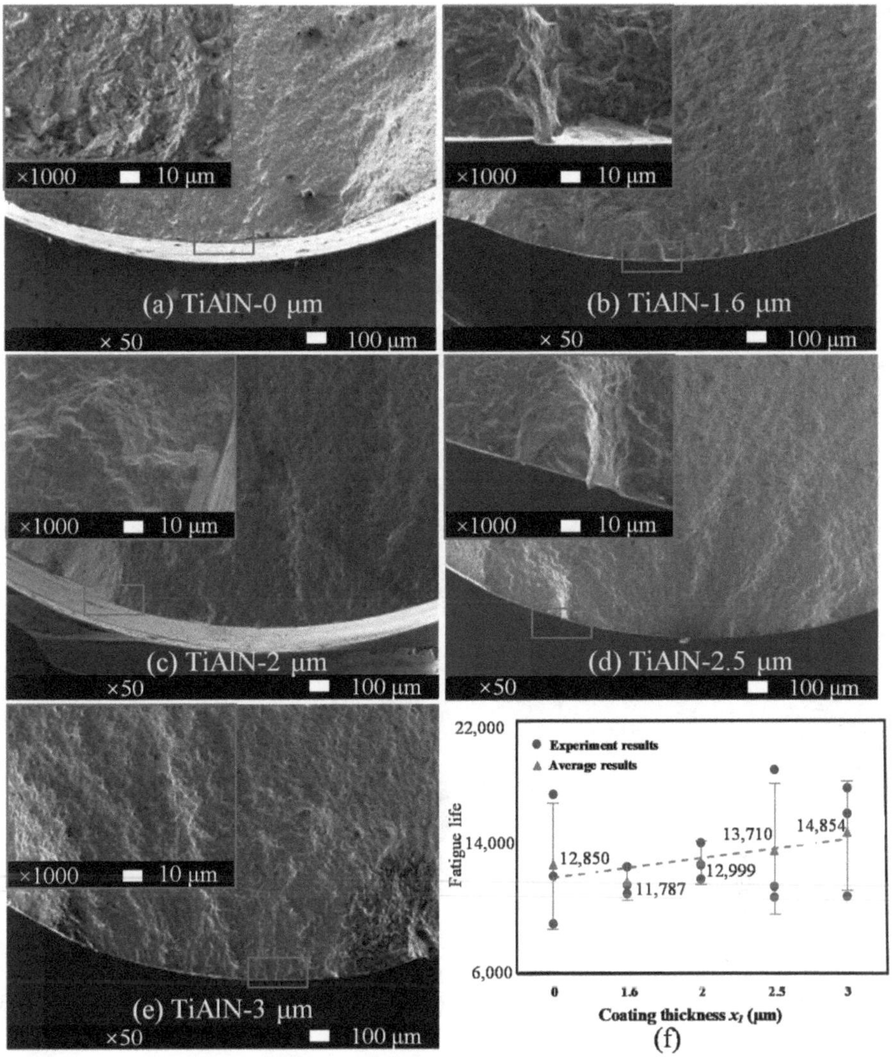

FIGURE 2.2 How cutting speed affects the tool-chip contact length on rake face of tool: (a–c) TiAlN-coated tool and (d–f) uncoated carbide tool. ([1] [Open access].)

in the tool-chip contact area at the cutting speed, 50 m/min, is shown in Figure 2.3. In addition, it shows an elemental scan result of the cross-sectional area. The Al_2O_3 oxide layer has a thickness, 0.8 μm, as shown in Figure 2.3a. According to the elemental scan result, as seen in Figure 2.3b, the oxygen distribution on the upper surface of TiAlN coating is noticeably larger than on the coating substrate. Figure 2.3c shows the atomic percentage and weight percentage of the oxygen element. These percentages have respective values of 11.43% and 17.28%. This further demonstrated the presence of oxide layer [1].

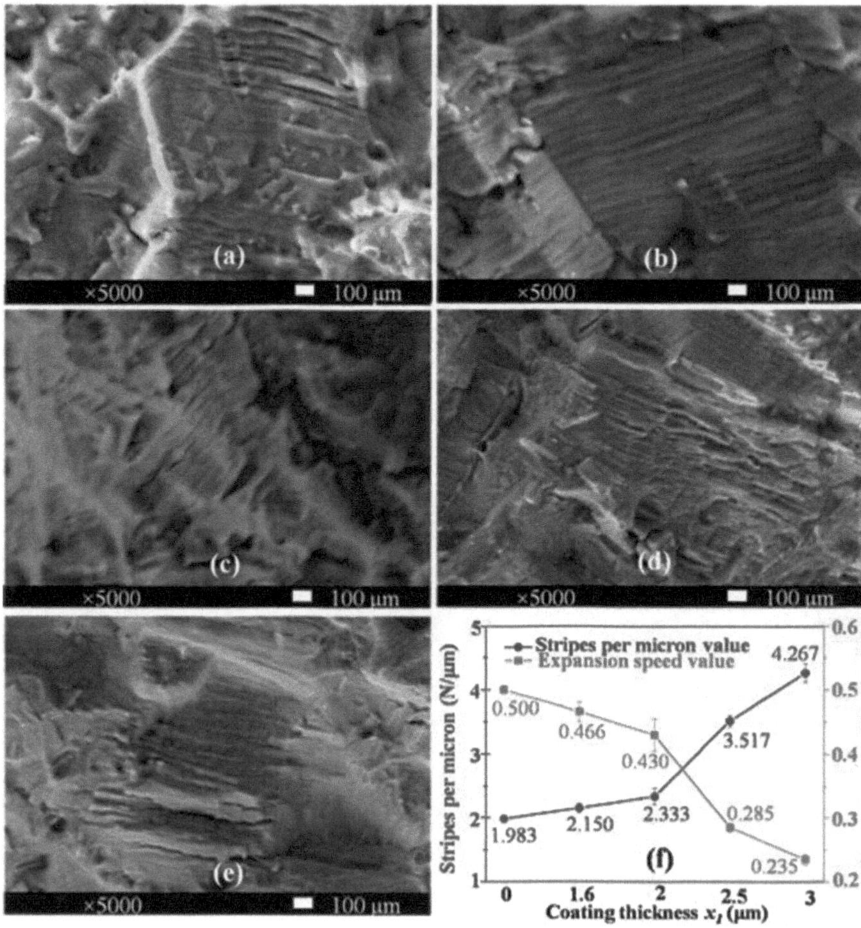

FIGURE 2.3 Elemental scanning and fracture surface morphology of the TiAlN-coated tool at a cutting speed, 50 m/min: (a) SEM image of fracture cross-section of the TiAlN-coated tool, (b) elemental oxygen scanning of the tool, and (c) weight and atomic percentage of elements. ([1] [Open access].)

2.1.1 Steel at High Speed

In general, high-speed steel is superior to other grades in terms of impact strength, hot hardness and reduced brittleness. It is also abrasion-resistant. Industrial cutting tools such as bits, cutters, saws, drills, taps, and broaches are the products that make the most use of these alloys. Compared to other carbon steels, HSS tools tend to stay sharp longer and the range of grades and surface finishes available opens up the possibility of specialized applications. These tools are used in everything from machining high-quality alloys to woodworking. HSS can be used to make punches, dies and other progressive forming components, although they are not generally considered cutting tools [1–3]. Due to their remarkable hardness and wear resistance, high-speed steels are well suited for a wide range of hand tools such as hand file blades, chisels, rasps, pocket knives, and kitchen knives. By combining additives at three different concentrations (1, 5, and 10 wt%), this study examines the composition of Madden-Julian Oscillation (MJO). The results, as shown in Figure 2.4, depict how the cutting forces and surface roughness change during minimum quantity lubrication (MQL) machining at specific feed rates, cutting speeds, and depths of cut. The corresponding tool failure scenarios are shown. According to results, MJO has lubricity properties better than other included ionic liquids, especially MJO + AIL10%, especially when it comes to viscosity. The improved performance increases its potential for using in improving tool life and working efficiency [11].

The main advantage of this type of tooling material is its flexural strength, which is directly related to toughness and is compared with the tooling materials. For all coatings, including multilayer coatings, solid lubricant coatings, TiN, TiAlN, TiCN, HSS, and high-speed steel powder metallurgy (HSS-PM). High-performance operations including milling, gear cutting, broaching, complex tapping, drilling, and reaming operations all benefit greatly from the use of HSS-PM. Rolls, knives, cold working tools, band saws, and disks use HSS-PM. However, tungsten carbide

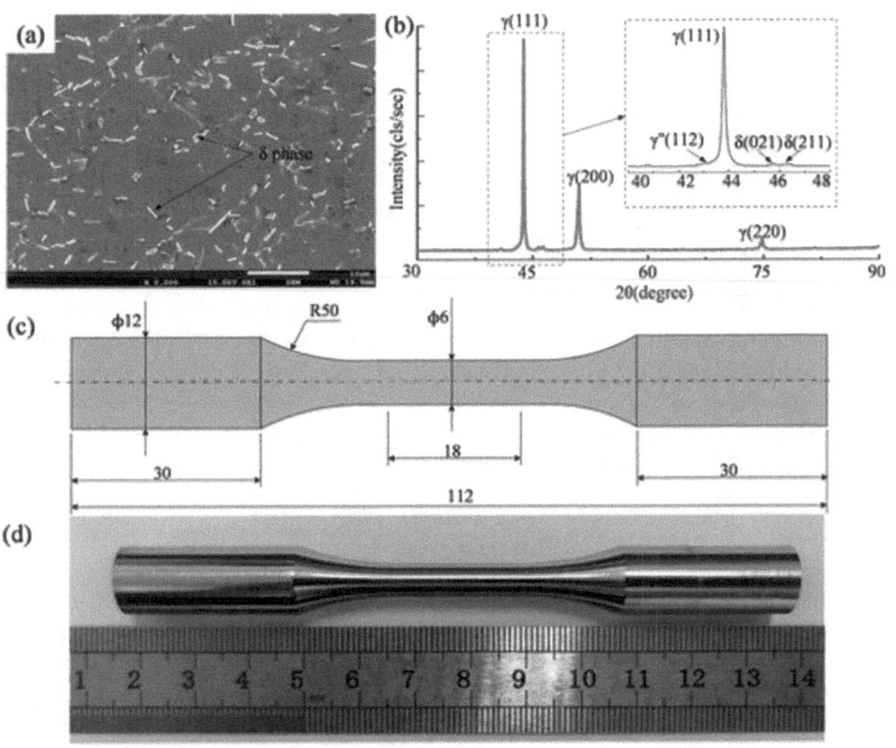

FIGURE 2.4 Typical tool failure modes in MQL turning when all specimens are lubricated. Most of the measured tool life is due to notch wear in acceptable tool wear mode according to ISO-3685 criteria. ([11] [Open access].)

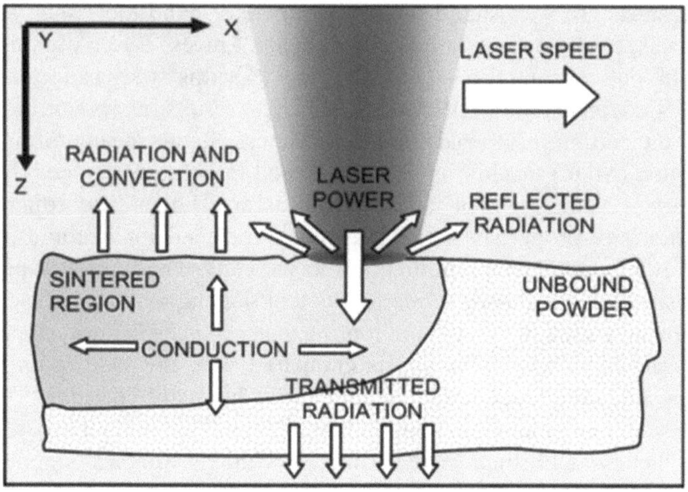

FIGURE 2.5 A comparison of internal phenomena occurring during various processing operations under different conditions. ([2] [Open access].)

is a more suitable tool material when it comes to machining hardened steels and extremely difficult-to-cut alloys (see Section 2.1.2). Numerous quality parameters are considered when evaluating a lubrication method for various machining operations. Figure 2.5 illustrates this concerning other internal phenomena. Section 2.1 shows the base oil. Section 2.2 demonstrates the base oil with bare nanoparticles. Section 2.3 illustrates the coated nanoparticle, sometimes called a hybrid nanoparticle, together with effects of all three circumstances. This figure shows that the hybrid nanoparticle and base oil machining procedure is the most effective. The process becomes much less effective when base oil is present during the same machining procedure. The latest technology and nanofluids are used in milling operations to solve the problem of thermal imbalance that occur during milling operations. Researchers have used several nanofluids in the milling process and found various improvements in the overall milling process. By preventing tool wear, these nanofluids do not improve surface quality and thermal balance but extend the tool's life [2].

2.1.2 Carbide Sinter (Hard Metal)

Tungsten carbide, often referred to as hard metal or cemented carbide, is a composite material composed primarily of tungsten carbide (WC) and metal binder, typically nickel (Ni) or cobalt (Co). This advanced material is known for its remarkable hardness, corrosion resistance and ability to continue performing at elevated temperatures and pressures, making it ideal for a wide range of machining and cutting applications. Tungsten carbide powder and metal binder are manufactured at high temperatures throughout the manufacturing process to create durable and strong material. The amount of binder affects the hardness and strength of the material, allowing it to be customized to meet specific application requirements.

In industries such as aerospace, tungsten carbide is commonly used to produce wear-resistant components, drills and cutting tools, and mining and automotive applications. Its unique combination of strength and hardness allows it to be used to efficiently machine even the hardest materials, increasing productivity and reducing tool wear. Recent advances in carbide sintering technology have improved formulations and processing methods, expanding the application range of technology and improving performance. Research continues into new bonding systems and improved material properties to improve sustainability and reduce environmental impact. In any case, carbide sintering remains an important component of modern manufacturing, driving advances in wear-resistant and cutting tool technology. Another important characteristic of solid metal is its elastic modulus. This value ranges from 500,000 to 640,000 N/mm^2 in solid metal and is 220,000 N/mm^2 in HSS. Thus, the grade of carbide is determined by two main description factors:

- Ratio of cobalt and tungsten carbide: The latter serves as a binder and usually varies from 6% to 12%. With a high melting point, 1493°C, cobalt helps minimize porosity by forming a soluble phase with tungsten carbide grains at 1275°C.
- Grain size: Finer grains correspond to harder metals; micrograined varieties contain particles less than 1 µm in size, while submicrograined varieties contain particles less than half a micron in size. Between 600 and 2100 HV, the hardness increases as binder content and tungsten carbide grain size decrease and vice versa.

Figure 2.6 shows that the use of cottonseed oil according to the MQL lubrication approach resulted in poor surface quality. When using 0.2 wt% nanofluid with the

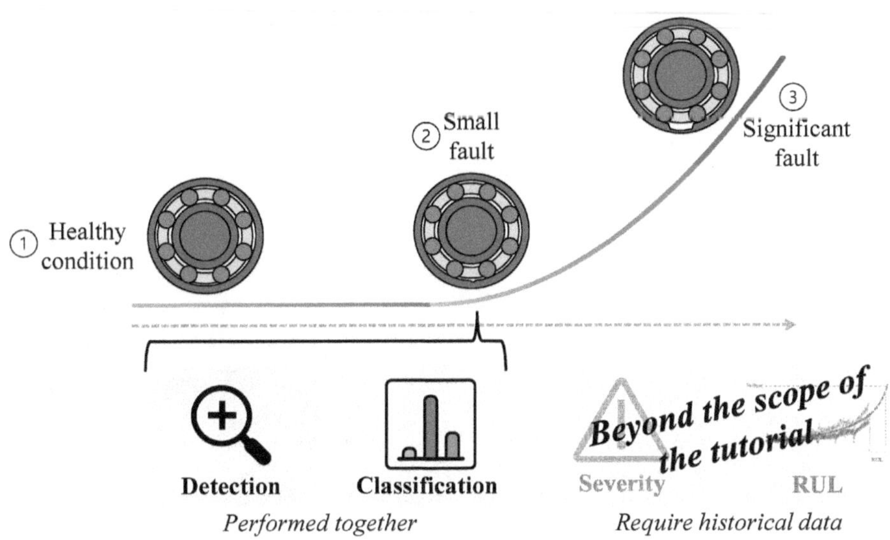

FIGURE 2.6 Surface quality of cottonseed oil. ([3] [Open access].)

same base fluid, the number of scratches is slightly less than 0 wt% nanofluid. The overall surface quality increases as the nanofluid concentration increases to 0.5 wt%. These results indicate that the nanofluid concentration has a significant effect on the overall milling process and balances the tribological properties and surface quality under excessive inclusion. There are various varieties of cemented carbide available today. Cemented carbides may also include varying amounts of tantalum carbide (TaC), niobium carbide (NbC), titanium carbide (TiC), and other carbides in addition to these basic WC–Co ratios. Cemented carbides can also be made with the cobalt binder phase completely replaced or alloyed with other metals such as nickel, chromium, iron, molybdenum, or alloys of these elements. This combination of materials makes them ideal for use as metal-cutting tools. When it comes to finishing equipment (especially those used for turning unique metals), the most commonly used steel bond or TiC-based cermet containing Ni and Ni (Mo) alloys are used. Fe–Ni iron alloys are attractive, although Ni and Ni alloys are the most commonly used binder materials for TiC cermets. Titanium dioxide-based ceramics are widely used due to their low density, good oxidation resistance, relatively high mechanical strength, and low (almost like steel) coefficient of thermal expansion [3].

The circulation times required to lower the bottom-hole temperature by 3°C for well depths of 5751 and 6251 m are 25 and 15 minutes, respectively, as shown in Figure 2.7. In theory, the flow rate should not be too high during circulation cooling, and the higher the temperature, the shorter the circulation time required to achieve the same temperature reduction [6].

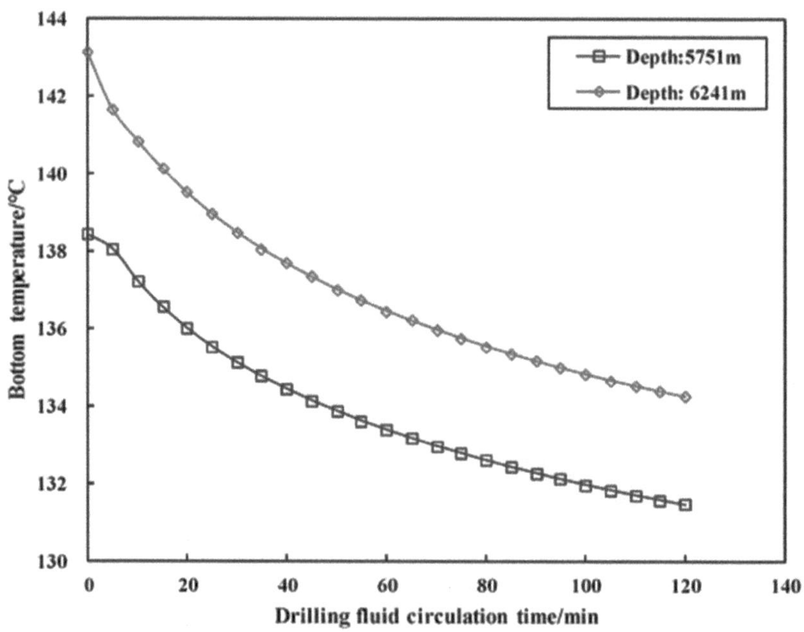

FIGURE 2.7 Relationship between circulation time and annular temperature in horizontal wells. ([6] [Open access].)

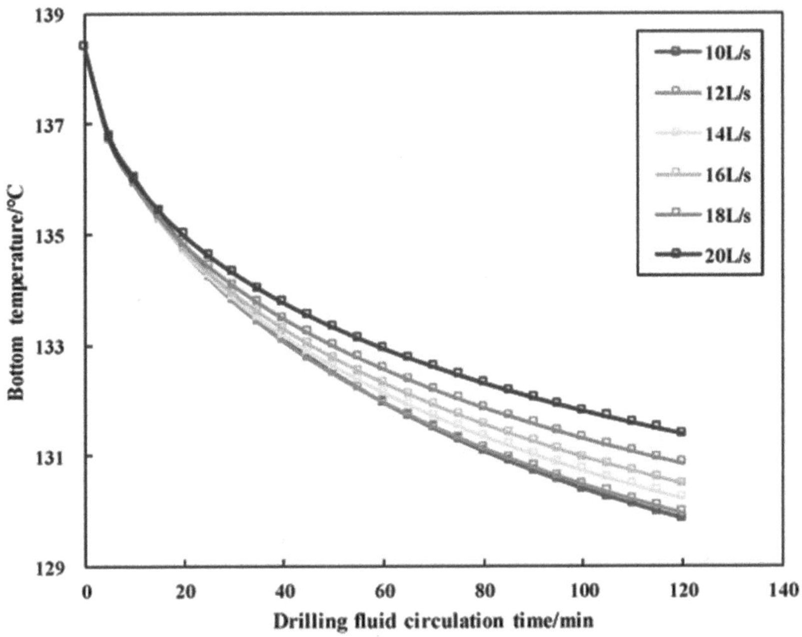

FIGURE 2.8 Variation of the bottom hole temperature with circulating time under different circulating flow rates. ([6] [Open access].)

As such, it acts as a barrier to the penetration of fluids. Thus, we conclude that drilling fluids are improved by using nanofluids. As a result, the filtration, rheology, wellbore and thermal aspects of drilling fluid are improved [4]. Figure 2.8 shows the relationship between different drilling fluid circulation flow rates and circulation times at a hole depth of 5751 m. For the same circulation time, the cooling efficiency decreases as the circulation flow rate increases [6]. This helps to improve the thermal properties of workpiece, which limits heat transfer and improves surface safety by reducing roughness and improving quality. The figure shows how applying nanofluid to the surface reduces the possibility of surface damage and increases surface roughness, both of which contribute to improving surface quality [5].

In real conditions, the interface formed by the liquid is ambiguous because plastic deformation of nanoparticles occurs there, but the overall friction in this scenario is reduced. The well temperature curve that changes with time during drilling is shown in Figure 2.9. It is clear that the well temperature shows a gradual upward trend with the increase of time. When there is no drilling, the drilling fluid is circulating [6]. Under all these circumstances, it is highly recommended to use nanofluids that improve the surface quality by reducing wear and roughness-like characteristics [6].

There are two ways to make carbide tools:

- Integrated tools are made by milling a raw carbide rod and creating a ball end mill or drill bit. The main advantage of these rotary tools is their impeccable balance; however, the main disadvantage is their high cost, given that

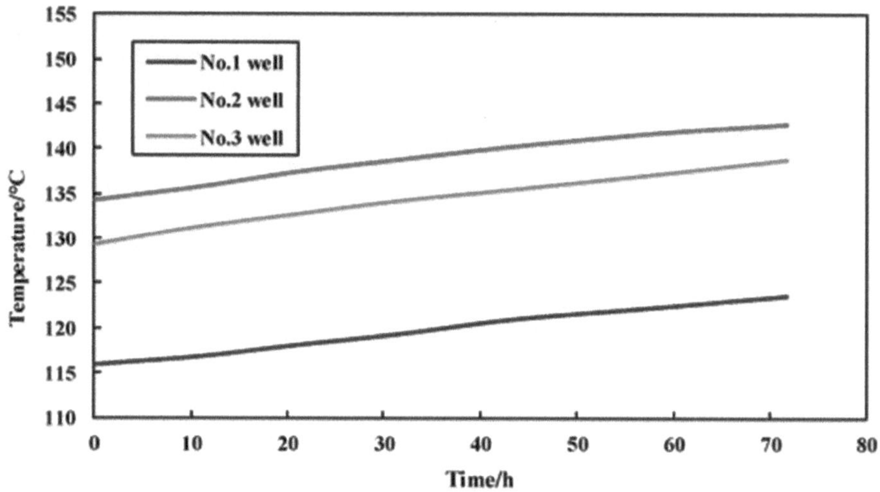

FIGURE 2.9 Relationship between drilling time and wellbore temperature in horizontal wells. ([6] [Open access].)

only a small and very precise part of the tool is worn during the cutting process. Each tool can be sharpened several times.

- Inserts: Tiny carbide inserts with unique geometries attached to steel tool holders. This arrangement is found in turning tools and large milling wheels and allows for rapid replacement of worn components.

The most common solid phase is tungsten carbide (WC), and most common binder phase is cobalt (Co) alloy. The basis of carbide structures and grades consists of these two components. Titanium carbide ceramics with a steel binder are used as blank, die materials and materials for high-cycle fatigue stamping. The ability of this material to change its composition to achieve the best physical and chemical properties, guaranteeing maximum resistance to oxidation, wear, deformation, degradation and corrosion, is one of its key features. Considering the cost of this material and its low tensile strength, it is typically applied in the form of ferrules that are attached or soldered to a steel shank. Typically, clamped attachments are used as disposable inserts. The solutions considered are organized based on the type of tool and workpiece material tested during machining. Figure 2.10 shows examples of AM cutting tools with customized coolant supplies.

2.1.3 Ceramics

Over the past 20 years, transparent ceramics have received much interest as an advanced optical ceramic material. Transparent ceramics have recently found a wide range of applications, influencing our daily lives in all areas including laser media, phosphors, scintillators, armored windows, infrared domes and electro-optical components [1–3]. Some of these applications are extremely complex. They are known

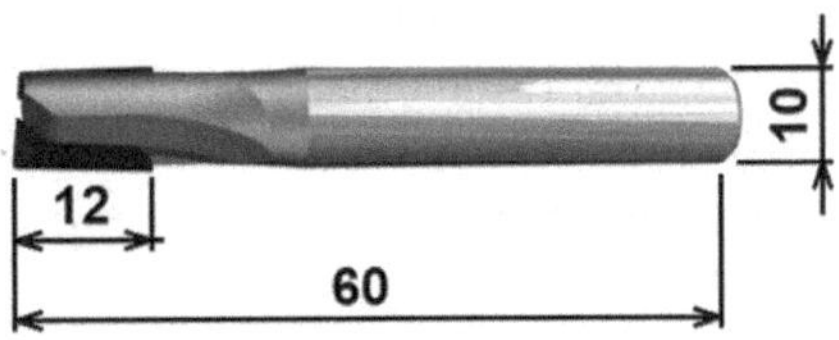

FIGURE 2.10 Cutting tools manufactured using additive manufacturing and customized coolant supply: (a) grooving tool with five nozzles; (b) drill with triangular channels; (c) threading tool with elliptical nozzles; (d) disk cutter with two nozzles, helical cutter with elliptical, and L-shaped nozzles; and (f) milling tool with L-shaped and round nozzles. ([7] [Open access].)

for their exceptional mechanical properties and optical clarity. Moreover, their use in photonics has gained prominence, especially when doped with rare earth elements. Ceramics are advanced materials characterized by their remarkable hardness, excellent temperature stability and resistance to wear and tear. Because they perform well at high temperatures and harsh operating conditions, ceramic materials such as alumina, silicon nitride and zirconia are increasingly being used in cutting tools. To achieve high density and structural integrity, fine ceramic powders are commonly used to create ceramic cutting tools. These techniques involve pressing and sintering. These tools can efficiently process hard and abrasive materials due to their high hardness compared to typical materials such as HSS and even carbide. Their chemical inertness reduces the likelihood of tool degradation and their low thermal conductivity helps maintain the integrity of cutting edge.

Ceramic cutting tools have a wide range of applications and used in a variety of industries, including manufacturing, automotive and aerospace for high-speed machining, finishing, and processing of materials such as composites and hardened steel. Improvements in ceramic compositions and production processes have increased their durability and reliability despite their brittleness. The goal of research into ceramic cutting tools is to increase their performance by creating new composite materials and coatings that will increase their utility and application. Ceramics represent a major improvement in cutting tool technology, enabling higher efficiency, longer tool life and superior machining outputs, as the need for high-performance machining grows. Transmission electron microscopy (TEM) characterization of nanopowder morphology and particle size of studied materials could be determined using high-resolution TEM images. Microphotographs of carbon monoxide (COM), commercial hydroxyapatite (HA), HA, and laboratory HA nanopowders reveal significant variations in size, shape, and uniformity. This small class of materials includes garnets ($Y_3Al_5O_{12}$ and $Lu_3Al_5O_{12}$), sesquioxides (Y_2O_3, Sc_2O_3, and Lu_2O_3), spinels ($MgAl_2O_4$), and fluorides (CaF_2). Hydroxyapatite (HA) has crystallites ranging in size from 20 to 80 nm and in various shapes such as spheres, rectangles and ovals. The image obtained at low magnification shows a high degree of homogeneity of material, clear visibility of grain boundaries and good separation of crystallites. However, there are also natural agglomerates that resemble spheres with a diameter of 200–500 nm. The shape of this nanopowder is described in detail in micrographs.

Tests are carried out on other commercially available powders to ensure that these agglomerates are not a fluke. The results that are achieved are the same. Energy dispersive spectroscopy (EDS) elemental analysis, which is not included here, was carried out but failed to reveal any differences in the chemical composition of larger and smaller grains. Correct conclusions are drawn about the presence of these large agglomerates in these powders that are sold commercially. However, the resulting crystallites have a clear structure and form uniform rods ranging in size from 20 to 100 nm. A more thorough study of particle size distribution was carried out. HA powder consists of particles with an average size of 40 nm, as can be seen in histograms of both types of powder (Figure 2.11). The aggregate size range is 175, 225, and 275 nm (indicated by red arrows).

Regarding LAB granules, the granular appearance (GA) reveals that the average size of the rod-shaped particles is approximately 40–55 nm in length and 19 nm in width. Without a doubt, linear alkylbenzene hydroxyapatite (LAB.HA) nanopowder demonstrates greater homogeneity [8].

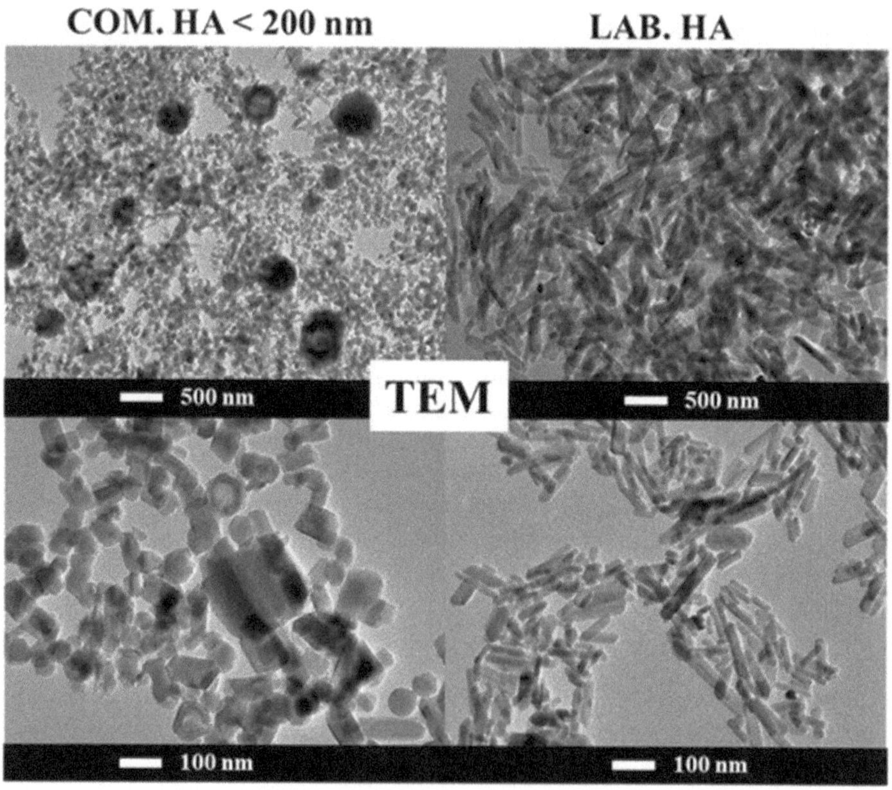

FIGURE 2.11 TEM of LAB and COM. Enhancing transparency in non-cubic calcium phosphate ceramics: effect of starting powder, LiF doping and spark plasma sintering parameters. ([8] [Open access].)

2.1.4 Hard Materials—Extra

PCD and polycrystalline CBN (PCBN) are particularly hard materials. The PCD and PCBN categories have multiple grades. PCD is appropriate for instruments designed to machine abrasive non-ferrous metals, polymers and composite materials. The hardest substance on Earth is diamond. The next hardest materials are ceramics and sapphire. In many industries, involving cutting and grinding tools, and bearings and machine parts, high hardness is a desirable quality. Hard materials are used in jewelry and watches because they can be polished to a stunning shine. Because sapphire and ceramics have good resistance to heat and corrosion, they are used in the production of crucibles, heater parts, medical equipment, and other related products. Industrial furnace viewing windows are made of sapphire, which is transparent, durable, and extremely heat-resistant. The term "cemented carbides" refers to nine different types of metal carbides: niobium (Nb), tungsten (W), zirconium (Zr), hafnium (Hf), vanadium (V), chromium (Cr), molybdenum (Mo), titanium (Ti), and tantalum (Ta); they are bonded (sintered) with iron (Fe), nickel (Ni), and cobalt (Co). They are as durable as solid materials and offer superior flexibility, strength, and wear resistance. They retain their hardness even at elevated temperatures and have minimal coefficients of thermal expansion. These qualities make them popular for use in molds, drills, and cutting tools. In addition, they have a high specific gravity and are extremely dense like gold [9].

By penetrating the grinding wheel's pores, the nanoparticle will form a layer of nanostructures with hydrophilic characteristics and contact angles less than 90°. A greater lubricated surface area is produced as a result of the particle droplets spreading and drawing themselves to the tool–workpiece surfaces (Figure 2.12) [10].

To determine which lubrication techniques result in the least wear during feed grinding, a comparative analysis of grinding wheel wear for different lubrication techniques was performed. After removing the material from each sample under appropriate grinding conditions, the grinding wheel wear is determined, as shown in Figure 2.13 [10].

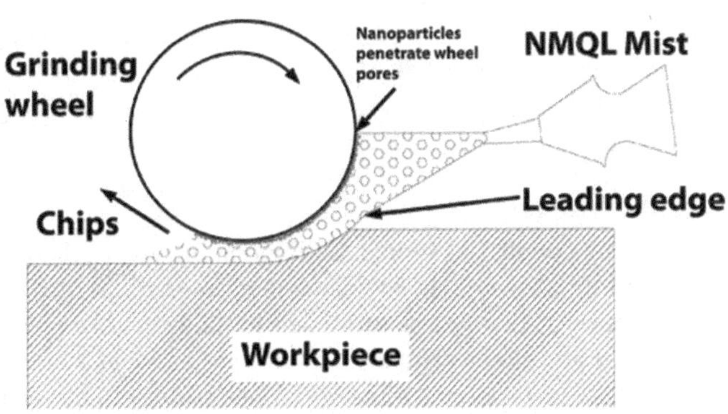

FIGURE 2.12 Schematic of nanostructures—wheel penetration while grinding operation is performed. ([10] [Open access].)

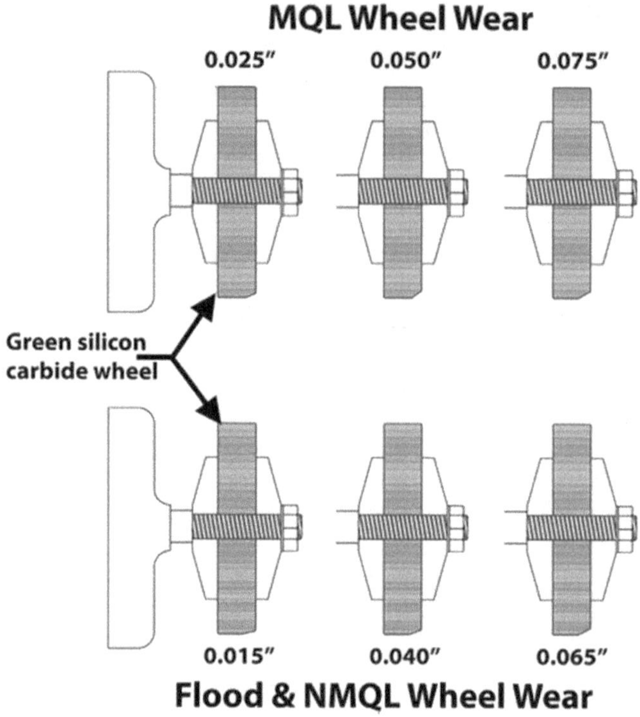

FIGURE 2.13 The wheel wear of the MQL test (top) and the wheel wear of the flood and NMQL tests (bottom). ([10] [Open access].)

This improves the overall efficiency of a certain machining process. As shown in Figure 2.14, hybrid nanofluids are generally used instead of single-component nanofluids to reduce cutting temperatures. Except for using these two types of nanoparticles with base fluids, these nanofluids are created in the same way. The composition of the hybrid nanofluid, which consists of two individual nanoparticles mixed with several base fluids that are either different or similar, is shown in Figure 2.14. After an appropriate homogenization process, they are converted into hybrid nanofluids, which are used for various applications, particularly as lubricants during machining processes [2].

2.1.4.1 PCD and Diamond

The process of sintering cobalt and synthetic diamond grains results in PCD plates, which are heated to high pressures. PCD comes in a range of grain sizes depending on the machining process. The hardest substance found in nature is diamond. As a result, it has the highest abrasion resistance. Diamond is an ideal material for cutting tools because of this special property, which also applies to synthetic diamonds such as PCDs. PCD tools are well known for their superior performance in machining non-ferrous metals such as composites and aluminum, as well as extended tool life and ability to operate at higher cutting speeds.

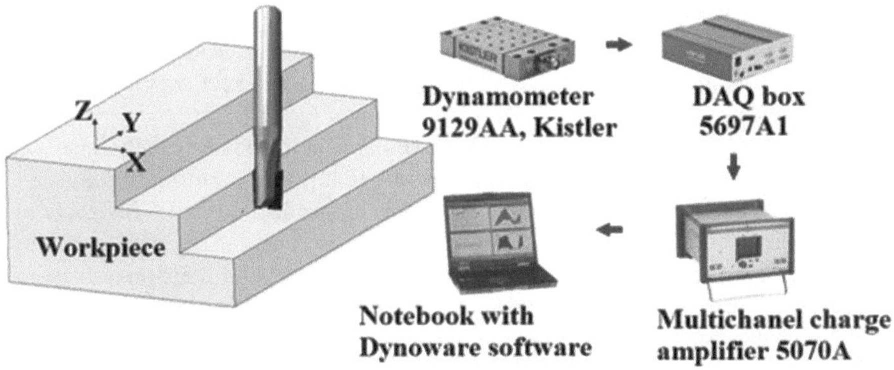

FIGURE 2.14 Potential use of hybrid nanofluids for solar, thermal, and photovoltaic energy as sustainable and renewable colloidal suspensions. ([2] [Open access].)

Randomly oriented diamond crystals bond to a tungsten carbide substrate to form PCD, an incredibly durable material. These man-made diamonds are created in laboratory using a high-temperature, high-pressure sintering method, in which the voids left by PCD grains are filled with a cobalt alloy. Additionally, because the individual diamond grains adhere to each other, the result is a very strong product that retains its shape and strength. The main advantages of PCD cutting tools include longer tool life and more reliable results. Depending on the level of production, carbide tools may have a shorter tool life of a few days or weeks. However, PCD tools have less downtime because they last for several months. With faster return on investment, fewer tool changes, high speed, and shorter cycle times, PCD tools are an excellent option for small shops to high-performance manufacturing plants. At their most productive and efficient, PCD tools can be up to 10 times faster than solid carbide tools. As a result, projects are completed faster, turnaround times are reduced, and the workload that can be completed increases.

Tighter tolerances, high time consumption, expensive and labor-intensive secondary finishing processes such as deburring and grinding can seriously hurt your profits. Compared to carbide and high-speed steel (HSS) tools, PCD tools are more accurate, have longer edge retention and produce more accurate shapes. This allows one to produce more parts faster without sacrificing quality or accuracy. Material versatility: Because PCD tools could be used with a limited range of materials, they are previously considered niche products. However, PCD tools can now be used for a wide range of materials and manufacturing processes, such as abrasive plastics, composites, and aluminum. Surface finishing tools can provide superior surface polishing without the need for post-processing. As a result, manufacturers can save money and time without sacrificing quality.

Depending on the specific requirements for cutting resources, surface finish quality, material type, and depth of cut, different classes of Principal Component Analysis (PCA) may be applied to optimize performance. For more information about our range of Polycrystalline Diamond (PCD) tools, please contact STF Precision, our trusted supplier of high-quality PCD tooling solutions.

2.1.4.2 CBN Polycrystalline

CBN is known as an industrial material with incredible hardness—second only to diamond—and outstanding thermal stability, making it an excellent choice for cutting and grinding tasks.

Recent advances in CBN technology focus on enhancing bonding techniques, creating new coatings and enhancing grain structure to improve performance. Research is ongoing to overcome sustainability issues in the production of PCBN tools and expand their uses. In any case, CBN is a breakthrough in cutting tool technology that gives manufacturers increased accuracy and efficiency in their manufacturing processes. However, since this layer is water-soluble, the use of water-based solutions results in faster wear of CBN compared to pure oil-based fluids. However, this does not preclude the successful using of CBN with water-based coolants. CBN wears five times faster than diamond when grinding aircraft-grade titanium alloys due to chemical-thermal degradation. In addition, the surface roughness and tool wear of different grades of CBN-I, CBN-II, and CBN-III inserts can be compared, as shown in Figure 2.15a–c. The erosion of CBN particles from the CBN matrix is the cause of wear of CBN-II inserts, as shown in Figure 2.15b. The metallic binder substrate or CBN-II tools have a reduced bond strength. The weaker bond between the CBN particles and binders has led to the separation of particles from the substrate [12].

Composed of closely packed boron and nitrogen atoms, CBN has special properties that make it easy to efficiently machine hard and abrasive materials such as ceramics, superalloys and hardened steels. High pressure and temperature are used during the manufacturing process to synthesize CBN, which produces a dense polycrystalline structure that improves the material's mechanical strength and wear resistance. The ability of CBN tools to withstand high cutting temperatures and speeds reduces tool wear and extends tool life—a critical advantage in high-production environments. CBN is widely used in a variety of applications, such as turning inserts and grinding wheels, especially in the tool, automotive, aerospace, milling, and decorative industries. Due to its inert chemical properties, it cannot react with the workpiece material, ensuring reliable performance and superior surface finishes. The main source of abrasive wear on the tool surface is these hard CBN particles, which have a hardness ranging from 2800 to 4200 HV. Since CBN-I has a lower thermal conductivity than other materials, more heat could be lost in the workpiece, resulting in thermal softening in the shear zone. The wear images of CBN-III tool are shown in Figure 2.15c. The increased cutting forces generated by CBN-III result in poor surface quality.

A popular commercial sintering method involves the using of high-purity micron cBN and addition of several binders (Co, Ti, and Al) [10–13]; however, the Vickers hardness can only reach 33–45 GPa [14,15]. According to recent research, starting with submicron CBN without any additives, the performance of PCBN can be significantly improved. It works well with nanomaterials when sintered at low temperatures [16,17]. They have some advantages, such as superior performance, easier extraction of raw materials and significantly lower sintering conditions. These studies show that using of cBN as a starting material has a potential to represent a significant advance in industrial production. In fact, submicron cBN has significantly smaller grains than commonly used cBN. The characteristics of samples are significantly affected by the grain size of starting materials. This is due to the pre-treatment of initial powder and

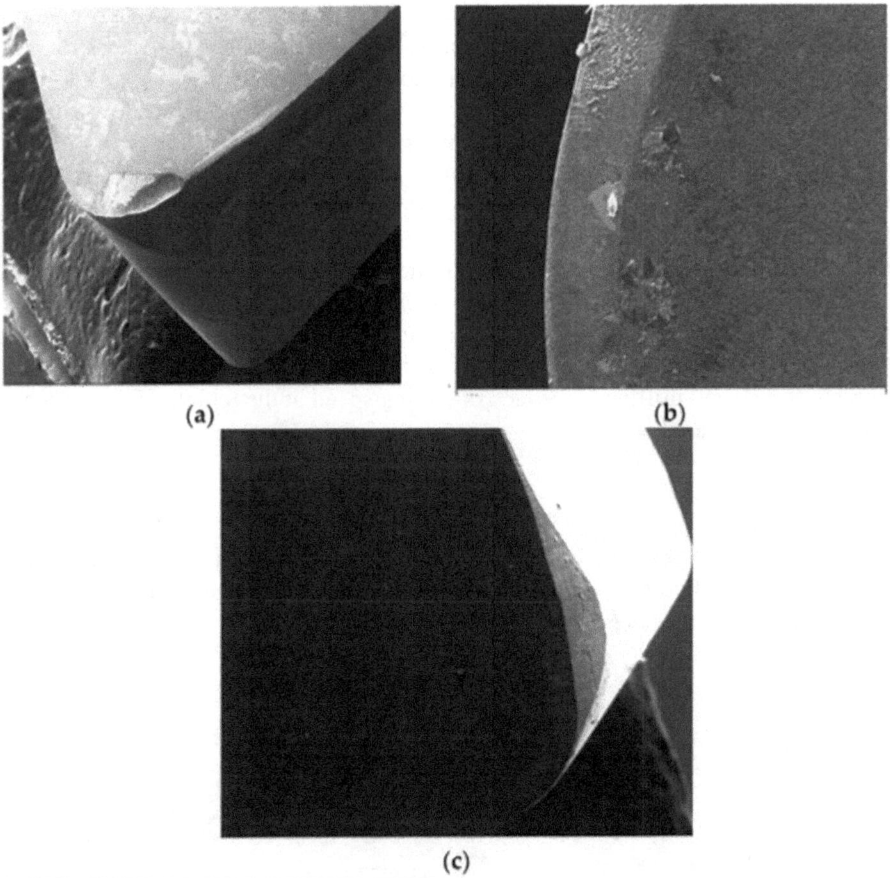

FIGURE 2.15 Abrasive arcs on rake face and VB on flank face are highlighted in tool wear images of (a) CBN-I, (b) CBN-II, and (c) CBN-III. ([12] [Open access].)

grain size of sintered samples. Therefore, it is worth exploring methods for sintering high-performance samples using micron-starting powders that do not contain additives.

Due to its incredible hardness, thermal stability, and chemical inertness, PCBN is a leading material for cutting and milling tools. Because of these properties, CBN is an invaluable tool for machining difficult materials where conventional cutting tools are inadequate, including superalloys and hardened steels. CBN is manufactured using advanced manufacturing techniques that ensure a robust, long-lasting tool that performs well at high temperatures and speeds, increasing productivity and tool life. CBN tools provide a reliable solution that meets the stringent requirements of industries, which increasingly demand higher performance and accuracy in machining applications.

The performance characteristics of CBN are being improved through ongoing research and technological developments, resulting in advances in grain structure, bonding methods and coatings. These developments expand the range of applications and enhance tool performance, making CBN a flexible choice across multiple industries, including manufacturing, automotive and aerospace.

In short, the distinctive qualities of polycrystalline boron nitride cubes reinforce their position as a cornerstone of contemporary cutting tool technology, enhancing the accuracy and efficiency of manufacturing processes while meeting the changing demands of industry. Therefore, these parameters are collected and evaluated after numerous milling tests on hardened steels to create shapes. Due to its special properties, PCBN is an ideal material for manufacturing continuous surfaces such as brake disks. Here's a closer look at why PCBN is suitable for this application. The main problem of ball end milling is the lack of cutting speed at the tool tip, which places a large mechanical load on PCBN.

2.2 COATINGS

Tool coatings can significantly extend tool life even in the absence of adequate lubrication or cooling. In addition, tool coatings continually improve the productivity of the machining process (see Figure 2.16).

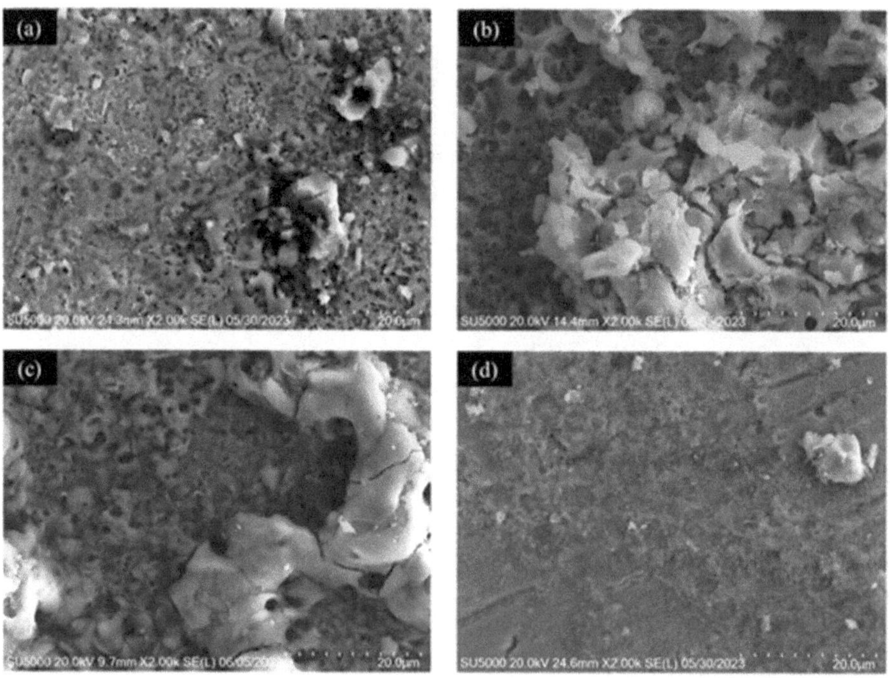

FIGURE 2.16 SEM-SE of superficial morphology of anodized in NaOH in 2000×: (a) Ti CP2, (b) Ti-6Al-2Sn-4Zr-2Mo, (c) Ti-6Al-4V, and (d) Ti Beta-C. ([13] [Open access].)

2.2.1 Physical Vapor Deposition (PVD) Coatings: A Historical Overview

The fundamental idea of PVD techniques is to grow a solid-state coating material on a base material through evaporation and/or spraying. The first known use of PVD coating dates back to 1857 when Faraday created the coating by vacuum-evaporating metal wire. Kunz has used this technique in 1888 to produce thin films, which are analyzed for their optical properties. For many years, this method is limited to physical research, but after the 1960s, it has begun to be used to treat the surfaces of cutting tools to delay tool wear.

PVD coating technologies do not obey the physical and chemical laws of hardening mechanisms. In addition, materials coated with thin films have elastic characteristics of the base substance. These factors make it possible to coat virtually any hard surface with virtually any material. PVD technique wear limits the service life of machine parts and life cycle of tools. Even as demand for some parts is increasing, downtime on production machines due to wear and tear increases costs. In this regard, the economic importance of wear delay increases every day. Improving the wear and friction characteristics of a part's surface offers a more practical and cost-effective solution to the problem than replacing older tooling and mold parts with more expensive ones. Thin film hard ceramic coatings have been very successful in this industry. The most important approach to applying hard ceramic coatings is PVD coating technology.

- Advantages of PVD coatings
- High hardness (~80–90 HRc) at 2400–4000 HV
- Increased abrasion resistance
- Low coefficient of friction and high lubricating surface
- Reduction in the thermal conductivity of the tool as a result
- Adaptability to the chemical environment
- The improved surface quality of machined parts
- Protection of workpieces from sticking to the surface of the tool mold
- Micron-level coating thickness does not result in dimensional change, and longer tool life results in lower tool cost per piece
- Reduction of downtime for workstations by reducing the frequency of tool changes
- Obtaining an even layer across the thickness of each area

Figure 2.16 shows the surface morphology of the alloys oxidized in NaOH as determined by scanning electron microscope (SEM-SE). This oxidation electrolyte is characterized by heterogeneous oxide distribution. The favorable oxide deposition is depicted in Figure 2.16b, where Ti-6Al-2Sn-4Zr-2Mo showed varying heights. In addition, there was cracking in that region of the oxidized metal. Figure 2.16c shows a similar morphology, indicating a tendency for oxide accumulation in certain areas. The decreased coating generation through oxidation for Ti Beta-C is demonstrated in Figure 2.16d [13].

To overcome this limitation, Platit has created nACRoTM, a new AlCrN coating containing silicon. This AlCrSiN-based final coating has been used effectively in drilling, milling and honing applications where high temperature and oxidation resistance of coating is required. As coating applications become more specialized than ever, it is difficult to explain current advances in hard machining coating technology in a broad sense; instead, alternative methods are being developed for similar processing procedures [10].

2.2.2 Optimizing Coating Selection for Hard Machining

A year after AlTiSiN coating was launched on the market, Oerlikon introduced a new Balinit Alcrona coating. This AlCrN coating is designed to enhance the performance of TiAlN coatings, especially in applications where high oxidation resistance is required. AlCrN coatings are comparable in hardness to TiAlN coatings, but they are distinguished by high oxidation resistance, and withstanding temperatures up to 1200°C (Figure 2.17) [14]. This can be explained by the presence of stable oxide $(Al,Cr)_2O_3$ formed during cutting, in contrast to the $TiO_2 + Al_2O_3$ oxides formed in TiAlN coatings. However, the main benefit of AlCrN coatings is limited to extending the life of carbide tools under typical cutting conditions. In most cases, carbide can be used successfully outside of harsh machining conditions. The last 5–6 years have seen remarkable growth in the machining sector, which is expected to reach USD100 billion in revenue by 2025 [11]. This is mainly because manufacturers can produce these high-quality and complex products at a faster pace thanks to computer numerical control (CNC) equipment and increased demand for higher-quality products. The six-axis CNC machine has attracted a lot of attention recently. Its ability to transform raw metal rods into a finished product makes it very suitable for the machining sector, especially for the production of extremely complex structures that can be completed completely without the need to stop [1,2].

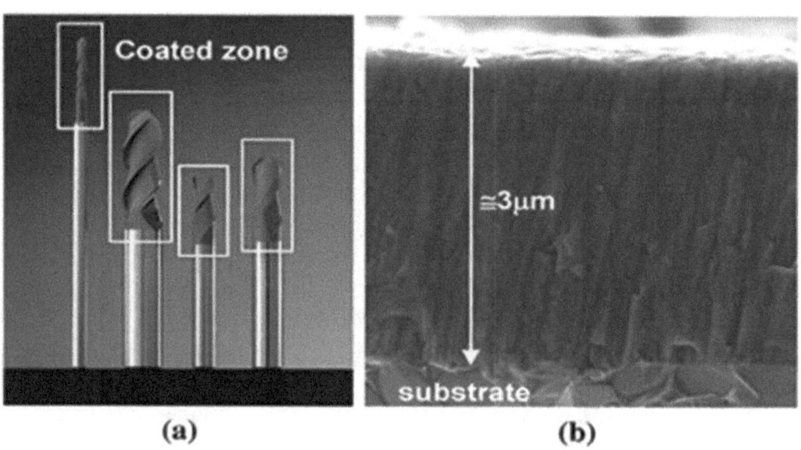

FIGURE 2.17 PVD-coated carbide tools (a) and AlTiN-coated cross-section (b). ([14] [Open access].)

2.2.3 Industrial Development of Various Compounds

The landscapes of the Boreal Forest region may undergo significant changes as a result of industrial growth and associated supporting infrastructure. Hydrological changes such as disturbances in levels, flows, and cycles can have a significant impact on emissions of organic carbon stored in peat and soils, as well as emissions of high-impact greenhouse gases, especially methane. Industrial growth, while once again releasing carbon stores that have existed for millennia, could lead to changes that accelerate the thawing of permafrost in some places. The goal of managing carbon storage in Boreal Forest biome should be to "not harm," as in medical industry. Following a precautionary measure would lead decision-makers to support keeping as much of Boreal Forest as possible free from human industrial activities that increase the likelihood of its carbon stocks being lost to the atmosphere, in addition to promoting global policies to reduce atmospheric carbon levels. Currently, 10% of Canada's Boreal Forest carbon stocks are located in protected areas. However, at least two-thirds of carbon in Canada's boreal forests is stored in intact forest areas. This means it is possible to create extremely large protected areas that include a significant portion of these vast carbon stocks within their boundaries, to protect carbon from loss or degradation as a result. Over the past few decades, industrial expansion and agricultural intensification have introduced hazardous waste into terrestrial ecosystems. Rapid and cost-effective cleanup of contaminated regions has become important due to toxic and carcinogenic properties of these substances, as well as their tendency to bioaccumulate in humans and other biological animals. The breakdown of resistant substances in soil can be influenced by rhizosphere organisms (Table 2.1). In addition, rhizosphere bacteria can help remove metal contaminants from the soil, including zinc, selenium and mercury. In the field of bioremediation, bacterial isolates and consortia have been the subject of numerous studies; however, degradability has been demonstrated by fungi, actinomycetes, mycorrhizal associations, and rhizosphere animals.

The application of plant-microbial systems in bioremediation, a technique that utilizes the inherent abilities of microbes and plants to transform, mineralize and complex organic and inorganic contaminants, has shown promising results. Compared to traditional soil remediation methods, bioremediation is less expensive, uses simpler technology and allows for minimal impact on the contaminated site. Rhizosphere microbes that mineralize pollutants into plant root exudates or plant root exudates have strong binding affinities for specific pollutants can aid in bioremediation, as can their innocuous counterparts.

Microbial communities associated with plant roots can contribute to mineralization of hazardous substances in the soil. There are various ways in which the "rhizosphere" can influence bioremediation. First, the increase in populations and activity of xenobiotic decomposers is thought to be stimulated by carbon substrates derived from roots. Second, high populations of different microorganisms in rhizosphere may promote bacterial conjugation, which may lead to emergence of new degradation pathways. Third, plants can become a substrate that enhances the synthesis of enzymes necessary for the mineralization of a dangerous substance.

The most significant obstacle in the field of bioremediation technologies is undoubtedly figuring out the processes that plants and creatures interact to clean up contaminated areas.

TABLE 2.1
Current Coatings for Tools

Coating	Color	Nanohardness (GPa)	Thickness (μm)	Friction (Fretting) Coefficient	Max. Usage Temperature (°C)
TiN TiCN	Gold	24	1–7	0.55	600
MP TiCN	Red-copper	32	1–4	0.20	400
CrN	Blue-gray	37	1–4	0.20	400
CBC	Metal-silver	18	1–7	0.30	700
AlTiN	Gray	20	0.5	0.15	400
μAlTiN	Black	38	1–4	0.70	900
TiAlCN	Black	38	1–4	0.30	900
Cromvic	Burgundy-violet	33	1–4	0.30	500
Gradvic	Gray	20	1–10	0.15	400
cVic ZrN	Gray	20/33	1.5–5	0.15	400
AlCrN	Gray	20/37	1–5	0.15	400
nACo	White-gold	20	1–4	0.40	550
nACRo	Blue-gray	32	1–4	0.60	900
nATCRo	Violet-blue	45	1–4	0.45	1200
nACo3	Blue-gray	40	1–7	0.35	1100
nACRo	Blue-gray	42	1–4	0.40	1150
nATCRo	Violet-blue	45/34	1–5	0.45	1200/900
	Blue-gray	40/34	1–5	0.35	1100/900
	Blue-gray	42/34	1–5	0.40	1150/900

Source: [13] [Open access].

Using this information, optimal combinations of plants and microbes can be selected and designed, and the restoration process can be optimized [7].

2.2.4 Current Developments in Hard Machining Coatings

The performance and durability of cutting tools in a range of industrial applications have greatly improved by recent developments in hard machining coatings. These coatings are essential for improving machining processes, particularly in high-speed machining and hard materials due to their primary design goals of improving wear resistance, reducing friction and enhancing thermal stability. Current developments focus on a range of new coating materials, including ceramic coatings, diamond-like carbon (DLC) and titanium-based coatings such as TiN, TiAlN and TiC. Tools can retain their integrity and sharpness in harsh environments, thanks to the exceptional hardness and chemical resistance of these materials. Furthermore, improvements in deposition methods such as chemical vapor deposition (CVD) and PVD have made it possible to create thinner, more uniform coatings that maximize efficiency while reducing tool weight.

To further improve the effectiveness of cutting tools, researchers are investigating multifunctional coatings that incorporate features such as self-lubrication and thermal barrier properties. By using less energy and producing less waste, these innovations aim to increase the sustainability of manufacturing processes as well as extend tool life. The development of hard manufacturing coatings has industrial demands for increased productivity and efficiency, securing their place as essential components of advanced manufacturing in the future. CVD TiAlN coatings have replaced the previous generation of hard coatings. As manufacturing processes become more efficient and sustainable, the market for hard machining coatings is changing rapidly. Cutting tools operating in harsh environments require coating materials that have undergone significant wear resistance, reduced friction and improved thermal stability. Examples of these materials include titanium-based, DLC and ceramic-based coatings.

PVD and CVD are two examples of innovative deposition processes that have made it possible to produce thinner, more uniform coatings that maximize tool performance while reducing weight. Additionally, additional features such as thermal barriers and self-lubrication are being introduced through the development of multifunctional coatings, improving overall machining efficiency and extending tool life. The industry's demands for increased efficiency and economy make it imperative to continually improve hard machining coatings. These developments not only help extend the life of cutting tools, but also promote environmentally friendly production methods by using less energy and producing less waste materials.

In brief, the continued exploration and innovation in hard machining coatings are essential to driving cutting tool technology forward, ensuring that manufacturers can effectively address the demands of today's manufacturing applications while increasing accuracy, economy, and sustainability [3]. In addition, cutters can be solid, coated or uncoated and may have inserts located along the edges of cutting blades. In most cases, these coated tools are considered for machining tasks. By extending tool life and improving the quality of machined parts, they improve overall process productivity. Moreover, when cutting materials with poor machinability, coated tools are an absolute necessity [14,15]. Combining the best qualities of the following structures constitutes the process of selecting a suitable coating structure:

- A monoblock coating or monolayer of same composition is applied in situations where exposure is low or cutting pressure is minimal.
- Double-layer coating is used when a hard coating and a top lubrication coating are required to improve chip control, to combine the benefits of an inner layer close to the base and top layer.
- Multilayer coating to prevent crack propagation through different layer materials and increased shear strength of coating.
- Adhesive layers: thin adhesion layer, with 0.05–0.2 μm thickness, is added to improve the adhesion of subsequent layer.
- Triple coating: Platit's innovative method for improving the structure of the coating, including a strong adhesive layer, durable, heat-resistant outer layer, and an inner layer.

The significant effect of cutting-edge radius on end mill life is shown in Figure 2.18. As can be seen, the ideal and no radius are significantly different. However, better edge stability can be achieved by improving the coating surface. Arc technology,

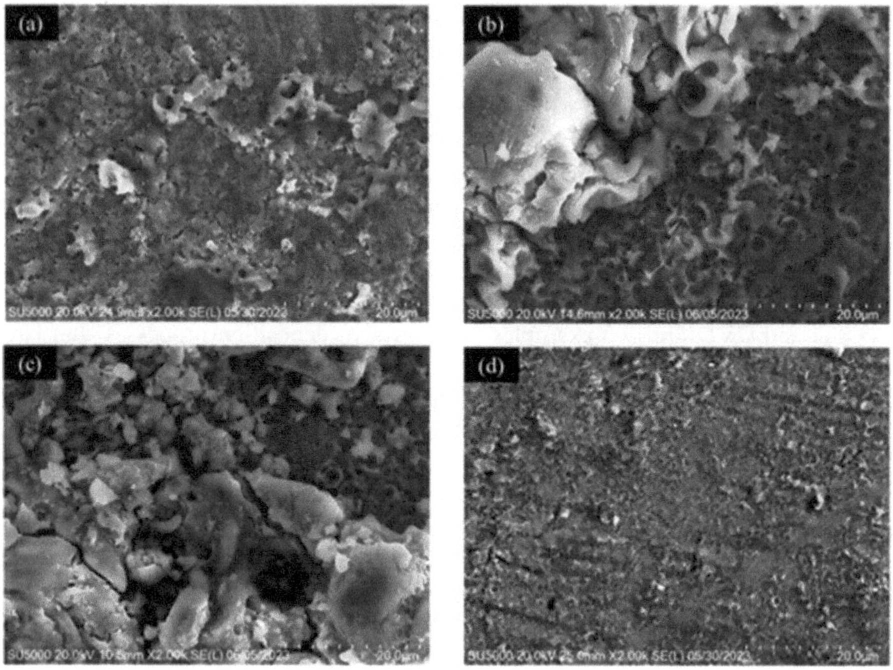

FIGURE 2.18 SEM-SE of superficial morphology of anodized in KOH in 2000×: (a) Ti CP2, (b) Ti-6Al-2Sn-4Zr-2Mo, (c) Ti-6Al-4V, and (d) Ti Beta-C. ([13] [Open access].)

which produces industrial PVD coatings is more economical and better suited to ensure consistent coating quality. The main disadvantage of this approach is the presence on surface of coating of droplets formed due to melting the target during arc burning [13].

Micrographs of the alloys oxidized in KOH are shown in Figure 2.18. The alloys oxidized in KOH showed greater variation than those oxidized in KOH; Figure 2.18a shows that the oxidized TiCP2 exhibits varying oxide heights. For Ti Beta-C, Figure 2.18 shows a similar pattern, except that the polishing lines are higher. A similar behavior is seen in Figure 2.18b and c, which are however inhomogeneous despite having higher oxidation generation [13].

Much of the roughness of coating is caused by these droplets bound to the surface of coating. Because edge roughness negatively impacts durability, drip removal procedures are typically performed on higher-quality tools. The lubricating effect of ultrasonic vibration-assisted turning (UVAT) and nanofluid minimum quantity lubrication (NMQL) processes can be further enhanced by adding diamond or graphene nanoflakes into the oil, which reduces the average surface roughness Sa compared with the CT processes. Conversely, as shown in Figure 2.19, the nanoparticles/nanofoils can be pressed into workpiece by vibrating cutting tool, creating many pits after removal. Therefore, the S_z value of UVAT and NMQL processes is higher than that of the CT and NMQL processes [16].

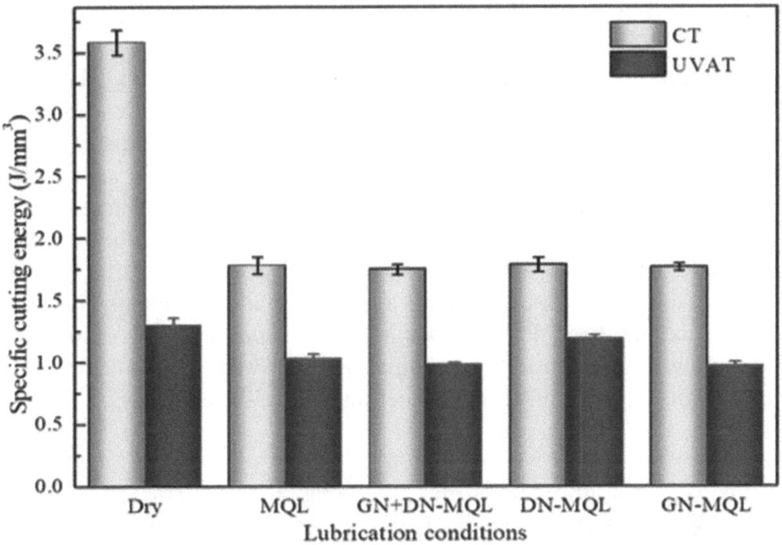

FIGURE 2.19 Surface morphology of UVAT processes under DN-MQL, GN-MQL, GN+DN-MQL, and MQL conditions is shown in 3D images (×1000). ([16] [Open access].)

2.3 WEAR OF TOOLS

Tool life, machining efficiency and surface quality are just a few of the many factors in metal cutting that are significantly affected by tool wear. Tool wear research is evolving as hot spots and trends to machining moves closer to intelligence. Tool wear significantly impacts the effectiveness, efficiency and feasibility of mechanical processes. It describes the gradual deterioration of cutting tools caused by a variety of wear mechanisms that occur during the manufacturing process, including abrasion, adhesion, fatigue and diffusion. Understanding the causes and types of tool wear is critical to increasing tool life and improving manufacturing outcomes.

Using uncoated carbide cutting tools (insert references TNMA160408 and TPUN160308, tool holder references DTFNR2020K16 and CTFPR2020-16), orthogonal cutting tests are performed in planing mode (Figure 2.20) [14]. The Ti-6Al-4V titanium workpiece has a parallelepiped shape and dimensions, $40 \times 20 \times 4\,mm^3$. The 40 mm length, which corresponded to direction X in Figure 2.20, is the cutting direction [14]. Using an Alicona InfiniteFocusSL microscope, the tool geometry is carefully measured in a tool-in-hand system (ISO 3002:1982 standard).

Identifying wear patterns and factors—such as cutting speed, feed rate, and tool material and workpiece properties—that contribute to tool wear have the subject of recent research. To facilitate preventive maintenance and reduce downtime, advanced monitoring techniques have been developed to predict tool wear in real-time. These methods include sensors and machine learning algorithms. Another key area for mitigating wear has been advances in tool coatings, materials and advances in composite materials and robust coatings that improve resistance to wear and thermal degradation. Furthermore, well-implemented lubricants and cooling techniques have been shown to reduce tool wear and enhance cutting efficiency.

All things considered, increasing productivity and maintaining the quality of manufactured products depends on effective tool wear control. Research into tool wear will be an essential to creating stronger tools and more efficient operating procedures as manufacturing technologies advance. However, there are currently no recognized best practices for tool wear suppression in tool wear research; signals are prone to disturbances, signal processing techniques are inefficient and model generalization is inadequate, among other things [14]. The largest difference between the results obtained using the algorithm and manual measurements is 4.98%. This demonstrates the improved detection accuracy of the proposed algorithm for online tool wear monitoring. Essentially, turning allows four wear mechanisms to occur:

When turning austenitic stainless steel and nickel-based alloys, notch wear is common.

- Crater wear: Metallurgical/chemical wear caused by adhesion and dispersion of fine particles from the rake surface of the tool onto the newly formed chips. An additional factor contributing to formation of scar on the rake surface,

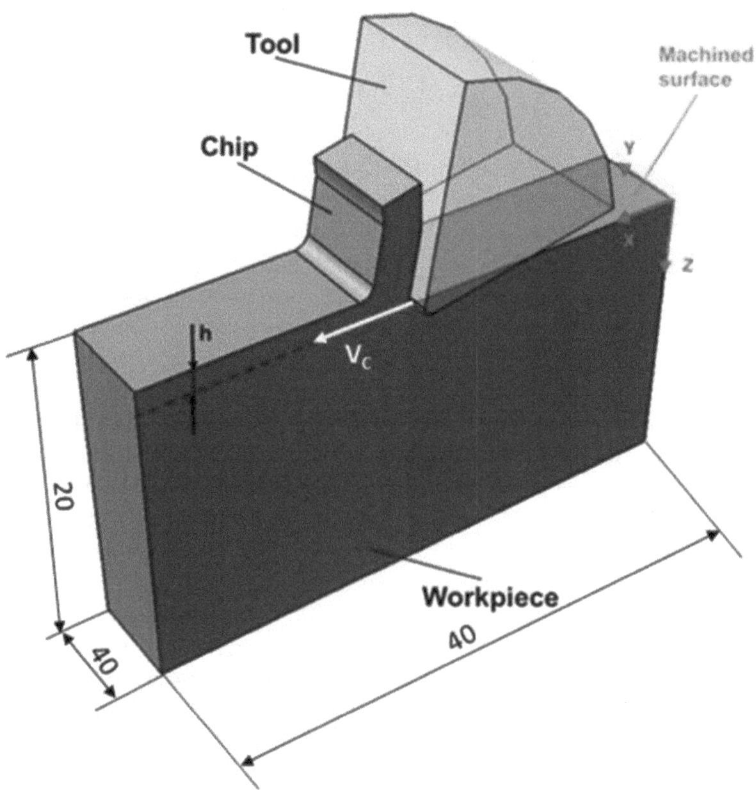

FIGURE 2.20 Schematic representation of orthogonal cutting estimates. ([14] [Open access].)

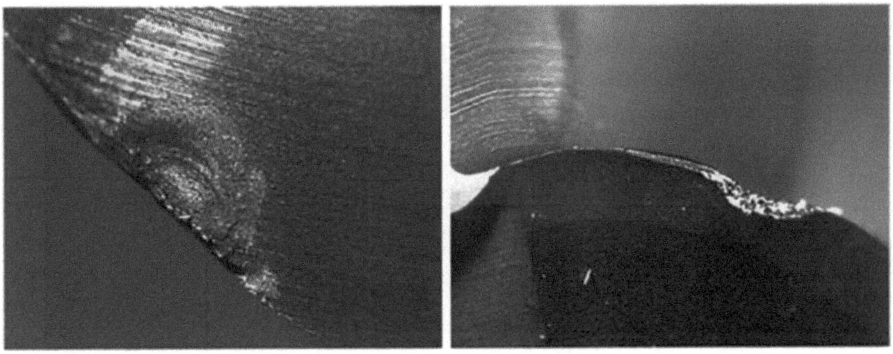

FIGURE 2.21 Double flank finishing of 50 HRC steel mold using ball milling. The worst VB (VB3) is shown as circle, while average VB (VB1) is shown as the rectangle. ([14] [Open access].)

which is often parallel to the main cutting edge is mechanical friction. When turning titanium alloys (see Figure 2.21) and other materials with low thermal conductivity, crater wear often occurs.

- Kerf wear: A combination of flank and rake wear, this wear occurs right at the major intersection point of the cutting edge of the work surface (which is also the depth of the cut line). This is a common occurrence when turning materials are used.
- Flap: As shown in Figure 2.21, this type of wear is applied on the lateral (relief) surface. Along the primary and secondary cutting edges of the tool, wear and chafing are not necessarily uniform. Because abrasion is the primary wear mechanism, it is more common when working with hard materials where there is no chemical affinity between the tool and the material.
- Adhesion: The tool rake and new chip surface are fused by high temperature and pressure. If the materials have a good metallurgical affinity, such welding will result in the formation of a thick adhesive layer, rear tearing of softer rubbing surface and rapid wear [14].

Adhesion is common in dry or nearly dry aluminum alloys but is rare in hard machining [14]. Like frictional wear of other mechanical components, the wear curve is divided into three stages:

- AB, where the area where the sharp edge wears out quickly. The wear size on the flank surface (VB) ranges from 0.05 to 0.1 mm..
- Zone BC, where the wear rate gradually increases and remains constant. This zone starts at 0.05 and continues to 0.6 mm.
- CD area with extremely high wear. Once this zone is reached, a worn tool must be replaced or sharpened before it breaks.

The type of wear that needs to be controlled in most machining operations is flank wear (VB) because it causes large differences in tool dimensions and therefore in the dimensions of machined parts. The maximum permissible value is 0.3–0.5 mm; the first is for finishing and the second for roughing. A typical variation of average VB over time at different cutting speeds. The main cause of deterioration is relative speed difference between the two contact surfaces, as in all friction situations [14].

2.3.1 Wear of Tools in Milling

Tool wear is closely related to tool life, workpiece surface quality and dimensional accuracy, intelligent control, and automated production maintenance. However, for many years, relevant researchers have mainly focused on studying the tool wear mechanism in the acute wear stage, which includes different processing methods, different superalloy materials, different tool materials and geometric characteristics and different coolant supply methods [15]. To guarantee the dimensional accuracy and surface quality of the workpiece, especially in the finishing stage, it is necessary to further investigate the morphology and wear mechanism of tool in initial and long-term wear stages. There are various benefits to using PVD of titanium aluminum nitride (PVD-TiAlN) coating in manufacturing especially for high-performance cutting tools.

In the study of milling tool wear, microtopography, tool wear curve, tool cutting force, spindle motor current, tool wear mechanism, chip topography and machined surface topography are investigated. SEM, EDS, and X-ray diffraction (XRD) are used to further investigate the tool wear mechanism during the acute wear stage [16].

This will help you fully understand the milling characteristics of GH4169 and test the cutting ability of tool. The results show that during the wear process of GH4169 superalloy tools, severe abrasive and adhesive wear occurs. In addition, the tool wear rate is high, cutting force increases, spindle motor current increases, and tool coating and substrate material are seriously worn. The main mechanisms responsible for tool wear are abrasive wear, adhesive wear, oxidative damage, and fracture. The rake and flank surfaces of cutter, as well as numerous hard abrasive carbides, present in the workpiece material are subject to severe abrasive wear. Fe, Cr, Al, and other elements are found on the front and side faces, and there is a severe adhesive wear. The front and side surfaces are covered with various oxides (Al_2O_3, Fe_2O_3, etc.), which indicates the presence of oxidative wear. Milling tool edge breakage is a very serious problem due to the interrupted cutting that occurs during the milling operation [17–20].

At the acute wear stage, as a result of extrusion and stretching, adhering particles or chips with different microscopic topographies stick to the wear zone of the tool. The microrelief of the machined surface of part subject to tool wear changed significantly throughout the entire wear period. VB observed and measured on the flank surface of end mills is the loss of particles along cutting edge, that is, where the rake and flank surfaces meet. Three measurements are possible:

- Uniform VB (VB1) means average wear along the axial cutting depth.
- Uneven VB (VB2): Uneven wear across several areas of the cutting edge.

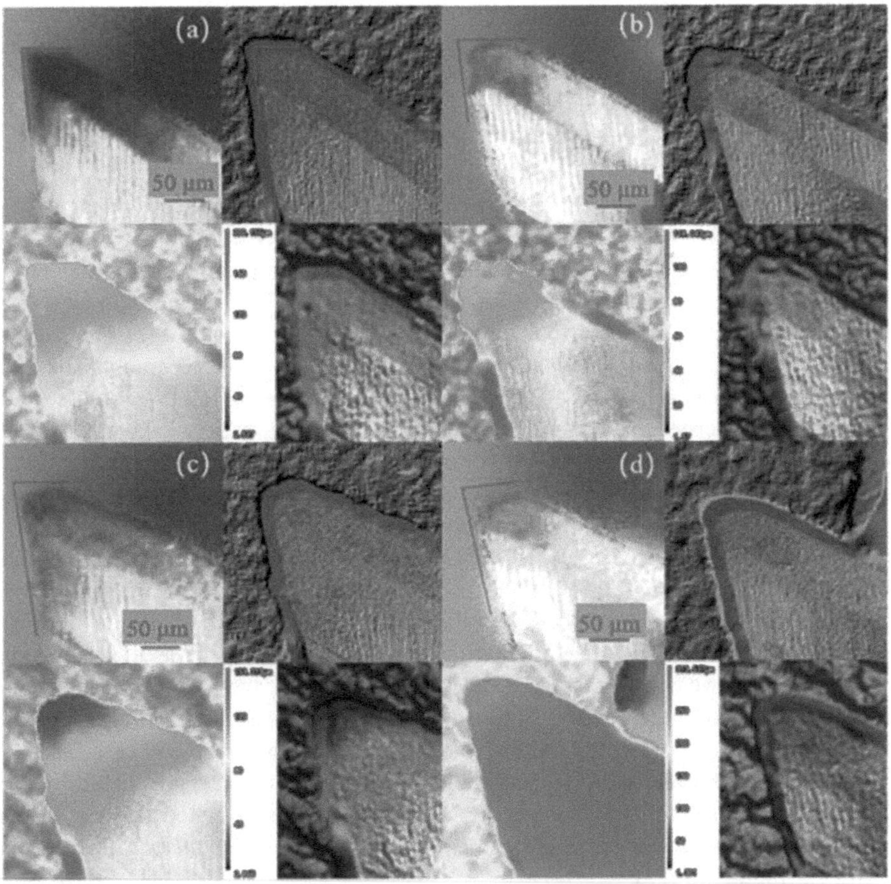

FIGURE 2.22 Study of wear topography of the 3D tool surface under different wear conditions. Confocal micrographs of initial, wear I, wear II, wear III and initial are shown in (a), (b), and (c), respectively. ([22] [Open access].)

- Localized VB (VB3): Wear is usually concentrated in certain areas. One type, known as notch wear, is located exactly at the depth of cut line and characteristic of materials that are prone to work hardening.
- Rake wear: This is the internal grooves of end mill that are found. The most common is crater wear (CT1), which is the progressive formation of a crater parallel to main cutting edge.
- Chipping (CH): Random, pinpoint flaking of cutting edge (see Figure 2.22). This is incredibly difficult to measure and avoid. It consists of tiny tool parts that break away from the cutting edge as a result of mechanical stress and short-term thermal stresses caused by heating and cooling cycles during intermittent machining operations.

- Uniform CH (CH1): A tiny edge of approximately the same size breaks along entire cutting edge upon contact with the material.
- Uneven CH (CH2): Sporadic, uneven CH, scattered across the entire cutting edge from one corner to the other.
- Flaking (FL): Loss of tool fragments, especially when using coated tools.

We focus on B4 because it has the most noticeable VB among the four cutting edges. Figure 2.22 shows the VB images of micro-milling cutter in different wear states using a confocal laser microscope. The symbols (a–d) represent the initial wear, first wear, second wear, and third wear, respectively [22]. Catastrophic failure is the rapid destruction and wear of tool. The standard measure of tool life is the average VB, as it indicates large changes in tool dimensions and workpiece dimensions. The maximum permissible values are 0.3–0.5 mm; the first is for finishing and second for roughing. A criterion for tool life is a chip larger than 0.5 mm. In metals with low machinability, several types of wear occur simultaneously, which aggravates and intensifies their harmful consequences [21,22]. The international standard for general milling specifies that the VB is the basis for determining the wear degree of cutter. For this reason, the study of VB is of great significance in the study of wear degree of micro-milling cutter. Figure 2.22 shows the contour and wear region of the tool. The new knife is smooth and flat, and contour of cutting edge is angular, as shown in Figure 2.22. The wear micro-milling cutter has a small wear zone, with very little wear at the tooltip. The cutting edge is aged and deformed, and micro-cracking of edge occurs. The wear II zone increases as machining process progresses. The 3D drawings show that additional delamination of blade material has further reduced the tooltip. More alarming is the blade material's delamination and edge's micro-cracking. The tooltip has completely delaminated at wear III, the most severe wear, and wear zone has spread along the side surface covering the entire cutting depth [23].

PROBLEMS

1. Describe the possibilities of materials needed to manufacture cutting tools.
2. List some limitations of sintered carbide metal.
3. Identify the problems arising from current developments in the manufacture of solid coatings.
4. Describe the challenges of optimizing coating selection for steel fabrication.
5. Explain PCD and diamond.

REFERENCES

1. Hao, G., Liu, Z., Liang, X., & Zhao, J. (2019). Influences of TiAlN coating on cutting temperature during orthogonal machining H13 hardened steel. *Coatings*, 9(6), 355.
2. Amin, A. R., Ali, A., & Ali, H. M. (2022). Application of nanofluids for machining processes: A comprehensive review. *Nanomaterials*, 12(23), 4214.

3. Duan, Z., Yin, Q., Li, C., Dong, L., Bai, X., Zhang, Y., … & Liu, Z. (2020). Milling force and surface morphology of 45 steel under different Al_2O_3 nanofluid concentrations. *Int. J. Adv. Manuf. Technol.,* 107, 1277–1296.
4. Seetharaman, G. R. & Sangwai, J. S. (2020). Effect of nanoparticles on the performance of drilling fluids. In *Nanotechnology for Energy and Environmental Engineering*; Ledwani, L., & Sangwai, J. S., Eds.; Springer: Berlin/Heidelberg, Germany, pp. 279–297.
5. Esfe, M. H., Bahiraei, M., & Mir, A. (2020). Application of conventional and hybrid nanofluids in different machining processes: A critical review. *Adv. Colloid Interface Sci.*, 282, 102199.
6. Zhang, S., Fu, J., Peng, C., Su, Y., Zhang, H., & Yang, M. (2024). The analysis of transient temperature in the wellbore of a deep shale gas horizontal well. *Processes*, 12(7), 1402.
7. Kelliger, T., Meurer, M., & Bergs, T. (2024). Potentials of additive manufacturing for cutting tools: A review of scientific and industrial applications. *Metals*, 14(9), 982.
8. Park, E., Kim, D. M., Park, H. W., Park, Y.-B., & Kim, N. (2020). Evaluation of tool life in the dry machining of Inconel 718 parts from additive manufacturing (AM). *Int. J. Precision Eng. Manuf.*, 21, 57–65.
9. Sharmin, I., Gafur, M. A., & Dhar, N. R. (2020). Preparation and evaluation of a stable CNT-water based nano cutting fluid for machining hard-to-cut material. *SN Appl. Sci.*, 2, 626.
10. Taha-Tijerina, J. J. & Edinbarough, I. A. (2023). Comparative cutting fluid study on optimum grinding parameters of Ti-6Al-4V alloy using flood, minimum quantity lubrication (MQL), and nanofluid MQL (NMQL). *Lubricants*, 11(6), 250.
11. Kazeem, R. A., Fadare, D. A., Ikumapayi, O. M., Adediran, A. A., Aliyu, S. J., Akinlabi, S. A., … & Akinlabi, E. T. (2022). Advances in the application of vegetable-oil-based cutting fluids to sustainable machining operations—A review. *Lubricants*, 10(4), 69.
12. Kumar, P., Chauhan, S. R., Pruncu, C. I., Gupta, M. K., Pimenov, D. Y., Mia, M., & Gill, H. S. (2019). Influence of different grades of CBN inserts on cutting force and surface roughness of AISI H13 die tool steel during hard turning operation. *Materials*, 12(1), 177.
13. Gaona-Tiburcio, C., Jáquez-Muñoz, J. M., Nieves-Mendoza, D., Maldonado-Bandala, E., Lara-Banda, M., Lira-Martinez, M. A., … & Almeraya-Calderon, F. (2024). Corrosion behavior of titanium alloys (Ti CP2, Ti-6Al-2Sn-4Zr-2Mo, Ti-6Al-4V and Ti Beta-C) with anodized and exposed in NaCl and H_2SO_4 solutions. *Metals*, 14(2), 160.
14. Outeiro, J., Cheng, W., Chinesta, F., & Ammar, A. (2022). Modelling and optimization of machining of Ti-6Al-4V titanium alloy using machine learning and design of experiments methods. *J. Manuf. Mater. Process.*, 6(3), 58.
15. Xue, C., Wang, D., & Zhang, J. (2022). Wear mechanisms and notch formation of whisker-reinforced alumina and sialon ceramic tools during high-speed turning of Inconel 718. *Materials*, 15(11), 3860.
16. Liu, G., Wang, J., Zheng, J., Ji, M., & Wang, X. (2023). An experimental study on ultrasonic vibration-assisted turning of aluminum alloy 6061 with vegetable oil-based nanofluid minimum quantity lubrication. *Lubricants*, 11(11), 470.
17. Sheeraz, M., Butt, Z., Khan, A. M., Mehmood, S., Ali, A., Azeem, M., … & Imtiaz, T. (2019). Design and optimization of piezoelectric transducer (PZT-5H stack). *J. Electron. Mater.*, 48, 6487–6502.
18. Ali, A., Sheeraz, M., Bibi, S., Khan, M. Z., Malik, M. S., & Ali, W. (2021). Artificial neural network (ANN)-based optimization of a numerically analyzed m-shaped piezoelectric energy harvester. *Funct. Mater. Lett.*, 14, 2151046.

19. Elahi, H., Mughal, M. R., Eugeni, M., Qayyum, F., Israr, A., Ali, A., … & Gaudenzi, P. (2020). Characterization and implementation of a piezoelectric energy harvester configuration: Analytical, numerical and experimental approach. *Integr. Ferroelectr.*, 212, 39–60.
20. Ali, A., Pasha, R. A., Sheeraz, M. A., Butt, Z., Elahi, H., & Khan, A. A. (2019). Investigation of electrical properties for cantilever-based piezoelectric energy harvester. *Adv. Sci. Technol. Res. J.*, 13, 76–85.
21. Naqvi, A., Ali, A., Altabey, W. A., & Kouritem, S. A. (2022). Energy harvesting from fluid flow using piezoelectric materials: A review. *Energies*, 15, 7424.
22. Zheng, T., Song, Q., Du, Y., & Liu, Z. (2022). Development of tool wear standards and wear mechanism for micro milling Ti-6Al-4V alloy. *Metals*, 12(5), 726.
23. Bejjani, R., Salame, C., & Olsson, M. (2021). An experimental and finite element approach for a better understanding of Ti-6Al-4V behavior when machining under cryogenic environment. *Materials*, 14(11), 2796.

3 Industrial Cutting Tools and Coating Deposition Applications

Hard Machining and Characteristics

3.1 INTRODUCTION

Parts with a hardness of more than 45 HRC are classified as "hard machined"; however, parts with hardness 58–68 HRC are typically machined using this technology. Materials used include hard chromium steel, tool steel, case-hardened steel, nitride cast iron, superalloys, heat-treated powder metallurgy components, and several types of hardened alloy steels. Any machine shop can provide the hardened metal cutting needed with sophisticated setups and specialized tool (holder) designs, as well as reliable equipment and extremely durable cutting materials. Although grinding is traditionally used to machine hardened steel parts, other methods can be used to machine hard parts. The cutting tool has several obvious advantages.

They perform well in many applications used today, especially those that involve milling, turning, and boring [1–3]. In modern manufacturing, this is no longer case, although many studies still adhere to the outdated idea that machining can completely replace grinding. Modern machine tools designed for heavy machining include grinding spindles, as the two operations complement each other well in a variety of useful industrial applications The three different kinds of tool coatings are considered in this study. The multilayer titanium aluminum nitride (TiN/TiAlN) coating known as "Firex coating" is used on carbide drills. It is better suited for dry machining and provides excellent wear resistance and thermal stability during drilling operations. Previous studies [4] show that Firex coating is superior to TiN coating and combines the advantages of titanium nitride (TiN), TiAlN, TiCN, and heat and fire resistance. For general metal and plastic cutting, TiN coating is used [12]. Due to its high elasticity, TiN coating can protect the cutting tool from adhesive and abrasive wear [12]. Particular attention is paid to tool matrix materials, tool coating materials [10–12] and recent advances [7–9]. Generally speaking, materials having a hardness greater than 45 HRC that are considered difficult to machine. The geometrical details of workpiece and drills are shown in Figure 3.1 [1].

A great place to start is a quick overview of the development of physical vapor deposition (PVD) coatings for cutting tools. Although the technology has been used since the 1970s, ultrathin TiN film was first commercially released as a tool coating

DOI: 10.1201/9781003604648-3

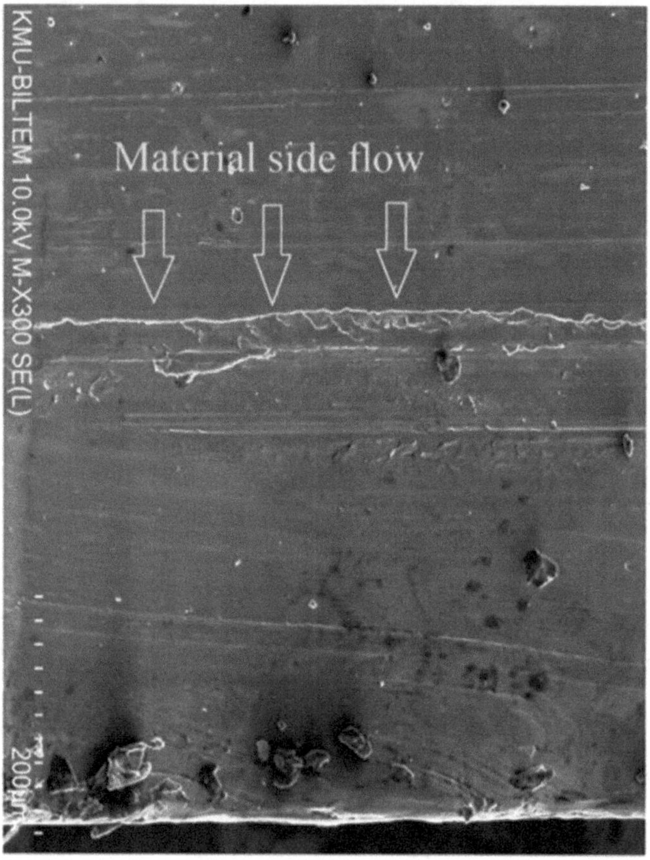

FIGURE 3.1 Information on (a) three different types of cutting tools used in study (side view) and (b) insert dimensions and hole pattern configuration. ([1] [Open access].)

in 1980s. Perhaps business influences have prevented its commercial use earlier. When this type of hard coating is used on the cutting edge of a machine tool, the tool life can be significantly increased. However, longer tool life leads to lower sales of new tools, which is bad news for tool manufacturers [4,5]. Although end users should be aware of them, research papers and promotional materials generally do not address the limitations and disadvantages of hard processing:

- Unit costs for tools for machining are significantly higher than for grinding.
- In some cases, the geometry or size of part makes hard turning impossible.

Generally speaking, unsupported parts should not have an *L*/*D* ratio greater than 4:1. Even though the tailstock is designed for long, thin workpieces, it vibrates when cutting under high pressure. Hard materials are difficult to process without specialized rigid equipment. The rigidity of machine determines the level of turning accuracy. The required part tolerances and surface finishes increase machine tool rigidity.

Coatings using nanostructures and nanolayers have become a subject of recent research. They have been tested where the results have been very promising, with tool life comparable to more modern nanocoatings [8,9]. When using PVD-coated TiN-cemented carbide cutting inserts to machine steel, cutting speed and feed have a substantial impact on tool life [10]. The built-up edge (BUE) is visible on all tools in this image, but it is comparatively larger on TiN, which is consistent with the results mentioned in [12] on drilling fiber metal laminates containing aluminum alloy sheets. This may be because dry drilling generates high temperatures, which reduces the thermal stability of TiN. In addition, some aluminum particles can be seen adhering to tool surface. The high speed causes these particles to melt and compress onto the tool's surface. Metal accumulation and adhesion at the tip are observed for each tool. In addition, pitting is not observed for each tool. However, the TiN tools have comparatively more on their surface. There are times when aluminum oxide particles filled these pits. The surface and edge of TiN tool are the most affected by working conditions in this study according to optical evaluation. However, the small number of holes drilled in this study did not allow conclusions to be drawn regarding the service life of each coated tool. The optical microscopic examination of tools after drilling is shown in Figure 3.2 [1].

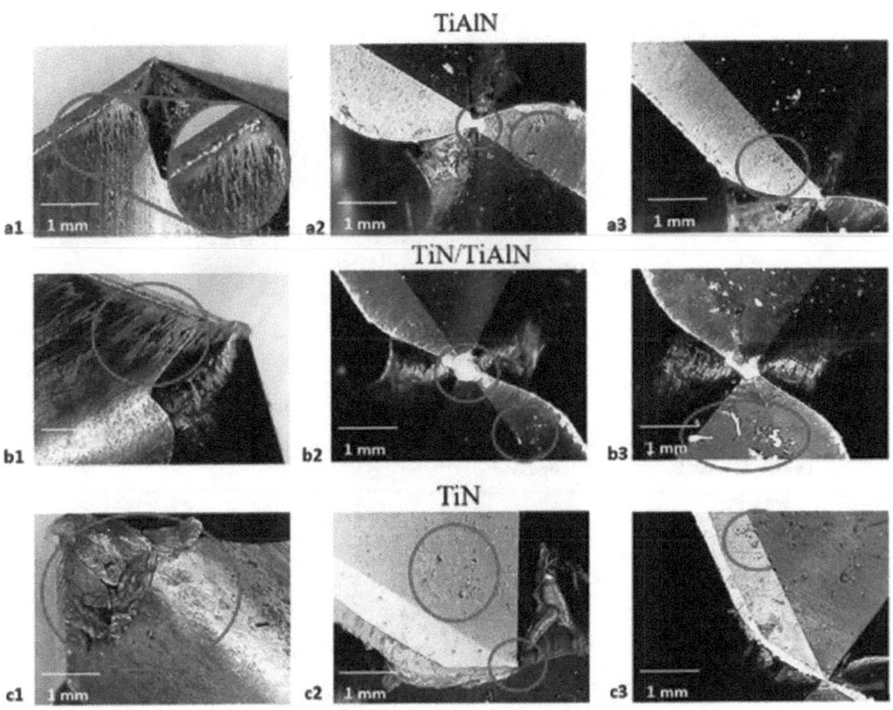

FIGURE 3.2 (a1,b1,c1) Aluminum adhesion and BUE, (a2,b2,c2) aluminum adhesion and surface corrosion on tool tip, and (a3,b3,c3) aluminum fills in surface corrosion. ([1] [Open access].)

The wear of PVD coatings is not impacted by the monolayer coating's thickness [11,12]. PVD coatings have much greater density, greater hardness, and greater stickiness. Nine 3D turning simulations are conducted using a finite element (FE) model to test the performance of generated models. This is a methodical overview of procedure and its importance. Due to the time-consuming procedure of validating the generated models, conventional parameters and settings are used in preparing the test configuration [7,8]. The material, cutting conditions, and cutting tool that are selected for this study have been effectively studied previously [3,4,6], as shown in Figure 3.3. To develop the FE model, the generated cutting tool sample with all its geometry (Figure 3.3a) is saved in the stereolithography (STL) file format and then

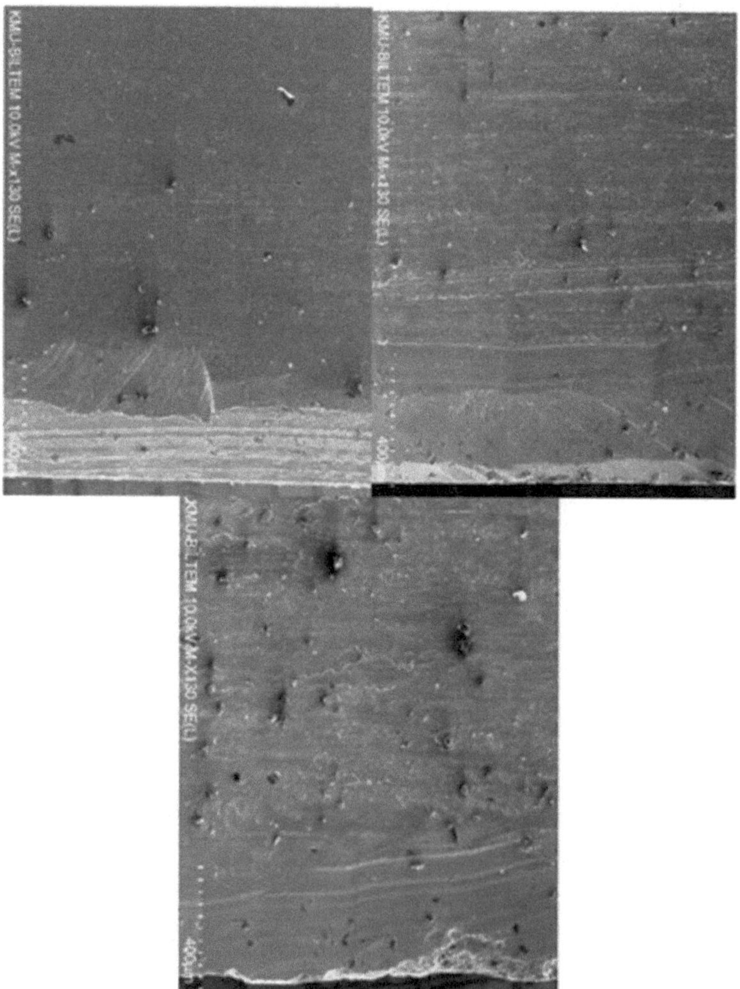

(a)

FIGURE 3.3 The model setup (a), the analysis domain (b), and the meshed tool (c). ([2] [Open access].)

loaded into DEFORMTM-3D version 12. It is meshed with approximately 50,000 tetrahedral parts and modeled as rigid. However, inversely, the mesh size of workpiece varied depending on feed rate. For each test, the minimum element size is set to 25% of the feed [2].

The biggest mystery is what coolant is appropriate for heavy machining operations. Dry milling is the best method for milling objects with interrupted cuts, such as gears. The reason for this is the possibility of thermal stress failure during the process of entering and leaving the grooves. The high tool tip temperatures during dry turning cause the pre-cut area to be annealed (softened) during continuous cutting, which lessens the material's hardness and facilitates cutting. It is useful to expedite the process during dry cutting because of this feature. There are definite economic benefits to dry cutting. On the other hand, it might be challenging to process, measure, and thermally deform parts during manufacture. Some applications are now required to use a coolant due to recent issues. When employed, a water-based coolant is required. It has been demonstrated that near-dry machining works well in demanding applications [9]. The dynamically formed BUE described above consists of several compounds formed during cutting, as well as highly polished and deformed particles of workpiece material [11–13]. The BUE and composite material are comparable in many respects [10,11]. These procedures apply to most commonly used carbide cutting tools, both coated and uncoated: The process involves five main steps; (a) material transfer from the counterface (chip and workpiece) to tool surface; (b) intense material adhesion resulting in BUE formation (Figure 3.4); (c) interaction between the adherent fragments of workpiece material and tool surface; (d) environment; and (e) complex phase transformations at tool/chip/workpiece interface [3].

For the self-lubricating TiB_2 PVD coating, in-depth wear resistance studies are conducted in this field [20]. The wear resistance data of multi-coated tools and untreated cutting tools in roughing of Ti alloy with severe BUE formation [2,8]. Severe BUE formation is observed on the uncoated tool according to the scanning electron microscopy (SEM) analysis, but the TiB_2-coated tool shows very little BUE formation. The different friction processes at the tool/chip interface are indicated by the chip types and their bottom surfaces. Figure 3.5 shows SEM images of chip types and bottom surfaces. The results show that the wear resistance of TiB_2 coating is significantly superior to that of untreated tool (Scheme 2). SEM analyses of worn surfaces confirm this (Figure 3.5) [3].

It is important to find solutions to the issues surrounding rigid equipment, industrial cutting tool properties and their applications. The intense processing results in a white covering that thickens with tool wear. It mostly affects parts like bearing strains, which are frequently formed on tools and exposed to high-contact pressures. Eventually, the white coating can come off, which would cause the tool to fail.

Most tool manufacturers have internal testing facilities to address this issue. The steel industry is able to carry out these tests. The attachment can precisely cut 400 parts; however after 300 parts, it can get so dull that a white coating forms. Figure 3.6 shows elements affecting cutting tool performance. The machining results are improved by using a coated tool [1].

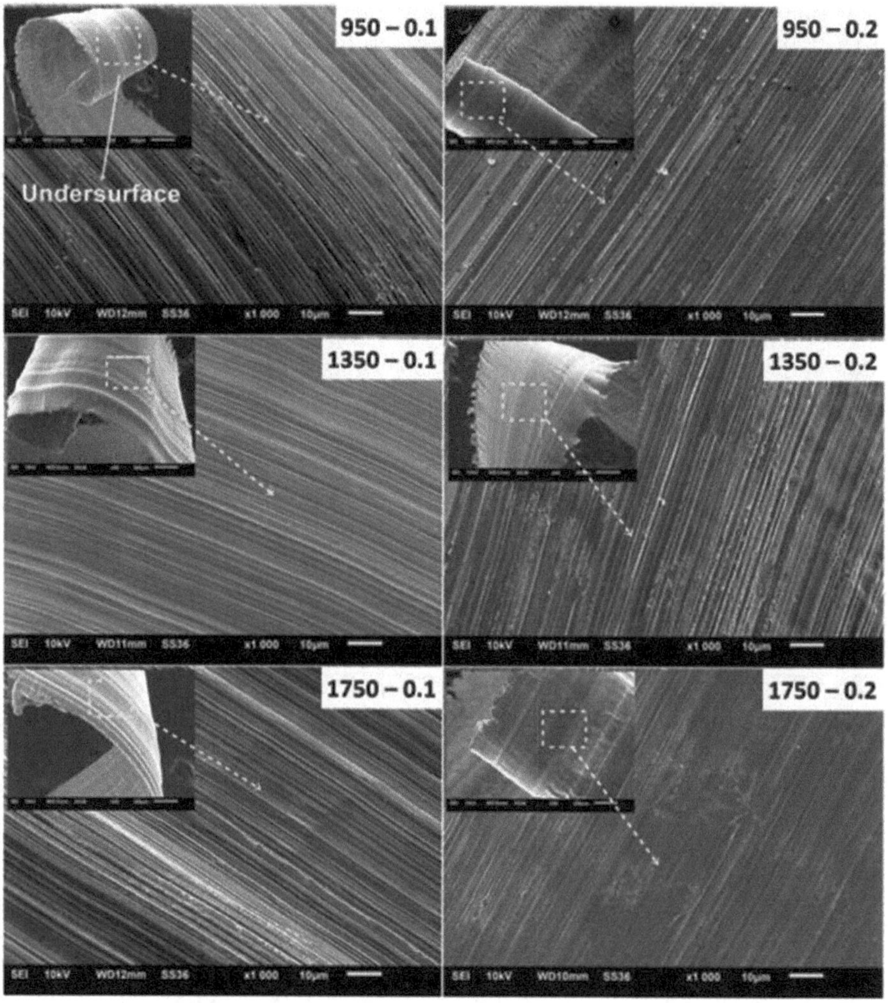

FIGURE 3.4 Schematic representation of machining of super duplex stainless steel (SDSS) with BUE formation using uncoated carbide cutting tool. Wear resistance data combined with scanning electron microscopy (SEM) studies of progressive BUE formation on the friction surface. (Adapted from [3] [Open access].)

3.2 ANALYSIS OF HARD MACHINING RESEARCH

Although there is a large amount of material in solid machines, a few studies have addressed this topic [9,10]. Traditional and intensive processing, while neglecting some basic truths, are well known in practice. As a result, the nature of hard machining discussed is understood very differently by leading tooling companies. Unfortunately, it seems that not enough machining researchers are "hearing" this point of view. The first detailed work that attempted to summarize the hard machining experience is published by [12]. Studying the technological characteristics of

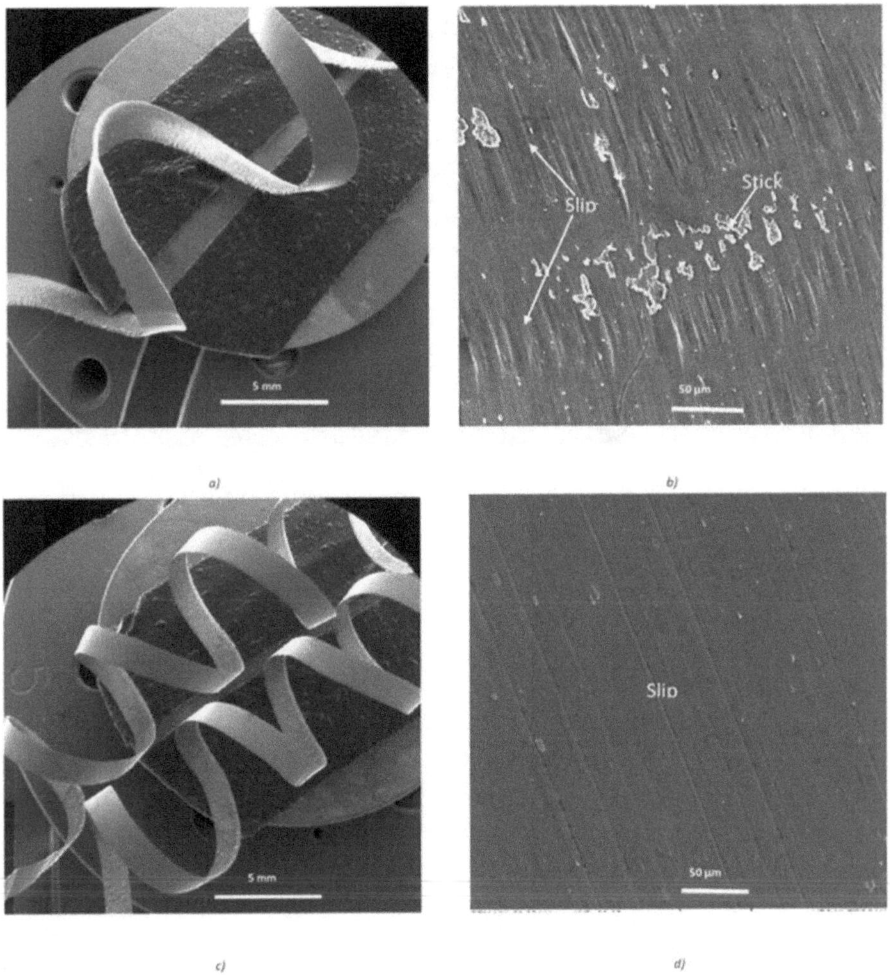

FIGURE 3.5 Chip type and bottom surface morphology of the chip for cutting tools that are (a, b) roughly machined, (c, d) unmachined, and (c, d) coated with TiB_2. (Adapted from [3] [Open access].)

these materials is necessary to create new technologies for processing hard metals. Using a polylactic acid fused deposition modeling template, the deformation behavior of the improved tool during the drawing process was investigated. All polymer forming tools have slight permanent deformations along their drawing diameters, which are caused by strong contact stresses. The operating parameters [20], lubrication parameters [21], properties of the deformed material [22], and the tool [23] play a complex role in the friction phenomenon occurring in selective melting fabrication (SMF) processes (Figure 3.7) [28].

Although "segmented chips" produced by adiabatic shear in difficult-to-machine materials have the same cross-section as saw tooth chips, the two types of chips produced by hard machining are different because they arise from different processes.

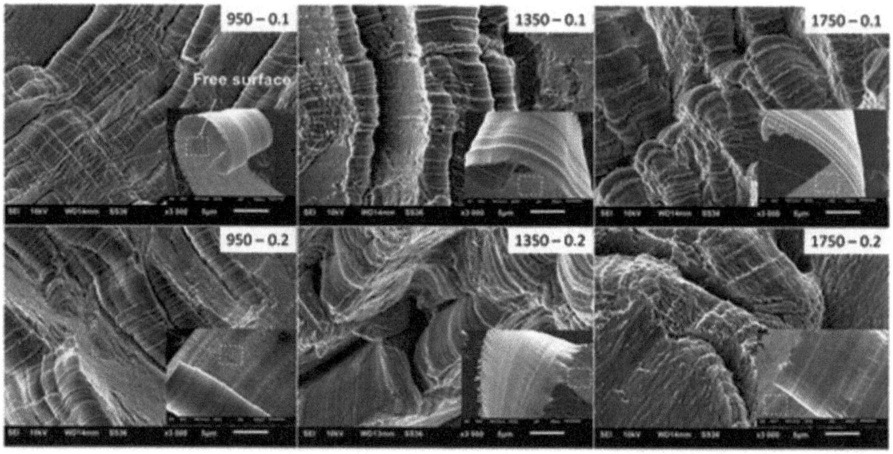

FIGURE 3.6 Factors influencing the cutting tool's performance. A coated tool yields better machining outcomes. ([1] [Open access].)

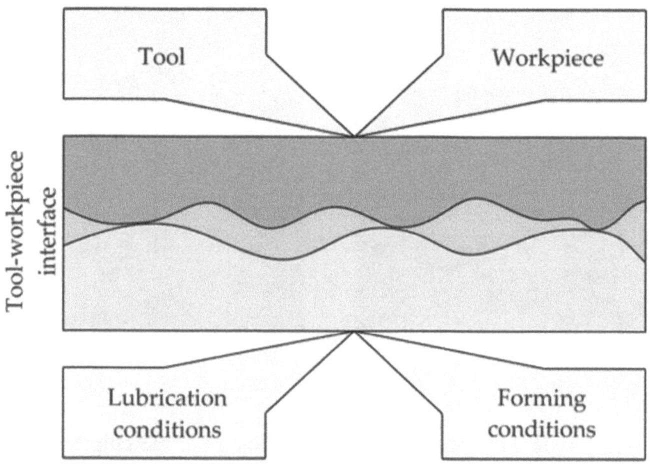

FIGURE 3.7 Scheme of the tribological system in SMF. ([28] [Open access].)

Compared to traditional machining, the shear angle is minimal. It increases sharply as hardness of material being processed increases.

- The disparity between the two components gets progressively larger. The working material's springing provides an explanation for this. In hard machining, a strong traction force is the primary cause of dimensional inaccuracies. For instance, the bearing steel AISI 52100, one of the materials used most frequently for severe machining and hence testing has a hardness range of 60–67 HRC. Gray cast iron has an elongation of greater than 7% and elongation at failure 5% in their testing. Additionally, neither the model nor its analysis account for the energy involved [11,14].

- Negative rake cutting appears to be the opposite of what is commonly practiced, with relatively low tangential (force) and radial (axial) cutting force components and a decreasing with increasing workpiece hardness.
- Neither the cause of temperature during complex machining nor its effect or its results is investigated.

Since SMF tools are subjected to high mechanical and frictional wear loads, their wear process is a complex phenomenon [7]. Tool wear is caused by oxidation, corrosion, and thermal and mechanical wear. Inadequate tool design may contribute to stress accumulation, which may lead to cracks when dynamic loads are applied. Inappropriate material selection (insufficient hardness, and in the case of hot SMF tools, limited resistance to high temperatures) is another factor affecting tool wear [8]. High frictional resistance may accelerate wear by causing undesirable microstructural changes in the tool material due to stress and high-temperature conditions [9]. The basic categories of wear consist of fatigue, abrasive wear, adhesion, and thermal wear (Figure 3.8) [28].

Pit wear and strong adhesive interaction at tool–chip interface, leading to development of BUE, where there are two main wear mechanisms seen in high-speed machining processes. Due to high cutting temperatures common at high speeds, pit wear mostly affects the rake surface of cutting tool. The thermal stresses and strains encountered during milling aggravate this wear mechanism. At the same time, the material accumulates on the cutting edge due to adhesive interactions at the tool–chip interface, affecting the surface finish and tool performance. Optimizing cutting parameters and selecting appropriate tool materials, which will ultimately increase tool life and machining efficiency, require an understanding of these wear mechanisms [4,7].

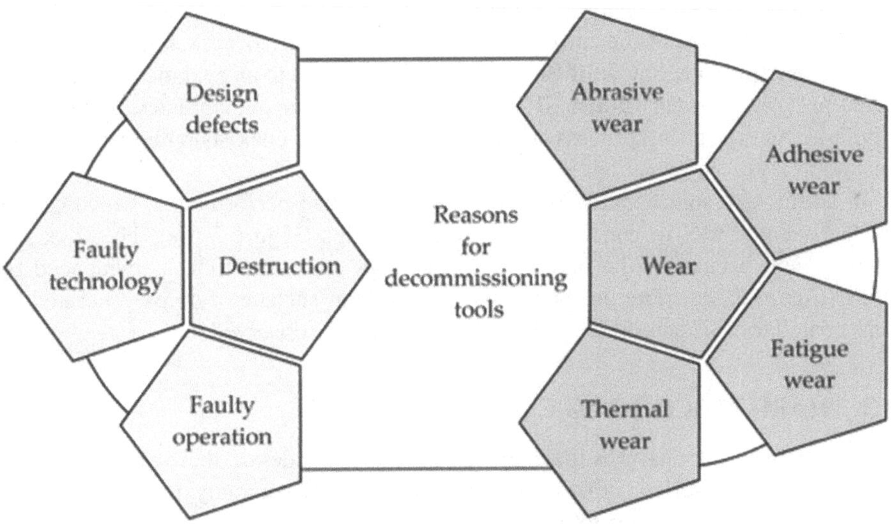

FIGURE 3.8 Causes of damage and wear in SMF tools. ([28] [Open access].)

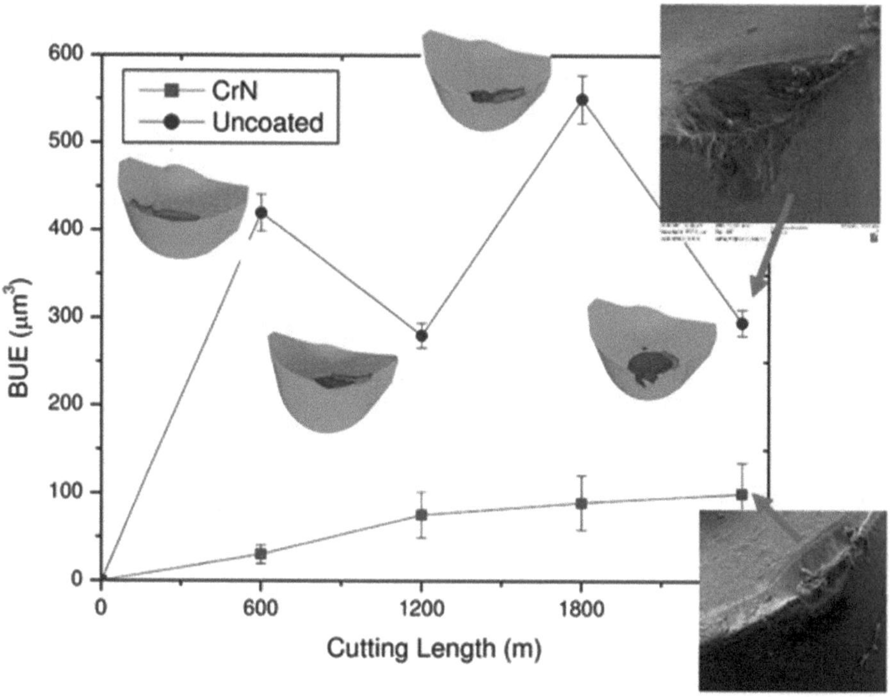

FIGURE 3.9 3D volumetric numerical data illustrating BUEs for coated and uncoated tools, taking into account the volume of peaks above the original tool-bearing surface during finishing of $TiAl_6V_4$ alloy. Blue areas represent crater formation areas and red areas represent BUEs. (Adapted from [3] [Open access].)

Over time, crater wear propagates throughout the cutting process, and when it reaches the side edge of the cutting tool, a rapid tool failure occurs. Figure 3.9 shows how the machining behavior of BUE formation compares to an avalanche. According to 3D progressive wear studies, BUE formation develops, accumulates, and eventually fails. As the cutting process continues, this behavior coexists with intense crater formation [19].

In addition, it significantly improves the tribological performance, which causes the CrN-coated tool to generate more swirling chips (Figure 3.10a). The stick-slip phenomenon occurs on the bottom surface of the uncoated tool, as evidenced by chips (Figure 3.10b). The morphology of the bottom surface of chips generated by CrN-coated tools is smoother, and the slip is primarily observed.

3.3 HARD MACHINING CHARACTERISTICS

Although this work analyzes the characteristics in great detail, there are several distinguishing characteristics that need to be considered when analyzing hard machining in order to set priorities for subsequent research and applications of this method. This can be achieved by comparing the energy balances of conventional and hard metal-cutting methods [13–17].

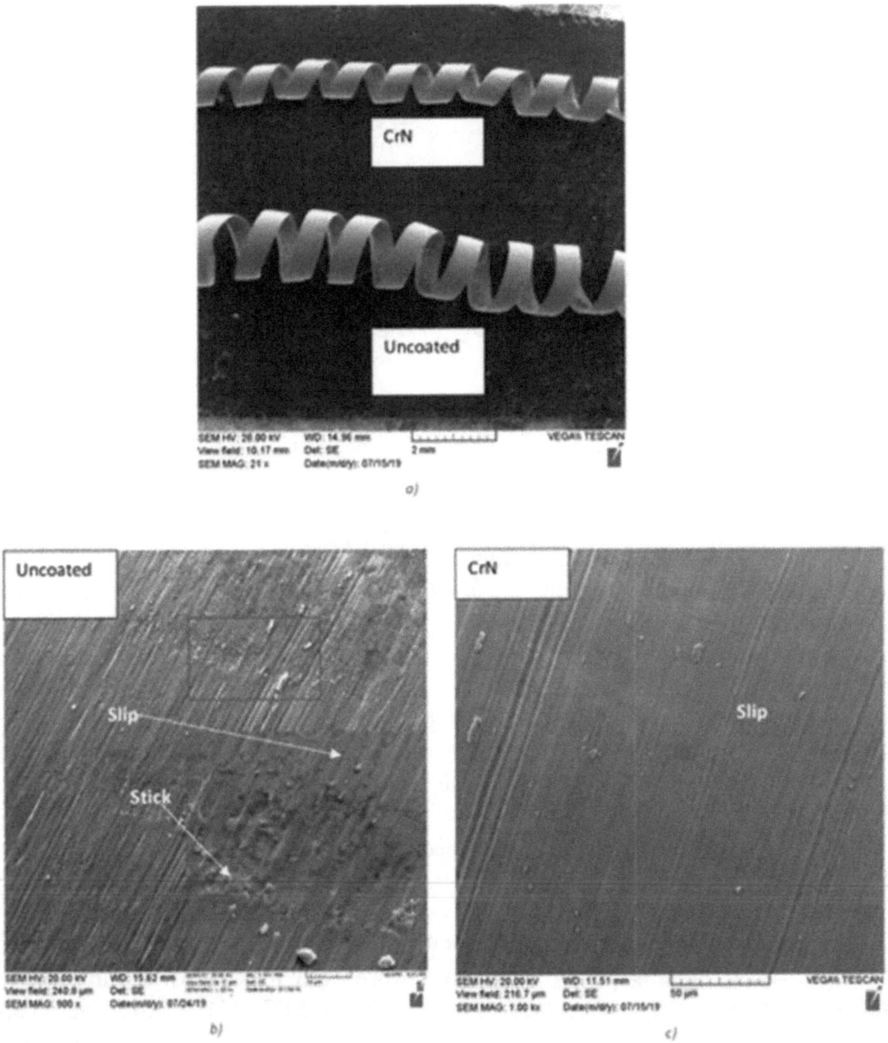

FIGURE 3.10 Chip appearance (a) and morphology of chip bottom surface for tools machining Ti_6Al_4V alloy that are (b) untreated and (c) CrN coated. ([3] [Open access].)

Several techniques, including force sensors, acceleration sensors, strain measurement, audio emissions [3], acoustic emissions [5], data-based monitoring approaches [5], image-based tool wear classification [6], artificial neural networks, and data-driven monitoring methods [7], are used to diagnose or monitor the condition of stamping tools. The impact of Seebeck coefficients on adhesive tool wear during the blanking process was examined [8]. They came up with a notion (Figure 3.11) for predicting tool wear and measuring thermoelectric currents. It was discovered that the development of adhesive wear is significantly impacted by microscopic processes that are influenced by thermoelectricity [28].

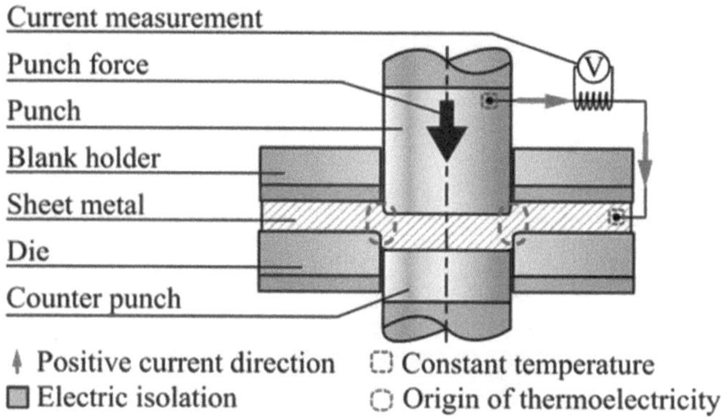

FIGURE 3.11 Tool used for the instantaneous measurement of thermoelectric currents. ([28] [Open access].)

An analysis of the tool life of compacted graphite iron (CGI) during machining under severe dry (no coolant) turning conditions is performed. The tool life values of a new low-stress multilayer super-thin cathodic (SFC) PVD coating (T2) with the same composition as a reference arc PVD coating Al60Ti40N (KC 5010) are shown in Figure 3.12 [3]. Due to the significantly reduced intensity of BUE formation in new coating compared to reference coating, the tool life is increased by 35% (see SEM images in Figure 3.12a). This can be attributed to the gradual peeling off of top surface of coating [5] rather than deep destruction of the entire coating layer (Figure 3.12b) [3].

After 400 m of cutting, a BUE is formed on the rake face of the coated turning inserts, as shown by the cross-sectional transmission electron microscopy (TEM) images of the worn area (Figure 3.13). This interaction resulted in the formation of a BUE directly on the tungsten carbide (WC)-cobalt (Co) (WC-Co) substrate and the entire AlTiN coating layer (reference) is worn away (destroyed) (Figure 3.13a). On the other hand, the coating layer is still visible in SFC–AlTiN multilayer coating (Figure 3.13b). The high-angle dark field (HAADF) image (Figure 3.13a) of the same area shows that the BUE is formed directly on the WC-Co substrate, where the visibly distorted coarse WC grains of reference coating are visible. The SFC–AlTiN multilayer coating does not contain any discernible WC grains, and the coating layer is located where the BUE is formed (Figure 3.13b). The sharpness of build-up/coating interface and absence of any interlayer formation are confirmed by the magnified TEM image (Figure 3.13b). In particular, there is no evidence of deformation or bending of columnar structure of AlTiN coating, which is typical for a coating that cannot withstand the strong frictional forces during cutting. Figure 3.13 illustrates how the adaptive response of self-organized critical systems (build-ups) and partial delamination, as shown in Figure 3.13, leads to a significant reduction in the size of energy dissipation spike. This prevents massive avalanches (Figure 3.13a) and significantly reduces surface damage (Figure 3.13b) [3].

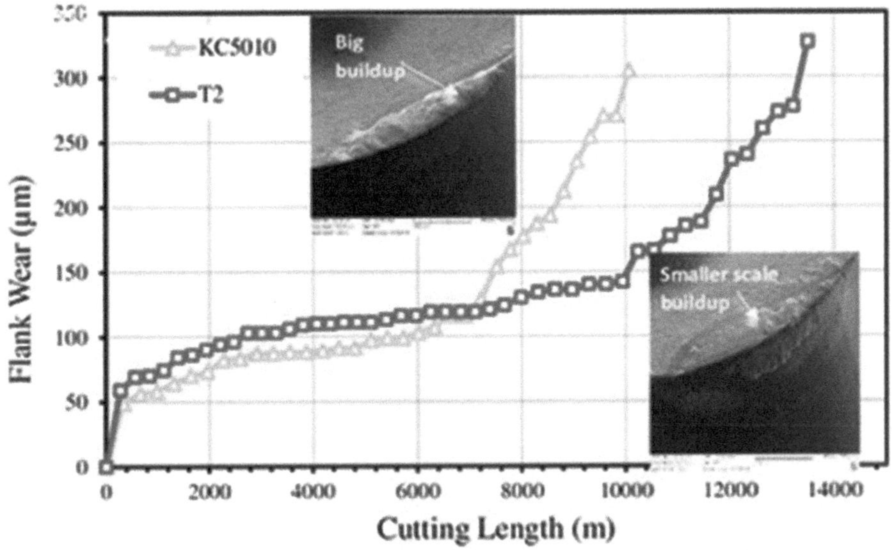

a)

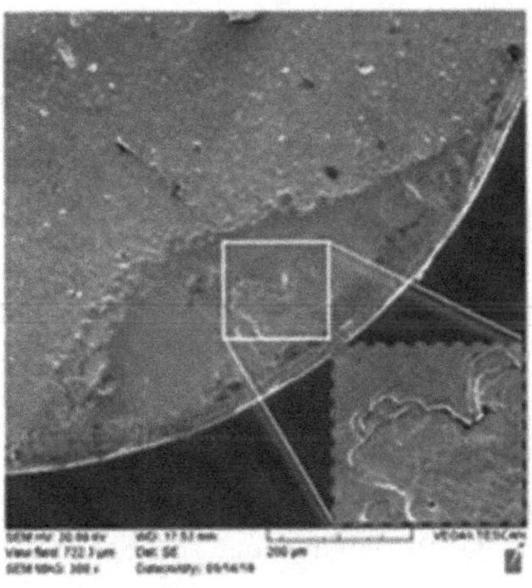

b)

FIGURE 3.12 During the machining of CGI, flank wear is shown as a function of cutting length for different PVD coatings. (a) Flank wear data combined with SEM images of BUEs; (b) SEM data after removal of BUEs by chemical etching. (Reproduced from [3] [Open access].)

3.3.1 Speed with Cutting Force

The Péclet criterion, a measure of similarity, is useful in understanding the role of thermal energy (heat) in hard machining [12]. The Péclet criterion for metal cutting is illustrated using the following processing parameters [7]:

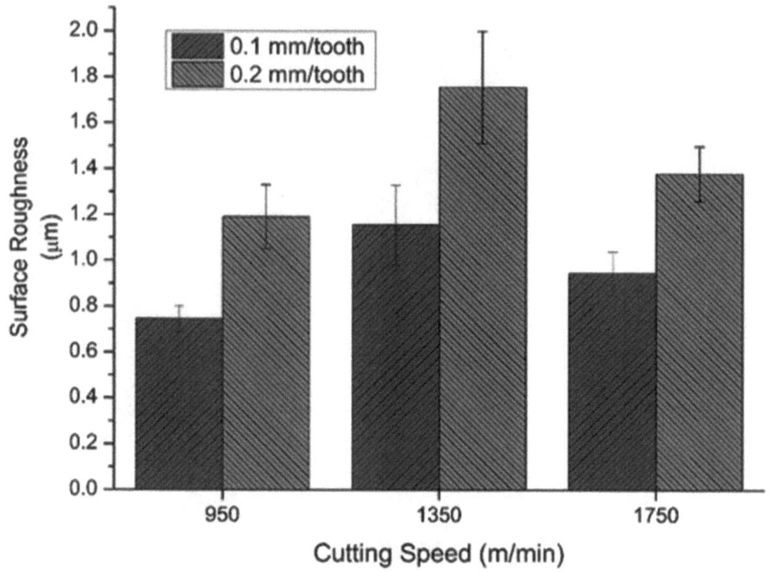

FIGURE 3.13 Transmission electron microscopy cross-sections of rake surface wafers: (a) Magnified high-energy annular dark field (HAADF) TEM image of reference AlTiN coating; (b) 10 μm SFC–AlTiN with a section length of 400 μm. (Adapted from [3] [Open access].)

$$\text{Pe} = \frac{vt_1}{w} \tag{3.1}$$

where w_w is thermal diffusivity of work material (m2/s), and v is speed.

$$w_w = \frac{k_w}{\left(c_p\rho\right)_w} \tag{3.2}$$

where $(c_p\rho)_w$ is specific volume heat (J/(m$^{3\circ}$C)) and (k_w) is thermos-conductivity of work material (J/(ms°C)). The number Péclet that expresses how much the cutting regime (vt_1) has an impact on material's thermal characteristics (w_w). Abrasive saw-tooth edges are formed due to high dynamic shear strength, creating notches on the tool [8]. The resistance of the processed material, which must be overcome during processing, is measured by the components of cutting force. According to test data, the titanium structure does not affect fundamental strength. Notable results are observed in the traction and pitching force scenarios. Regarding nitriding process NK, these forces for titanium are much less than that for MC Ti. The cutting force component is calculated both tangentially and axially using geometric relationships and dynamometer measurements (F_X, F_Y, and F_Z). When using a fiber optic or other optical beam delivery, the device is placed in front of the tool to direct a laser beam onto a workpiece. The workpiece is softened by laser heat, which facilitates cutting. The heat is concentrated around the cut itself due to laser beam that is so precisely focused [2]. The chips dissipate heat, yet the physical characteristics of material being sliced remain unchanged as a result of heat.

3.3.2 The Cutting Forces Major Axial (Thrust) Components

Currently, the automotive and aerospace industries are the main consumers of titanium and its alloys. Sports and recreation industry, bioengineering, shipbuilding industry, aerospace, and architecture also use it. In addition, nanocrystalline materials (NCs) exhibit unique physical and mechanical properties due to their distinctive structure [2]. They are used in cutting tools and electronic devices to improve performance and protection, as well as in aviation for thermal barrier coatings and in automotive industry for anti-corrosion coatings on engine components. The analysis was performed on different tool surface concepts and compositions using CrAlN PVD coatings (Figure 3.14). When compared to an unmodified tool, it was found that the surface features and PVD coating significantly extended the tool life (84%) [28].

Arc plasma deposition is the preferred coating method in many industries due to its versatility and efficiency [9]. These properties provide them with a great potential for energy applications and make attractive for using in biological applications [1], such as bone and dental implants. All these applications must be considered using manufacturing procedures to produce the appropriate parts. The most common industrial technology is mechanical processing of useful and practical metals. However, the same characteristics that are considered difficult to work with AISI 52100 steels are presented in Figure 3.14. If the applied stress is greater than the elastic limit, it is referred to the yield limit, which, for annealed 52100 steel, is 370 MPa; the material will display both elastic and plastic deformation.

The total elongation after the failure of a flexible component, as shown in the distance diagram, in the case under consideration is less than 20%. This makes it easy to determine where the point draws a line parallel to line from point 0. As a result, the distance is sometimes called elastic recovery or elastic return in materials processing. The pressure at point of contact between the tool and workpiece increases sharply because this friction creates a relief, which leads to wear on the flank surface. This explains the significant radial (axial) component of the cutting force, and severe wear on the flank surface happens when cutting something hard. However, prolonged heat reduces its flexibility [3].

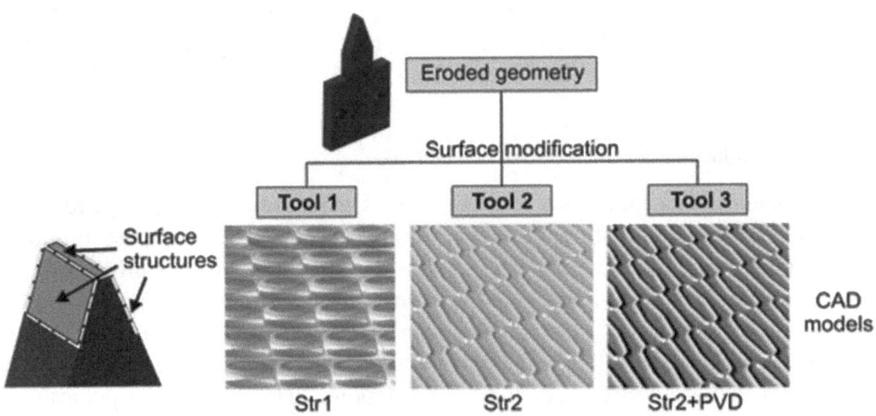

FIGURE 3.14 Surface structures (Str1, Str2) of the forming tool. ([28] [Open access].)

3.3.3 Power Invested in Creation of Surfaces

The ability of solid materials and objects to restore their previous shape after removing loads or external forces causing a deformation is called elasticity. When an object regains its original size and shape after the load is removed, it is called elastic. Different materials behave differently for different physical reasons, which are influenced by their microscopic structure. For example, the stretching of the polymer chains under influence of applied force provides the elasticity of rubbers and polymers. Conversely, under the influence of external pressure, the crystal lattice cells that make up the structure of metal material undergo changes in size and shape, which determine the elasticity of the metal. The two factors that determine the elasticity of the material are the elastic limit and elastic modulus. Materials that require high loads to create large deformations and are difficult to deform have a high modulus of elasticity, one example being steel strips. Materials that easily deform under load, such as elastic bands, have a low modulus of elasticity. The material deforms permanently if the stress under load is too great, at which point it takes on a different shape and size rather than returning to its original state. The stress value above which a material ceases to behave elastically and instead permanently deforms is known as elastic limit. The material with the lowest elastic modulus is the most elastic among materials with same elastic limitations [12–18]. Surface engineering processes are constantly evolving, providing tremendous opportunities for a wide range of applications across multiple industries. From consumer products to electronics, medical devices, aerospace and automotive components, product performance and durability are greatly influenced by the adaptability of metals to their surface quality and process. Subsurface chip analysis is performed for the cutting tools with low-stress coating and control sample (Figure 3.15). For the control coating, a strong slip phenomenon with severe surface deterioration is often observed (Figure 3.15a). The severity of this catastrophic phenomenon is significantly reduced in the new low-stress coating; first of all, slip spots are evident (Figure 3.15b) [3].

Research into new materials, improved creative processes and integration of surface technology into other manufacturing processes are enhanced by ongoing research and development in this field. This work's proportional influence on heavy machining becomes much more substantial because it rises with the material's strength. In any model of chip creation, the labor necessary to generate a new surface must be taken into account in hard machining while balancing forces and/or energy.

3.3.4 Requirements for Rigid Processing

Models created using multibody tools in a real-time virtual environment require a trade-off between computational efficiency and accuracy. When complex multibody models come into contact with other parts of the system or with external surfaces, they are deformed to some degree. Different modeling assumptions may be applied, resulting in different types of models depending on the intended use of the virtual environment. Although finite segments and methods offer computationally efficient ways to represent both large and small deformations, they are unable to convey the stresses and strains experienced by deformable components. FE analysis is a reliable method for describing the deformation of a system, including stress and strain fields,

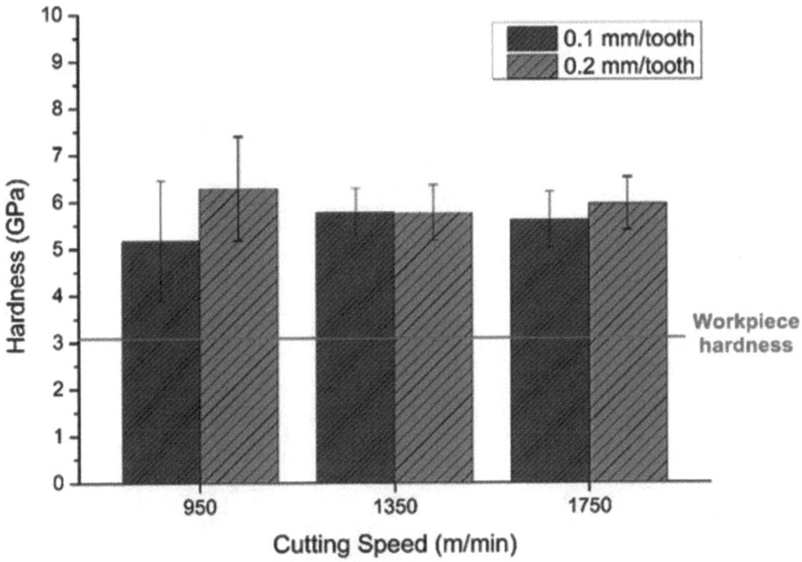

FIGURE 3.15 SEM images of combined texture of bottom surface of chip are obtained by conventional arc PVD coating (reference) and low-voltage multilayer TiAlN coating using an Infinite Focus G5 variable focus microscope (Alicona). ([3] [Open access].)

but requires more expensive computations. Appropriate contact models must be included in systems where touch or impact scenarios occur [3]. However, the driving force (friction changes) in the majority of sliding systems is frequently unknown or needs to be discovered by modeling. Frictional vibrations can, at most, be minimized its frequency can be raised out of the undesirable range by altering the dynamics of the system [1–4].

3.4 OPERATIONS FOR HARD MACHINING

The final manufacturing procedures of parts and components typically involve complex manufacturing such as boring, drilling and the required dimensions and surface finishes typically require stringent manufacturing standards and conditions.

3.4.1 Hard Turning

Hard turning (Figure 3.16) has the obvious benefit of eliminating grinding processes. The same machine, which "softly" spins the part before hardening it, enables this productivity and maximizes the equipment's effectiveness. However, some stores commit the error of initially selecting the incorrect tool insert for the job (one that is less expensive). Tool steel, bearing steel, and case-hardened steel are examples of commonly used hard materials [4]; this is untrue since these materials have a hardness that is substantially lower than 47 HRC, which means that the mechanism for how chips form during the operation, including the materials used for tools, differs greatly the best process predictability, from a metallurgical perspective.

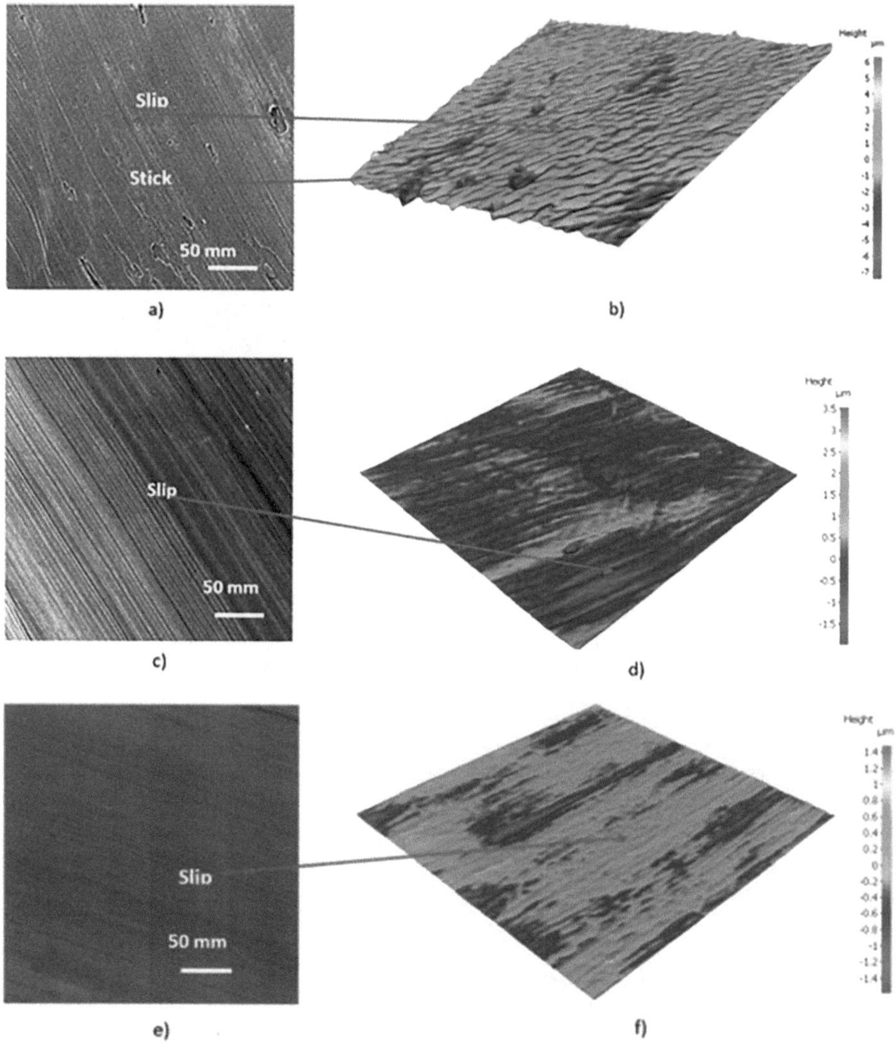

FIGURE 3.16 SEM images of combined surface texture of underside of the chip obtained using an Infinite Focus G5 variable focus microscope (Alicona) with tools that are (a,b) untreated, (c,d) PVD coated, and (e,f) CVD coated. ([3] [Open access].)

As previously stated, the *L*/*D* ratio for unsupported workpieces should generally not exceed 4:1. When cutting long and thin pieces, the tailstock is supported but excessive cutting pressure can create vibration.

The level of turning accuracy is determined by the rigidity of the machine. Most modern machines have the rigidity needed to handle tasks that require heavy turning. In many cases, the overall condition of a car is more important. However, machine tool rigidity becomes an increasing concern as required part tolerances and surface finishes increase. The best option is to use specialized equipment and turning centers designed for serious machining. However, the secret to success in turning is not the rigidity of the machine, but system. Unfortunately, the industry does not fully understand this simple concept. To maximize system rigidity, all overhang, tool, and part extensions must be minimized such as eliminating spacers and shims [3]. The uncoated tool left sliding wear marks visible on the underside of the chip (Figure 3.16a and b). The underside of the chip produced by a PVD-coated tool has showed significant sliding but no sticking (Figure 3.16c and d). A weak sliding mark is visible on a chemical vapor deposition (CVD)-coated tool (Figure 3.16e and f). The conclusion is that a CVD-coated carbide material with an optimal combination of micromechanical properties significantly improves the friction conditions at the tool/chip insert interface and controls self-organized critical process (SOC). This directly demonstrates the behavior of the integrative system [3].

It therefore makes sense that any coolant used should be water-based and of low concentration. The fact that chips break almost completely when hand-milled indicates that they generate the necessary amount of heat. In Figure 3.17, the flaking wear is seen in combination with the wear morphology on the rake face. The reason for this may be the embrittlement of the tool in an extremely low environment. Consequently, the tool strength is significantly reduced due to flaking wear. Overall, however, the trend indicates that cutting with LN2 reduces tool wear for both low and high speeds significantly better than other two circumstances [23].

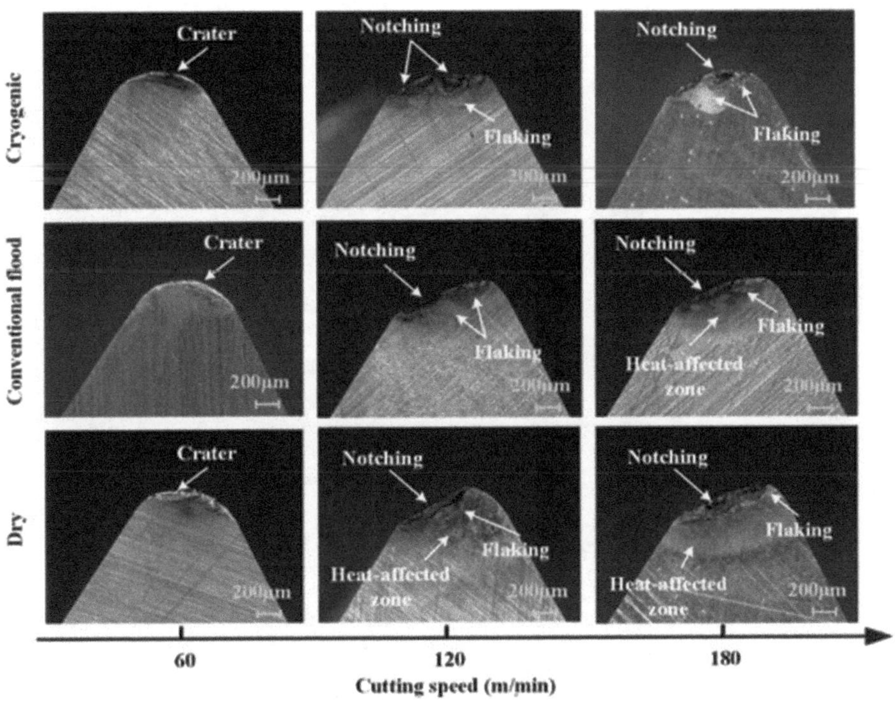

FIGURE 3.17 The shape of rake wear. ([23] [Open access].)

3.4.2 Reaming and Hard Boring

Reaming is a widely used method for making holes in manufacturing. When it comes to machining smaller holes, reaming is more practical and cost-effective than internal grinding and fine boring [4]. Boring is a machining technique in which a pre-made hole is enlarged using a cutting tool. In situations where deformation and dimensional changes caused by heat treatment create undue constraints in "soft stage" of machining, hard boring will be a more economical alternative that reduces waste [10–14].

A brief description of hole machining includes boring, reaming, drilling, and honing. In this chapter, we report every step of the metal-cutting process. In addition to being important surfaces on components such as disks, rings, couplings, brackets, and bushings, holes are common surfaces encountered in the machining process. A blisk with 20 channels is demonstrated in Figure 3.18 [29]. The machine tool life cycle ceramic, metal-ceramic, or private cellular networks (PSBN) plates is mounted on the tool body directly or in parts. Multi-bladed tools are used to achieve high performance. However, single-edge tools with backing plates are used when the hole being machined must have a high degree of accuracy. Clamping tiny inserts onto tiny boring bars becomes more challenging for holes smaller than 12 mm in diameter. To solve this problem, a soldering tool is used [27].

3.4.3 Hard Milling

Businesses frequently concentrate on creating dies, and dies are used to cut materials like P20, H13, W5 and S7. Solid rods and cavities have traditionally been produced from these materials using electrical discharge devices. These materials can now be

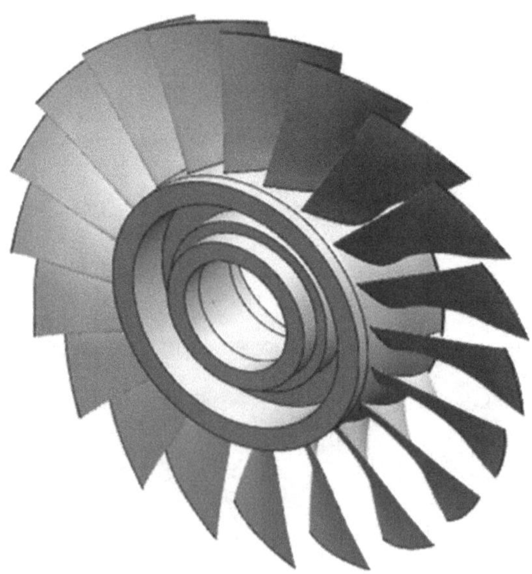

(a)

FIGURE 3.18 (a) A blisk with 20 channels. ([29] [Open access].)

produced directly into hardened materials, thanks to the development of new technologies over time. This is accomplished by utilizing new tool-path techniques and hard milling processes. Prominent mold manufacturers have realized that, in an increasingly competitive global environment, adjusting to new technologies may be essential to their survival. The deployment of the system involves the system's most important element, the machine. The machine needs to be made for hard milling and has some high-speed machining center features [9]. Hard milling demands a machine's fundamental design as well as specific machine parts like the motor, spindle, and computer numerical control (CNC). It is a crucial to have a study base with adequate vibration-dampening properties. Because polymer concrete bases often have vibration dampening that is 6 to 10 times better than cast iron, they are a viable option for high-speed, heavy-duty milling. Digital drives' quick acceleration and deceleration capabilities and superior contouring accuracy aid in cutting down on cutting tool wear. Depending on the application, each of these tools has different strengths and limitations. End mills made of solid carbide are of highly expensive, precisely ground and coated. Cutter sets with interchangeable carbide inserts are a second category of carbide cutters. They are not ideal for harsh milling tasks. Ceramic replacement inserts are a third type of strengthened ceramic inserts with whiskers. Faster cycle times and fewer operations per item are advantages of employing a milling system with changeable ceramic inserts. Using interchangeable ceramic inserts, ceramic cutters can safely and quickly grind everything from big face mills to small diameter end mills. Modern whisker-reinforced ceramics may run at speeds much above the point at which they melt since they have a melting temperature of over 2000°C that fails with carbide tools [5].

3.4.4 Hard Broaching

The broaching operation is the most cost-effective way for mass-producing parts with complex features in the field of metal cutting where high production rates and great component stability are required. One motion can be used to draw the part's profile. While exterior broaching is intended to machine the surface profile, internal broaching begins from a hole that has already been pre-machined. Both horizontal and vertical directions are open to firmware. Both linear and helical cutting movements are possible [6–8]. It is feasible to use one of the two rigid broaching techniques with this tool configuration:

- The cutting speed during typical broaching procedures is known to let between 0.5 and 6 m/min. Since that time, researchers and professionals have focused on rigid broaching's effectiveness and dependability, as well as productivity and surface quality. The outcomes thus far are outlined below:
- When it comes to the quality and dependability of broaching operations, coated carbide broaches, intended as side-cutting tools, perform well. It obtains an IT profile quality of 7–8.
- When utilizing carbide broaches, an efficient cutting speed is identified as a crucial component in the broaching process. When compared to green (soft) feed, the cycle time at this speed is shorter since the feed stroke lasts between 1 and 2 seconds.

3.4.5 Operations for Manufacturing Hard Gear

Cutting, forming, grinding and heat treating are the four main steps in the gear manufacturing process required to produce gears that meet these requirements. Using techniques such as hobbing, forming, milling, broaching, and gear cutting produces precise tooth shapes on gears. One of the most important technologies in the production of gears is abrasive processing. Over the past 20 years, among the methods of abrasive processing of hardened gears, operations with cutting speeds 0.3–5 m/s have gained importance. Much work has been done to make these low-speed grinding operations more productive. The development of new abrasive tools is an integral part of this study, requiring a large amount of empirical research and, therefore, a significant investment of time and money [3,4]. The method allows you to change the geometric surface of side surfaces of teeth and improve the surface quality. In addition, the noise level of the gearbox is reduced due to characteristic surface roughness of final gears. The most common process and wear pattern of milling tools are the development and propagation of cracks caused by thermal fatigue (high processing temperatures). The origin and wear patterns of coated tools are still the subject of recent research as they have a direct impact on tool life. Coating wear studies can help develop new coatings and improve processes.

Recent research has shown that using these methods can extend the life of coatings and improve surface quality and overall process efficiency. However, machining with rigid gear machines places unique demands on the tool, machine, and overall process. Understanding these prerequisites is necessary for the successful implementation of this process. High-speed steel is the material most commonly used in the industry today to make cooktops. In addition, final machining cannot be done on a high-speed steel milling machine. Even when these relatively soft materials are used, this can only go so far before additional finishing with other dental instruments such as razors is required. The main reason is the incorrect angle of the lateral surface along the tooth profile. To improve the characteristics of the finished surface, the cutting wedges of form-cutting tool are created with an inclination angle of 3°–4° [17]. Since the tooth profile must be maintained within a limited tolerance when a cutter is sharpened repeatedly, substantially smaller flank angles are used in hob tools. Due to the high yield strength and high elasticity of processed materials, when working with hard materials, excessive friction occurs on the side surface. The workpiece and flange stick together due to high temperatures caused by this friction, reducing surface quality and shortening tool life. In addition to inserts used for small units, there are carbide inserts with replaceable carbide for complex machining and design of drilling and boring tools.

3.4.6 Skiving

3.4.6.1 Hard Turning

Turning is a very effective and simple machining method that produces high-quality surfaces for rotating body parts made from a variety of materials. The efficiency of processing is determined for each specific case, the best combination of workpiece, tool materials, and cutting parameters. The conditions, geometric characteristics of cutting tool, the state of machine, and other technical data are taken into account. Machine parts consisting of hardened steel annealed and special cast iron and other difficult-to-cut materials are successfully machined. The use of cubic boron nitride

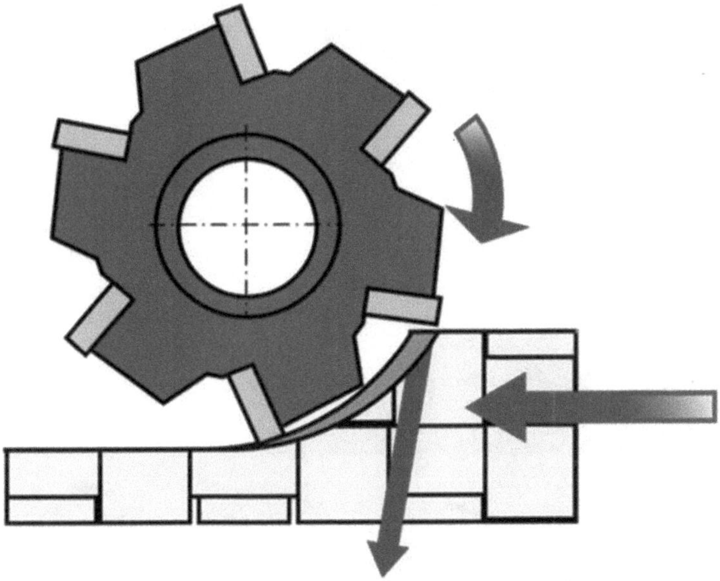

FIGURE 3.19 The arrangement of warp and weft yarns from the point of view of delamination during down milling. ([30] [Open access].)

(CBN) tools for fine turning of workpieces composed of these materials avoids the need for grinding and traditional method of producing surfaces with a roughness less than $R_a = 1.25$ µm. Currently known as "hard turning", this processing method involves machining materials with a hardness greater than 47 HRC [11–14]. The arrangement of warp and weft threads from the point of view of separation during drawing is shown in Figure 3.19 [30].

3.4.6.2 Skiving of Gear

Polishing hardened gears are known as gear hobbing. Using this method, errors in the gear ends and profiles associated with heat treatment are eliminated. The end result is increased transmission accuracy. Hardened gears with a hardness up to 63 HRC can be processed with carbide cutters [4]. The development of extremely sophisticated automated machine tools has made this possible, making the approach efficient and practical [5].

Figure 3.20 illustrates the hobbling of hardened spur gear with a diameter of 16 mm. The body of the worm gear tool is made of high-speed steel, with carbide cutting inserts soldered onto it. Each side surface of gear is sharpened by removing 0.5 mm thick allowance in one pass. The advantages of this technology include the capability to modify gear flanks such as cone and crown shapes over distances exceeding 100 meters in a single pass, along with significant reductions in tooling costs and cycle times. The character has achieved DIN 8 quality, which is the most accurate indicator of process quality. The carbide cutter appears to be protected from impact and sudden overload by a negative rake angle. However, this is not entirely true as process modeling is sometimes too confusing and important variables are

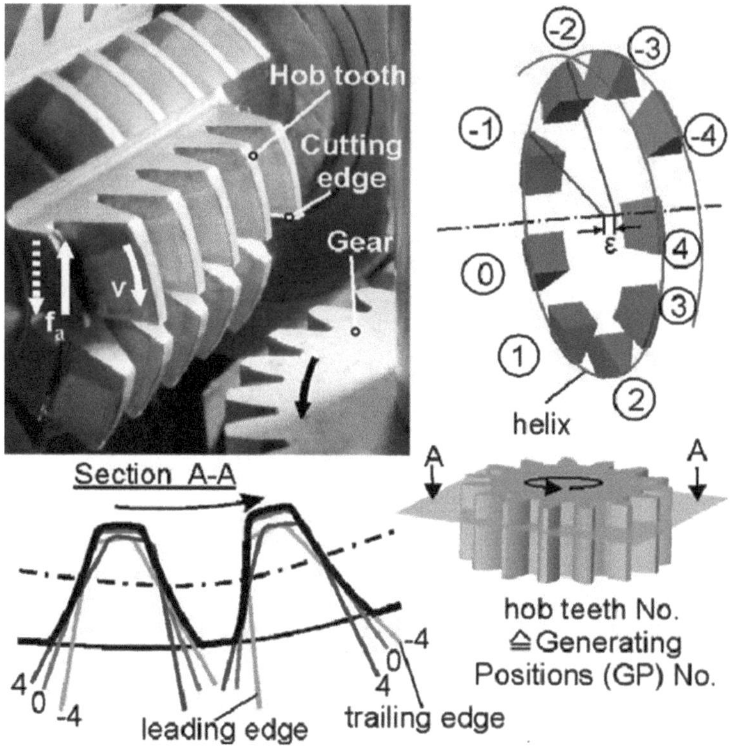

FIGURE 3.20 Tooth geometry of traditional and skiving hobs is compared. ([27] [Open access].)

ignored due to the geometry of tool teeth in machine tool is not taken into account [7–9]. Figure 3.20 illustrates the unique tooth geometry of the cutter. Many scientists believe that the main difference between sharpening teeth with worms and cutting teeth is the negative rake angle of the tool in hand, caused by the displacement of the front surface of the tooth [27].

3.5 TECHNOLOGY OF COATING

High-performance carbide-coated tools are the result of significant advances in coating technology. These exhibit improved qualities such as wear resistance, toughness, and hardness. Chemical coating techniques have advanced of new tactics and solutions for the selective coating of complicated surfaces. To enhance the performance of sol-gel coatings, new materials have been created and hybrid organic-inorganic coatings investigated [3,4]. Developments in etchant formulation, process control, and masking techniques have proven advantageous in achieving high accuracy and reproducibility in chemical etching procedures.

3.5.1 Backdrop

Cutting tools with hard protective coatings are several micrometers thick. Until the 1980s, TiN, TiCN, and Al_2O_3 produced by CVD are used to make carbide tools. However, TiN is first PVD-coated and used by cutting factories, followed by TiAlN [4]. Currently, much work is being done to improve cutting performance (CVD coatings) [5]. These layers can serve as a bridge between the desired characteristics of top layer and interfacial adhesion mechanism. Due to the increased friction created by this hardening property between the insert tip and workpiece, higher temperatures are encouraged, resulting in diffusion and oxidation [4,5].

3.5.2 Significance of the Coating Layer

PVD and CVD coatings are applied using multiple cathodic arc deposition [6]. PVD instruments are more fragile than CVD instruments. PVD-coated tools offer more finishing options. On the other hand, CVD tools show better potential for roughing [7]. Substrates and layers of PVD coatings are being studied [8]. Tools coated with CVD (moderate temperature) materials are more durable and less likely to break [9]. Numerous studies have been conducted on different types of coatings and their application in specific situations. Figure 3.21 shows the characteristics of various coatings. Mechanical properties are determined by the properties of substrates of various coatings [9]. TiN is a durable substance that resists corrosion and wear well [5].

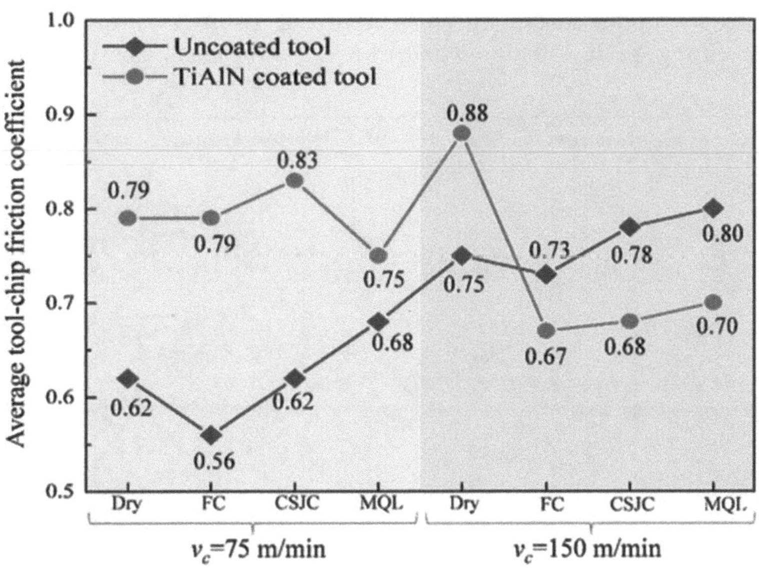

FIGURE 3.21 Surface topography of Ti-6Al-4V after turning with an uncoated carbide tool under dry conditions with different cooling rates at cutting speeds of 75 and 150 m/min. ([26] [Open access].)

For a better finishing, a number of coatings are applied to the tool surface, including TiN, TiCN, and TiAlN [15]. The impact of tool coatings on process efficiency is significant. Coated carbide tools can significantly reduce friction [12]. Tools used for dry cutting are usually made of carbide [14–17]. The production rate of CVD is higher than TiCN (single layer) [7,9]. Laser surface engineering methods take advantage of advances in laser technology, making surface modification more accurate and efficient [13]. Electroplating procedures have been improved using chemical methods, resulting in improved coating quality, higher deposition rates, and reduced environmental impact [18].

The 3D surface topography of Ti-6Al-4V after turning with bare tools at different cutting speeds is shown in Figures 3.19 and 3.20. Obvious reactions are observed on the cut surface of Ti-6Al-4V under both dry and alternating cooling. Figure 3.22 illustrates the surface topography parameters Sa and Sdr. With the increasing in speed from 75 to 150 m/min for different cutting pairs, Sa increases by at least 12.88% after the initial turning of Ti-6Al-4V. It can be demonstrated that high cutting speeds can cause significant deformation of top layer material of Ti-6Al-4V [23–26].

When machining a titanium alloy (Ti-6Al-4V) using both uncoated and titanium aluminum nitride coated (TiAlN) tools under different cooling conditions conventional flooding (CF), compressed super jet cooling (CSJC), and Minimum Quantity Lubrication (MQL) a higher surface roughness (Sa) is observed compared to machining under dry conditions. For the lowest surface roughness when machining Ti-6Al-4V with bare and TiAlN tools, respectively, MQL and flood cooling (FC) are suitable for cooling conditions. When machining Ti-6Al-4V with the TiAlN tool at a high cutting speed, 150 m/min under MQL cooling conditions, the mechanical

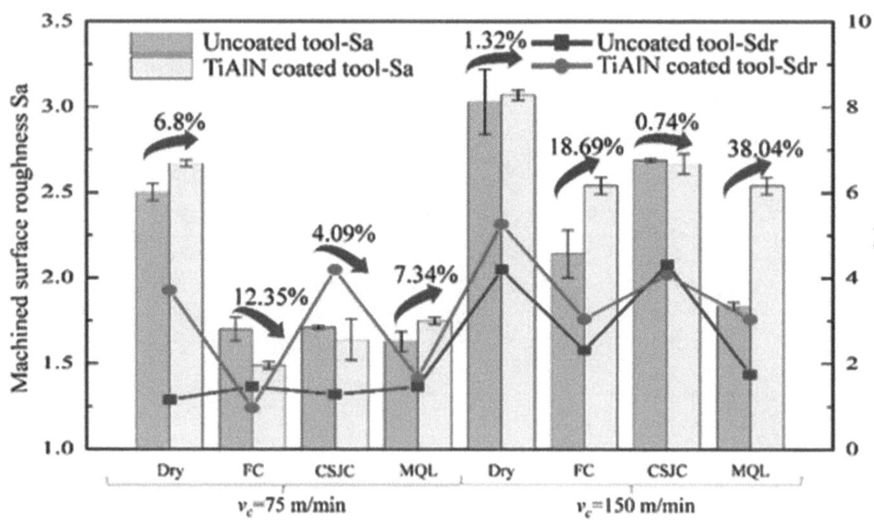

FIGURE 3.22 Surface topography of Ti-6Al-4V alloy after turning with TiAlN-coated carbide tool under dry conditions and different cooling conditions, at cutting speeds, 75 and 150 m/min. ([26] [Open access].)

surface roughness is similar to that underflow cooling conditions. Figure 3.22 shows that the TiAlN coating deposited under dry and MQL cooling conditions increase the mechanical surface roughness of wTi-6Al-4V alloy by 6.80% and 7.34% at a low cutting speed, 75 m/min [26,27].

The high speed (150 m/min) applied to the cutting tools may result in worse mechanical and thermal loads due to the harsh interaction characteristics between the tool and the workpiece. Compared with the CSJC and MQL cooling settings, the use of flow cutting can provide a superior cooling effect and longer tool life. After removing $1.5 \times 104\,mm^2$ Ti-6Al-4V, Figure 3.23 shows the tool wear topography of selected tools under different dry cutting and cooling conditions at the cutting speeds 75 and 150 m/min. Turning Ti-6Al-4V at a higher speed revealed relatively obvious

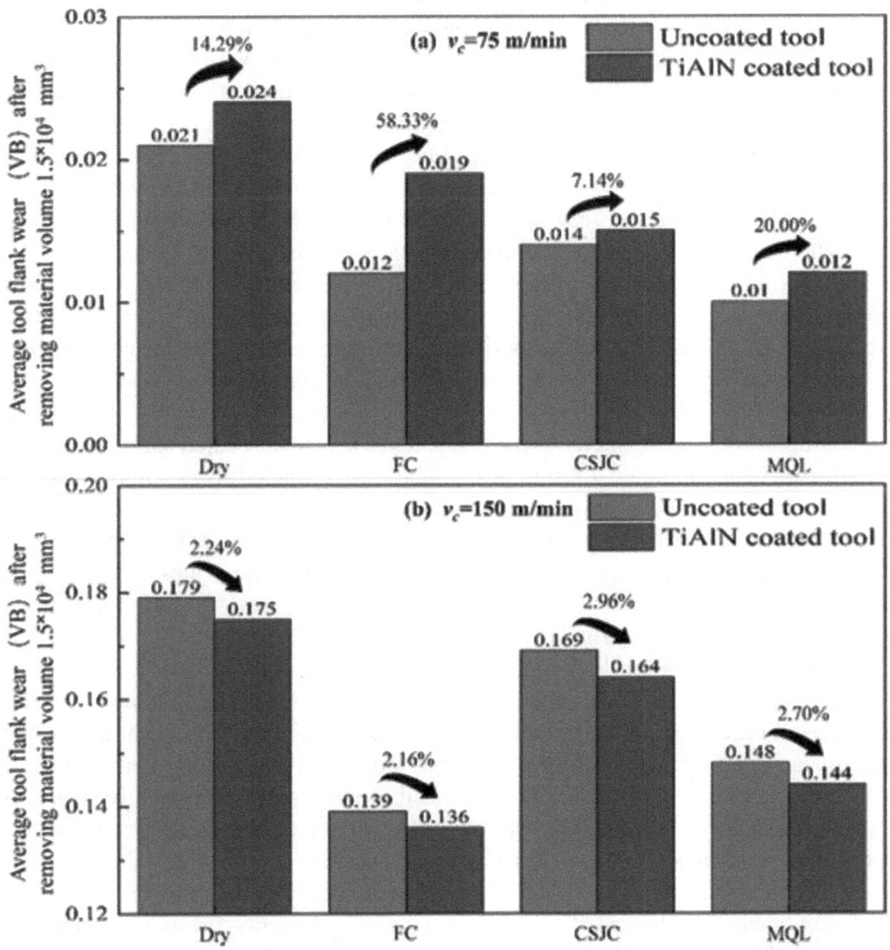

FIGURE 3.23 Tool wear topography after removing $1.5 \times 104\,mm^2$ of Ti-6Al-4V alloy using uncoated and TiAlN-coated tools under dry conditions with different cooling rates at cutting speeds of 75 and 150 m/min. ([26] [Open access].)

pitting and flank wear. When turning Ti-6Al-4V at high speed, 150 m/min, the average flank wear of tool is an order of magnitude higher than at low speed, 75 m/min, after removing the same amount of material from the workpiece [26].

3.6 FUTURE AND CURRENT EVENTS

The release of innovative types of tools will occur much faster than in the past, as the time required to produce new tooling material will become shorter and shorter. However, the main obstacles in tool development continue to be hardness, wear resistance and a balance of strength and toughness, which have been difficult to achieve. As a result, the future tool industry is expected to favor tool materials that combine the hardness and wear resistance of ultra-hard materials with the strength and toughness of high-speed steel and cemented carbide. This trend continues the tradition of creating and studying instruments. The most important point to note is that various coated instruments and composite structures can partially or completely overcome the above difficulties. Thus, the tools coated with a composite structure have the best opportunity for development. Tooling materials will be in high demand and will face increasing challenges in the future regarding workpieces and production processes. In addition, the future tool industry will likely be characterized by the emergence of new types, shifts in factors that determine the market share of each type and situations where complementary and competing products exist.

The cutting tools market will focus on meeting the constant need for innovative tooling ideas. It is an important to remember that cutting tool wear characteristics, tribological evaluation, tool life experiment, and mechanical behavior are the main factors when using a cutting tool to machine high-strength low-alloy steel. With the current state of coating technology, choosing the right coating structure and mixture can be important. Cutting factories have begun to use cement-coated carbide tools due to their excellent tribological properties, which have attracted the attention of scientists. Research is aimed at developing new coating methods and materials for cutting tools. In the manufacturing sector, economic and environmental factors are leading to ever-increasing cutting speeds, as well as decreasing of using coolants and lubricants. At the same time, the need to extend tool life requires a constant further development of carbide technology. A crucial manufacturing procedure that boosts productivity is dry cutting. The tool will function more dynamically due to the continual trend toward better cutting processes and technological features. In general, the mechanical engineering relies heavily on the utilization of nanostructured PVD coatings. When turning hardened steel, a modified tool design may be advantageous to provide better finishes on improved tool materials. Advanced tooling materials typically consist of matrix base material and coating material applied on top of matrix. As a result, extensive research is carried out to select materials suitable for tool bodies and coatings. Since tool wear is a key indicator of tool life, longer tool life and improved machining productivity can be achieved. The materials from which metal-cutting tools are made to have the following qualities; heat resistance, wear resistance, strength, and toughness. Among them, the most important are hot hardness and strength. As a result, the best tool materials are expected to combine strength and hot hardness, another important goal of tool development in industry. Additionally, different tool materials can be used for different purposes due to their different qualities.

PROBLEMS

1. Explain several basic production characteristics of cutting tool manufacturing.
2. List some of the forces involved in creating surfaces.
3. Determine the requirements for developments in the field of processing of solid materials in production.
4. Characterize the cutting forces of the main axial components (thrust).
5. Explanation of the skiving of gear.

REFERENCES

1. Al-Tameemi, H. A., Al-Dulaimi, T., Awe, M. O., Sharma, S., Pimenov, D. Y., Koklu, U., & Giasin, K. (2021). Evaluation of cutting-tool coating on the surface roughness and hole dimensional tolerances during drilling of Al6061-T651 alloy. *Materials*, 14(7), 1783.
2. Tzotzis, A., García-Hernández, C., Huertas-Talón, J. L., & Kyratsis, P. (2020). CAD-based automated design of FEA-ready cutting tools. *J. Manuf. Mater. Process.*, 4(4), 104.
3. Fox-Rabinovich, G., Gershman, I. S., Yamamoto, K., Dosbaeva, J., & Veldhuis, S. (2020). Effect of the adaptive response on the wear behavior of PVD and CVD coated cutting tools during machining with built-up edge formation. *Nanomaterials*, 10(12), 2489.
4. Qiu, J., Zhang, S., Li, B., Li, Y., & Wang, L. (2023). Research on tool wear and surface integrity of CFRPs with mild milling parameters. *Coatings*, 13, 207.
5. Petrovic, A., Lukic, L., Ivanovic, S., & Pavlovic, A. (2016). Optimisation of tool path for wood machining on CNC machines. *Proc. Inst. Mech. Eng. Part C J. Mech. Eng. Sci.*, 231, 72–87.
6. Mughal, K., Mughal, M. P., Farooq, M. U., Anwar, S., & Ammarullah, M. I. (2023). Using Nano-fluids minimum quantity lubrication (NF-MQL) to improve tool wear characteristics for efficient machining of CFRP/Ti_6Al_4V aeronautical structural composite. *Processes*, 11, 1540.
7. Doluk, E., Rudawska, A., Kuczmaszewski, J., & Miturska-Barańska, I. (2021). Surface roughness after milling of the Al/CFRP stacks with a diamond tool. *Materials*, 14, 6835.
8. Ozkan, D., Panjan, P., Gok, M. S., & Karaoglanli, A. C. (2020). Experimental study on tool wear and delamination in milling CFRPs with TiAlN- and TiN-coated tools. *Coatings*, 10, 623.
9. Xu, Z. & Wang, Y. (2022). Study on milling force and surface quality during slot milling of plain-woven CFRP with PCD tools. *Materials*, 15, 3862.
10. Wang, B., Wang, Y., Zhao, H., Wang, M., & Sun, L. (2021). Mechanisms and evaluation of the influence of cutting temperature on the damage of CFRP by Helical Milling. *Int. J. Adv. Manuf. Technol.*, 113, 1887–1897.
11. Landini, P. & Volkert, M. R. (2000). Regulatory responses of the adaptive response to alkylation damage: A simple regulon with complex regulatory features. *J. Bacteriol.*, 182, 6543–6549.
12. Wang, I., Bisoyi, H. K., Zheng, Z., & Gutierrez-Cuevas, K. G. (2017). Stimuli-directed self-organized chiral superstructures for adaptive windows enabled by mesogen-functionalized graphene. *Mater. Today*, 20, 230–237.
13. Fox-Rabinovich, G. S., Gershman, I. S., & Veldhuis, S. (2020). Thin-film PVD coating metamaterials exhibiting similarities to natural processes under extreme tribological conditions. *Nanomaterials*, 10, 1720.
14. Gershman, I. S., Gershman, E. I., Mironov, A. E., German, S., Fox-Rabinovich, G. S., & Veldhuis, S. C. (2016). Application of the self-organization phenomenon in the development of wear-resistant materials—A review. *Entropy*, 18, 385.

15. Fox-Rabinovich, G. & Totten, G. E. (2006). *Self-Organization during Friction: Advanced Surface-Engineered Materials and Systems Design*. CRC Press: Boca Raton, FL, USA.
16. Yau, B.-S., Huang, J.-L., Lu, H.-H., & Sajgalik, P. (2005). Investigation of nanocrystal-(Ti1−xAlx)Ny/amorphous-Si_{3N4} nanolaminate films. *Surf. Coat. Technol.*, 194, 119.
17. Wang, Z., Wang, C., Zhao, Y. L., Hsu, Y. C., & Li, C. L. (2020). High hardness and fatigue resistance of CoCrFeMnNi high entropy alloy films with ultrahigh-density nanotwins. *Int. J. Plast.*, 131, 102726.
18. He, Q., Paiva, J. M., Kohlscheen, J., Beake, B. D., & Veldhuis, S. C. (2020). An integrative approach to coating/carbide substrate design of CVD and PVD coated cutting tools during the machining of austenitic stainless steel. *Ceram. Int.*, 46, 5149–5158.
19. Qiu, J., Zhang, S., Li, B., Li, Y., & Wang, L. (2023). Research on tool wear and surface integrity of CFRPs with mild milling parameters. *Coatings*, 13, 207.
20. Petrovic, A., Lukic, L., Ivanovic, S., & Pavlovic, A. (2016). Optimisation of tool path for wood machining on CNC machines. *Proc. Inst. Mech. Eng. Part C J. Mech. Eng. Sci.*, 231, 72–87.
21. Mughal, K., Mughal, M. P., Farooq, M. U., Anwar, S., & Ammarullah, M. I. (2023). Using nano-fluids minimum quantity lubrication (NF-MQL) to improve tool wear characteristics for efficient machining of CFRP/Ti_6Al_4V aeronautical structural composite. *Processes*, 11, 1540.
22. Doluk, E., Rudawska, A., Kuczmaszewski, J., & Miturska-Barańska, I. (2021). Surface Roughness after milling of the Al/CFRP stacks with a diamond tool. *Materials*, 14, 6835.
23. Wang, X., Zhang, X., Pan, D., Niu, J., Fu, X., & Qiao, Y. (2023). Tool wear and surface integrity of γ-TiAl cryogenic coolant machining at various cutting speed levels. *Lubricants*, 11(6), 238.
24. Xu, Z. & Wang, Y. (2022). Study on milling force and surface quality during slot milling of plain-woven CFRP with PCD tools. *Materials*, 15, 3862.
25. Wang, B., Wang, Y., Zhao, H., Wang, M., & Sun, L. (2021). Mechanisms and evaluation of the influence of cutting temperature on the damage of CFRP by Helical Milling. *Int. J. Adv. Manuf. Technol.*, 113, 1887–1897.
26. Zhao, J., Liu, Z., Gong, Z., Liu, A., Cai, Y., Wang, B., … & Ji, H. (2023). Comparison of surface morphology and tool wear in the machining of Ti-6Al-4V alloys with uncoated and TiAlN tools under dry, minimum quantity lubrication, flood cooling, and low-temperature spray jet cooling conditions. *Micromachines*, 14(6), 1263.
27. Daniyan, I., Mpofu, K., Ramatsetse, B., & Gupta, M. (2021). Review of life cycle models for enhancing machine tools sustainability: Lessons, trends and future directions. *Heliyon*, 7(4), e06790.
28. Trzepieciński, T. (2023). Approaches for preventing tool wear in sheet metal forming processes. *Machines*, 11(6), 616.
29. Zhao, J. & Chang, Z. (2022). A new approach of modelling bottom edge cutting in 4-axis rough milling of complex parts and its application on feed rate optimization. *Micromachines*, 13(12), 2071.
30. Knap, A., Dvořáčková, Š., & Knápek, T. (2022). Study of the machining process of GFRP materials by milling technology with coated tools. *Coatings*, 12(9), 1354.

4 Machining of Hard Materials

4.1 FUNDAMENTAL HARD MACHINING (HM) FEATURES

The process of machining materials with hardness greater than 45 HRC (Rockwell C hardness), which required grinding, is known as hard machining (HM). High precision and surface finish can be achieved in hard materials using this technique, which offers significant benefits in terms of productivity, economy, and flexibility. It uses advanced cutting tools and machining strategies. The use of high-performance cutting tools specifically designed to survive the harsh conditions involved in machining hard materials such as carbide, ceramic, or CBN is one of the fundamental principles of HM. Higher cutting speeds, lower cutting forces, and less heat impact on the workpiece are characteristics of the process, which increases surface integrity and dimensional accuracy.

Another critical element of HM is the process of maximizing tool life and minimizing wear by optimizing machining parameters such as feed rates, depths of cut, and tool geometry. Efficiency is further increased by allowing real-time adjustments to machining processes through the integration of advanced monitoring and control systems. Solid-state machining, which is replacing traditional milling techniques, signals a major move toward more efficient manufacturing practices as industries demand higher performance and shorter lead times. The fundamentals of solid-state machining continue to be refined through ongoing research and technological developments, making it a critical step in modern manufacturing.

4.1.1 HM Definition

Solid machining is the process of using specially designed cutting tools to withstand the harsh conditions of materials with a hardness greater than 45 HRC. By using advanced tool materials including high-speed steel (HSS), ceramics, and CBN, this process makes it possible to precisely and efficiently form solid components without the need for traditional grinding methods. Solid machining has many advantages over traditional processes, including shorter machining times, better surface finishes, and increased dimensional accuracy. Solid machining is more affordable. Solid machining increases productivity by reducing thermal distortion and tool wear through the use of high cutting speeds and finely tuned machining parameters. This strategy is increasingly being used in some sectors such as manufacturing, automotive, and aerospace, where high-performance parts are essential. Solid machining continues to evolve as a result of technological advances including new developments in tool materials, coatings, and monitoring systems to increase its efficiency and expand its uses in contemporary manufacturing processes. Values of 1 μm Rz (or 0.1 μm Ra)

DOI: 10.1201/9781003604648-4

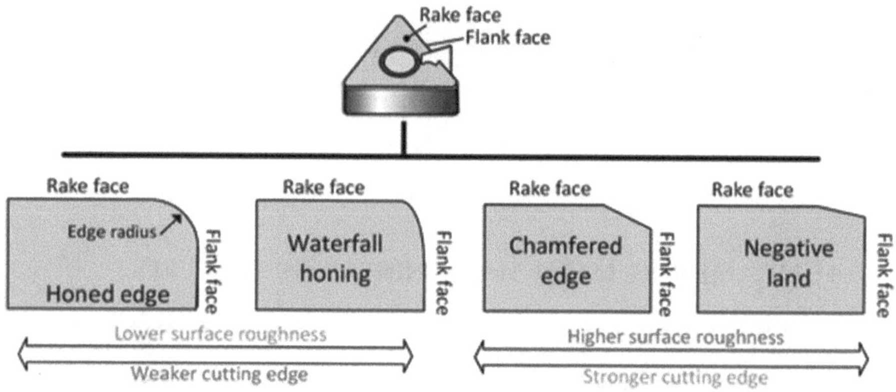

FIGURE 4.1 Typical edge preparations for tools. ([20] [Open access].)

in high-precision CBN machining and corresponding dimensional tolerance IT3 are achievable using space fillers with beveled, sharp, and polished cutting edges for grinding the surface of SAE 1070 high carbon steel. Figure 4.1 shows the improved surface quality achieved by space fillers with beveled, sharp, and polished cutting edges compared to other edge preparation combinations including beveled, sharp, and polished cutting edges [20].

On the other hand, deep turning followed by finish grinding can also be a useful pre-finishing procedure for parts with exceptionally tight tolerances. Their use has expanded to some important industrial areas, including die and mold manufacturing, automotive, hydraulics, and roller bearings. Gears, gear shafts, bearing races, and other transmission parts are typically machined by turning, while high-speed milling dominates die and mold manufacturing.

The macroscopic morphology and schematic structure of bare TC4, single-layer Ta, and multilayer Ta/Ti/Zr/Ta coatings are shown in Figure 4.2a–c. All the surfaces of the samples are smooth as can be seen; however, the single-layer and multilayer coatings have a slightly yellower appearance than the parent TC4 alloy. α and β phases in TC4 alloy have a roughly equiaxed structure (Figure 4.2d). Figure 4.2e and f shows the cross-sectional scanning electron microscopy (SEM) images of coating samples, which allow their thickness and structure to be observed. Zr and Ti layers are black, while Ta coating layer is comparatively shiny. Ta and Ta/Ti/Zr/Ta coatings have a thickness of approximately 3.28 and 3.10 μm, respectively, as shown in Figure 4.2b and c. In addition, the thickness of Ta/Ti/Zr/Ta coating layers is determined using the first layer or Ta, which is 0.70 μm, followed by the second layer or Ti, which is 0.93 μm, third layer or Zr, which is 0.75 μm, and fourth layer or Ta, which is 0.72 μm. In addition, as shown in Figure 4.2e and f, a columnar structure is observed in all the coatings. In particular, the single-layer Ta coating in Figure 4.2e has shown a clear columnar structure that is perpendicular to the surface. In comparison, the Ta/Ti/Zr/Ta multilayer coating has showed a less columnar structure [2].

The lubricating effect of material can be measured by its friction coefficient, although this does not necessarily indicate its wear resistance. For this purpose, further studies of wear scars are carried out. SEM and optical microscopy (OEM)

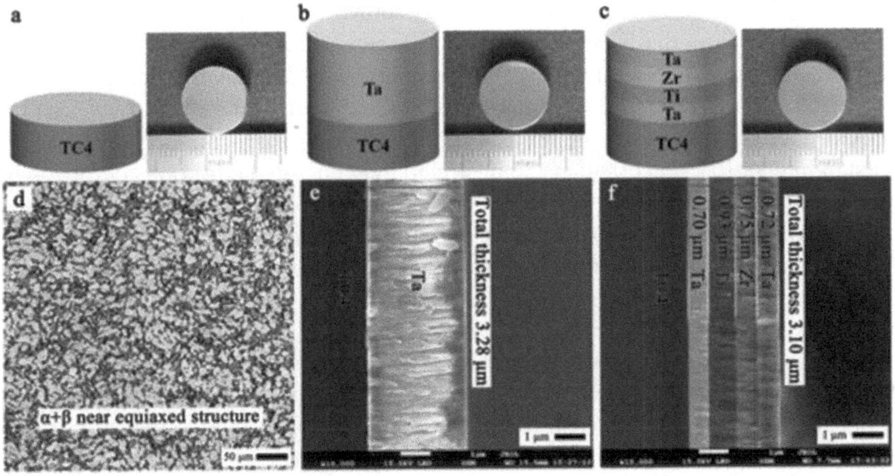

FIGURE 4.2 Structure of TC4 alloy, Ta coating, and Ta/Ti/Zr/Ta coatings shown schematically with the naked eye. (a–c) Optical micrographs of the TC4 alloy, single-layer Ta coating, and multilayer Ta/Ti/Zr/Ta coating, respectively. (d) Optical micrograph of the TC4 alloy; (e, f) Cross-sectional scanning electron micrographs of the Ta and Ta/Ti/Zr/Ta coatings, respectively [2]. [Open access].

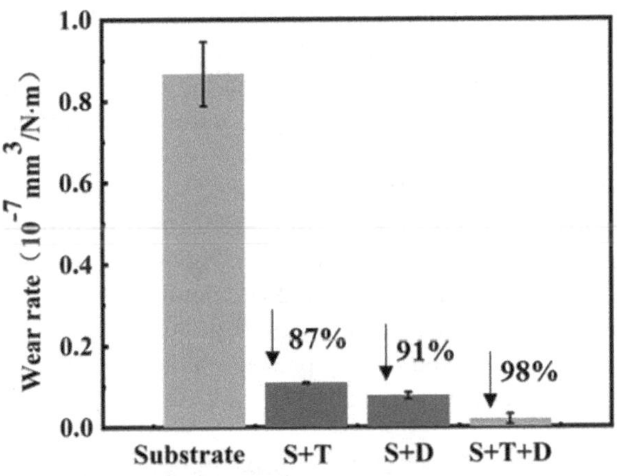

FIGURE 4.3 Schematic diagram of the friction and wear mechanism of the textured H-DLC coating on the $CuAl_{10}Fe_5Ni_5$ surface. ([1] [Open access].)

images of wear scars on coated and uncoated TC4 specimens sliding on SiC balls under a normal force of 0.5N are shown in Figure 4.2a–c. When subjected to a normal load, 0.5N, the surface showed obvious wear scars. To be more precise, parallel grooves containing a large number of adhered black particles are visible on uncoated TC4 alloy. H-Diamond-like carbon (DLC) coatings were deposited on the $CuAl_{10}Fe_5Ni_5$ substrate (S+D) and the E90 elliptical texture surface (S+T+D) to further improve the tribological performance of the $CuAl_{10}Fe_5Ni_5$ surface. Figure 4.3 shows the temperature and friction coefficient. The friction coefficient of the

$CuAl_{10}Fe_5Ni_5$ substrate is relatively high, as shown in Figure 4.3a, and it changes significantly during the friction process and averages about 0.17. Due to the ability of the texture to absorb wear debris and retain lubricant and the excellent lubricating property of the H-DLC coatings, the E90 elliptical texture and the H-DLC coating help to stabilize and reduce the friction coefficient compared with the substrate. The outstanding antifriction effect is demonstrated by reducing the friction coefficient by 95% and 94%. The friction coefficient is approximately 0.08, which is the difference between the textured (S+T), coated (S+D) surface, and the textured surface of the H-DLC coating (S+T+D). On the other hand, the friction coefficient of the textured coating increases due to the edge scraping. The ideal size of the elliptical texture reduces wear and friction on the surfaces of GCr15 and $CuAl_{10}Fe_5Ni_5$ by promoting the trapping of wear particles and the storage of lubricating oil on the surface texture. The surface texture effectively reduces the rise in friction temperature, which is more noticeable for $CuAl_{10}Fe_5Ni_5$ than for GCr15 substrate, by increasing the contact area between $CuAl_{10}Fe_5Ni_5$ and lubricating oil. Finally, Figure 4.3 suggests and illustrates the friction mechanism [1].

In addition, typical wide-wear grooves and parallel ridges with a certain amount of wear particles are observed on all surfaces of specimens, indicating that abrasive wear is the main wear mechanism [2]. The wear scar results have showed that uncoated TC4 has the worst wear resistance, and Ta coating has the lowest width and depth values, 624.4 and 6.7 μm, respectively. The equivalent values of Ta/Ti/Zr/Ta multilayer coating are in the middle, 629.8 μm in width and 10.0 μm in depth, respectively. It is worth noting that the scar depth of Ta/Ti/Zr/Ta coating increases. This may be due to the poor adhesion strength of coating to substrate under a high normal load. This may be due to the insufficient adhesion of several layers and smaller thickness of each layer [23]. In addition, SEM and energy dispersive X-ray spectroscopy (EDS) scanning of the GCr15 counter bodies sliding on the substrate and the $CuAl_{10}Fe_5Ni_5$ samples with E90 texture were performed to further investigate the wear morphology and mechanism, as shown in Figure 4.4. The surface of the top GCr15 sample is severely eroded and covered with numerous grooves after sliding on the $CuAl_{10}Fe_5Ni_5$ substrate, as shown in Figure 4.4a. According to the corresponding EDS mapping, O and Cu are abundantly present on the worn surface. According to these results, GCr15 experiences oxidative and adhesive wear due to the transfer of

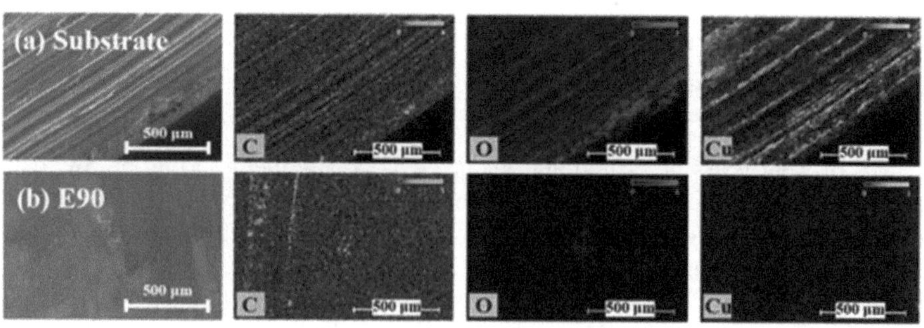

FIGURE 4.4 GCr15 counterfaces sliding against the substrate and E90-textured $CuAl_{10}Fe_5Ni_5$ are shown in SEM pictures and EDS maps. ([30] [Open access].)

$CuAl_{10}Fe_5Ni_5$ substrate to its surface. The worn surface of GCr15 was significantly less grooved when it slid on the elliptical textured surface of E90 [2].

Typically, dry machining is performed using CBN cutting materials. This becomes possible due to the high heat resistance of cutting material, as well as the positive influence of high temperature in the chip formation zone. The insert structure experiences severe thermal stress due to lack of coolant or interruptions in cutting. This can cause structural cracks and ultimately lead to failure of the insert. Steel in the shear zone is heated to high temperatures during HM, which softens it. Most of the heat generated in shear zone goes into the chip rather than the workpiece. This indicates that the edge of the plate is not exposed to heat during the process. Although carbide is subjected to severe testing, the hardness of CBN does not change until it reaches 1200°C, with a decreasing in hardness occurring at about 800°. Chemical resistance is another deciding factor, especially given current temperatures [5–7].

4.1.2 Grinding Operations with Comparison

Precision machining relies heavily on grinding processes to provide excellent surface finishes and tight tolerances on a variety of materials. This process involves removing material using abrasive wheels or belts, works particularly well on hard materials that are difficult to machine using conventional techniques. Many variables such as type of grinding technique, the properties of abrasive and specific application requirements can greatly affect the efficiency and effectiveness of grinding processes. Surface grinding, cylindrical grinding, internal grinding and centerless grinding are examples of common grinding processes. Each technique is unique and suitable for a specific set of applications. For example, cylindrical grinding is used for circular components, while surface grinding is best for flat surfaces. Additionally, the performance and results of grinding process are influenced by the choice of grinding wheels such as silicon carbide, diamond or aluminum oxide.

Recent advances in grinding technology, such as the creation of ultra-fine materials and high-speed grinding, have increased precision and productivity. Furthermore, the integration of intelligent monitoring systems facilitates immediate feedback and adjustments, thus optimizing grinding parameters to enhance efficiency. Grinding provides better surface finishes and excellent geometric accuracy compared to other manufacturing methods. However, compared to processes such as grinding or turning, it typically entails slower material removal rates. Material hardness, desired surface smoothness, and production volume play an important role in decision between the grinding and other manufacturing techniques.

In short, knowledge of different grinding processes and their relative benefits is vital for manufacturers looking to improve their manufacturing processes. Grinding continues to be a vital part of contemporary manufacturing, driving advances in materials processing and precision engineering as technology evolves further. A flexible grinding operation means that a single, autonomously controlled machine can flexibly produce a family of components, or possibly multiple families, For example, one machine that can be used is designed to produce cylindrical parts with varying diameters and shoulders. Peel sanding is a word that many sander manufacturers are using these days. Peel grinding allows the grinding wheel to be used like hard-turning tool by combining high-speed grinding techniques with computer numerical control (CNC). A 5 mm wide grinding wheel can often produce a shape or several diameters by following a pre-programmed path. The variety of parts or operations that can be performed on a single machine is expanded by the peel-grinding machine. The following are a few fundamental differences between the various processing methods [2]:

- Hard turning can be completed in one setup and done under dry conditions, which greatly speeds up the operation.
- Lathes provide greater production flexibility.
- The CNC lathe can perform roughing and finishing operations in one clamping. Changing tools at a turning center or turning station makes it easier to automate multiple turning operations.
- There are no coolant, maintenance, or disposal costs because hardening is dry.

Two interactions between the abrasive material and the grinding surface produce frictional effects when grinding hard, brittle materials on the surface. The flow of the workpiece over the contact surface creates the first effect, while the friction between the grinding surface and the granular wear ridge creates the second effect (see Figure 4.5) [32].

The roughness of the ground surface is directly affected by surface grinding process parameters, including depth of cut, lateral feed, and linear feed. Elevated levels of either will result in an increased rate of thermal energy release due to material removal. As already explained, this will result in surface roughness and possibly thermal damage. It is important to remember that obtaining the best surface quality, the cutting direction should be parallel to direction of grain in surface layer. The expenses of turning and grinding are compared in Figure 4.6 [4].

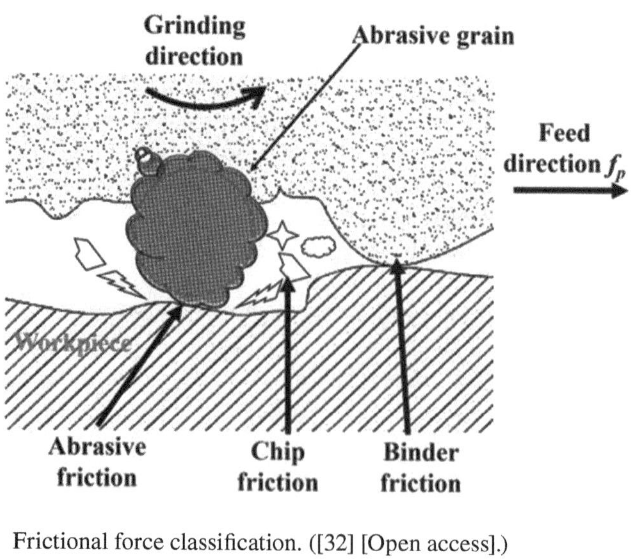

FIGURE 4.5 Frictional force classification. ([32] [Open access].)

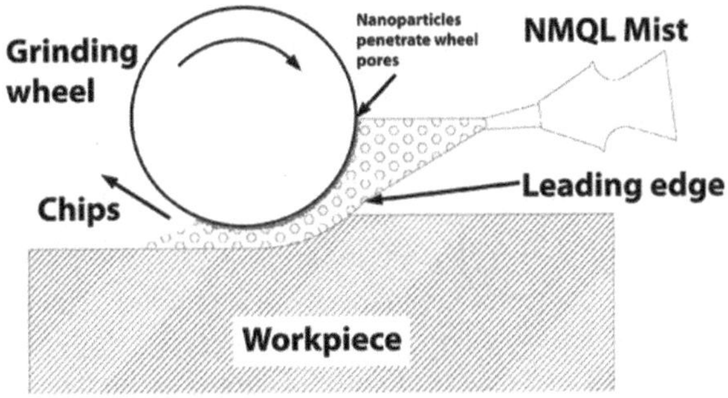

FIGURE 4.6 Comparing the costs of grinding and turning [4].

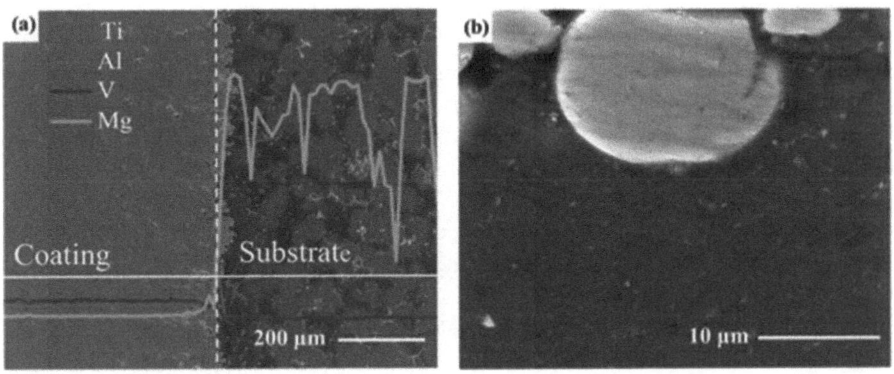

FIGURE 4.7 TC4 alloy at 0.5 N, Ta coating at 0.5 N, Ta/Ti/Zr/Ta coating at 0.5 N, TC4 alloy at 2 N, Ta coating at 2 N, and Ta/Ti/Zr/Ta coating at 2 N as SEM images with enlarged areas of worn surfaces. ([2] [Open access].)

Figure 4.7 shows high-magnification microscopic images of wear scars on coated and untreated TC4 alloy specimens sliding on SiC balls under normal loads of 0.5 and 2 N to further investigate wear behavior. The typical morphologies of worn surface obtained under normal load (0.5 N) are shown in Figure 4.7a–c. As for TC4 alloy, a large amount of adhesion region and wear debris are observed on wear scar (high-magnification image; rectangular area in Figure 4.7a), indicating that adhesive wear is a part of wear process.

The detached wear debris created a "cold weld" effect on the contact surface of specimen due to harder friction pair continuously wearing the surface of TC4 alloy to form adhesion nodes. However, this bonding condition is created by "adhesion". The resulting shear force acts on the wear scar path due to repeated movement of friction pairs along the path on specimen surface. This causes some of weakly bonded debris to form nodes, break away from them, and migrate back, creating additional accumulation nodes [2]. As for the single-layer Ta coating, it is obvious that the coating cracked after wear experiment under the standard stress (0.5 N, Figure 4.7b). At a higher magnification, it is found that the coating gradually peeled off has a broken region, which is the main source of abrasive particles. In contrast, the multilayer Ta/Ti/Zr/Ta coating did not degrade over time; rather, the entire part aged due to insufficient adhesion (Figure 4.7c).

4.1.3 Technology-Related Procedures: Hard Machining

Rigid machining with a geometrically defined cutting edge offers several advantages, including fewer grinding operations, enhanced surface quality, and shorter processing times. PHORN has extensive experience in tool technology and offers a wide range of tools for the efficient machining of steels with a hardness greater than 56 HRC. CBN has proven to be an effective cutting material for grooving applications. After diamond, CBN is the second hardest substance known. When used correctly, CBN tools wear much slower than tools made from conventional cutting materials. As a result, greater shape and dimensional accuracy can be achieved, and hard materials—steel with a hardness of up to 70 HRC—can be processed with consistency.

Peripheral (end) milling and face milling are the two main milling procedures, as shown in Figure 4.8. Face milling is used to create smooth surfaces on a workpiece, while peripheral milling is typically used to cut slots or profiles. In this study, an optimization strategy using milling operations was tested [28].

There are no different grades of CBN. CBN volume fraction, fillers, particle size, and ceramic/metallic binder phase (cobalt/nickel) contribute to differentiation. Rotary milling cutters are used in milling procedures to remove material from a workpiece to create precise shapes or surface features. CBN is a cutting material that can be used with several Horn tooling systems. These are usually tools specifically designed for their intended use. However, Horn offers conventional tools equipped with this extremely durable cutting material. Built-up edge (BUE): As shown in Figure 4.9, the BUE can form at the tip of the tool, changing the tool nose radius and creating higher surface roughness values. The adhesion of workpiece material to tool cutting edge during machining operations is the primary source of BUE in this process. This adhesion is caused by the high temperatures and pressures at cutting edges. The micromechanics of BUE can be influenced by many factors, including the workpiece material, cutting conditions and tool material [28].

According to DMG Mori (often referred to simply as DMG), which is a well-known German company known for its expertise in the production of machine tools, particularly CNC machines, machining hard parts can bring significant financial benefits

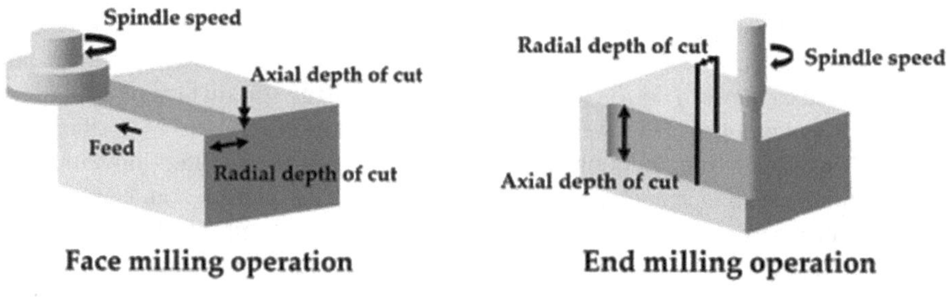

FIGURE 4.8 Milling operations. ([28] [Open access].)

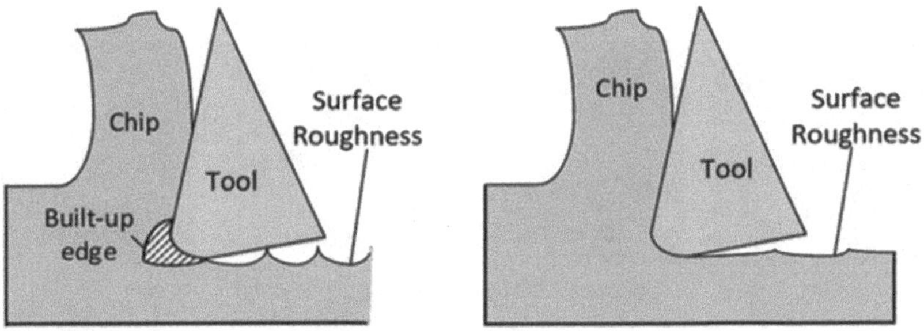

FIGURE 4.9 Comparison of high-speed machining (HSM) and traditional die and mold-making methods. ([20] [Open access].)

through a combination of laser finishing and high-speed cutting (HSC) rough milling, which can operate at speeds up to 42,000 rpm. Production cycle time is reduced by almost six times through the use of laser processing and elimination of electrical discharge machining procedures [2]. The complete machining of tiny, precise parts made from both metallic and non-metallic materials is a special application of this method. High-speed machining (HSM) and conventional die and mold-making techniques are contrasted in Figure 4.9 [20].

Nearly all of the cells cultured in the four groups exhibit normal morphology, lamellar expansion, and a uniformly distributed spindle shape. Moreover, it is observed that no dead cells are found and cell number in each group doubled during the incubation period. However, the non-TC4 coated group had relatively fewer cells compared with other groups. After 24 and 96 hours of incubation, the cell numbers in both the monolayer and multilayer groups are higher than the control group [15].

Figure 4.10 illustrates the fluorescence microscopy images of live (green) and dead (red) cells in control group, non-TC4 coated group, Ta monolayer and Ta/Ti/Zr/Ta multilayer groups after 24, 96 and 168 hours of incubation [2]. The cell morphologies in the three experimental groups are similar to those in the control group as shown in Figure 4.10 [2].

4.2 TOOLS AND EQUIPMENT

4.2.1 Machine Tools

Every machine has a way to store tools and work, as well as to accurately control the depth of cut. Cutting speed is the relative movement of cutting edge of tool relative to workpiece. The feed motion is the speed at which the uncut material comes into contact with tool. As a result, four bionic tools are created, as shown in Figure 4.11, which combine traditional tools with good mechanical properties of projectiles. To facilitate comparison and analysis of performance of different types of tools, the tool is presented as shown in Figure 4.11. Tool 1 is a curved, convex tool that extends from the tip of blade inward at a certain angle. Tool 2 is a bionic tool with a concave, curved surface that is depressed at an angle from the tip of knife to inside of blade. Tool 3 is a flat, convex bionic tool with a protrusion at a certain angle from its tip.

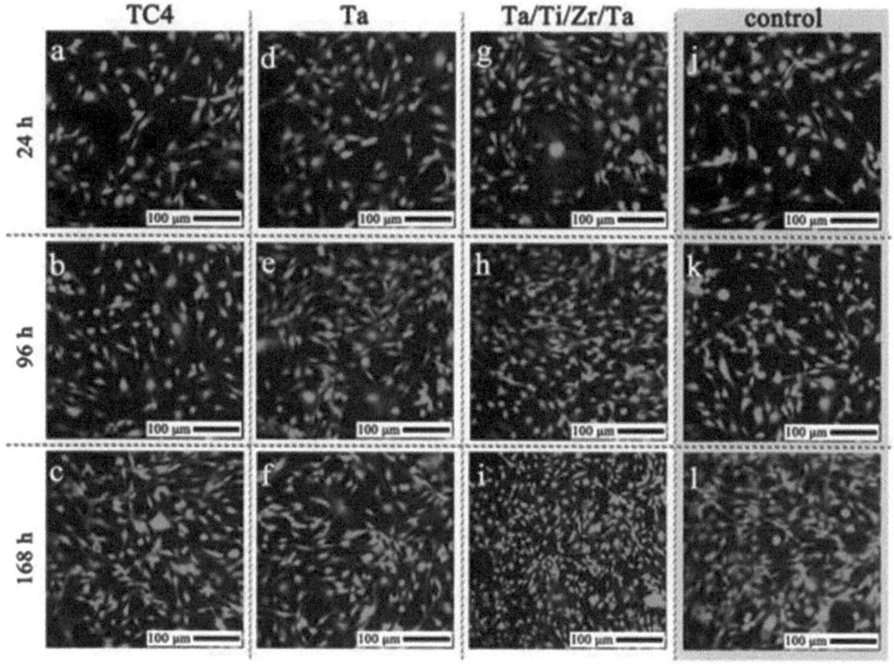

FIGURE 4.10 Fluorescence micrographs showing live (grey) and dead (black) cells in different samples after 24, 96 and 168 hours of incubation for (A,D,G,Y), (B,E,H,K), and (C,F,I,L). ([2] [Open access].)

The final aluminum parts are rectangular with dimensions, 160 × 100 mm. Eighteen panels are cut and ready for testing due to selected specifications. Since all conditions are applied three times and showed average values in this study, this shape is chosen. The welding tool has a flat shoulder and cylindrical pin is made of W302 steel. The dimensions of tool are 15 mm for shoulder, 5 mm for the pin and 4.5 mm for length. According to some assumptions, the ideal pin length is 85% of base metal thickness [19]. Previous studies have shown that the ideal length is 4.5 mm. Figure 4.11a and b illustrates a schematic diagram and photograph of tool, respectively [29].

Since overheated tools can reduce their cutting performance, this study highlights the importance of temperature control in metal-cutting operations. Since cutting speed and shear force affect heat generation during machining, their interaction is critical. The material being cut and thermal resistance of cutting tool affect the optimum cutting speed. The shape of workpiece, hardness of machine tool, and depth of cut are important factors that affect machining efficiency. There are two main types of metal-cutting tools: multi-point tools such as drills and milling cutters, which have multiple cutting edges, and single-point tools that are used for tasks such as drilling and turning. This describes two working methods; the workpiece moves against a fixed tool as in the diagrams, where the tool moves against the workpiece in a linear motion. To improve machining performance and reduce cutting-edge wear, it is necessary to properly adjust the edge angles or clearances. To reduce heat and friction,

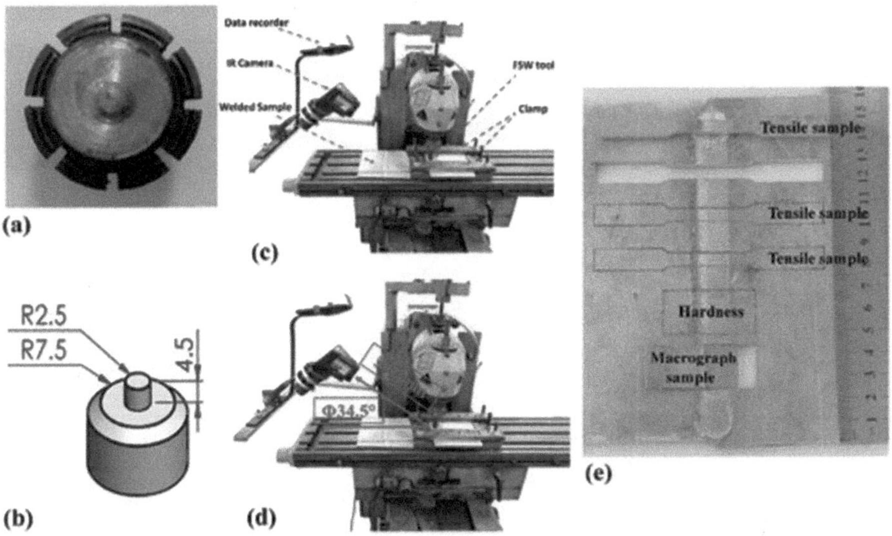

FIGURE 4.11 (a) Schematic diagram of the friction stir welding (FSW) tool, (b) schematic representation of tool dimensions, (c) experimental setup of the FSW process, (d) viewing angle of an infrared camera, and (e) test specimens for microstructure and mechanics. ([29] [Open access].)

and to cause wedging motion when chips are formed, cutting tools are marked with rake angles. When using convex, convex, concave and convex-flat tools in the test, the main cutting force increased significantly. According to an engineering study on tool tilt angle (TTA), the maximum contact area between the workpiece and welding tool occurs at a TTA of 1.5°, or approximately 263.22 mm^2. On the other hand, the smallest contact area, which is approximately 225.1 mm^2, is observed at TTA 3°. After controlling for variables such as surface tension and flash generation, these measurements are obtained by examining the tool contact surfaces and aluminum alloys. The predicted contact area is shown in Figure 4.12, which gives a visual representation of how TTA affects the tool–workpiece interaction [29].

The cutting tool must be harder than workpiece and maintain the temperature of its cutting edge created by cutting motion to remove chips from the workpiece. The first steel used in machine tools has a carbon percentage of 1–2%. Although these carbon steel tools are reasonably priced, their cutting performance tends to decrease at temperatures above 400°F (205°C). The invention of high-speed steel made it possible to operate tools at speeds two or three times higher than those allowed by carbon steel, doubling or tripling the productivity of machine shops around the world. Eighteen percent of one of the most popular varieties of HSS consists of tungsten, 4% chromium, 1% vanadium, and only 5%–8% carbon. In addition, due to the risk that temperature fluctuations in the machine shop and heat generated by machining process itself may affect productivity, it has a thermal distortion stabilization mechanism [22].

By circulating temperature-controlled fluid through key machine components, this technology reduces the amount of thermal deformation that the machine structure is

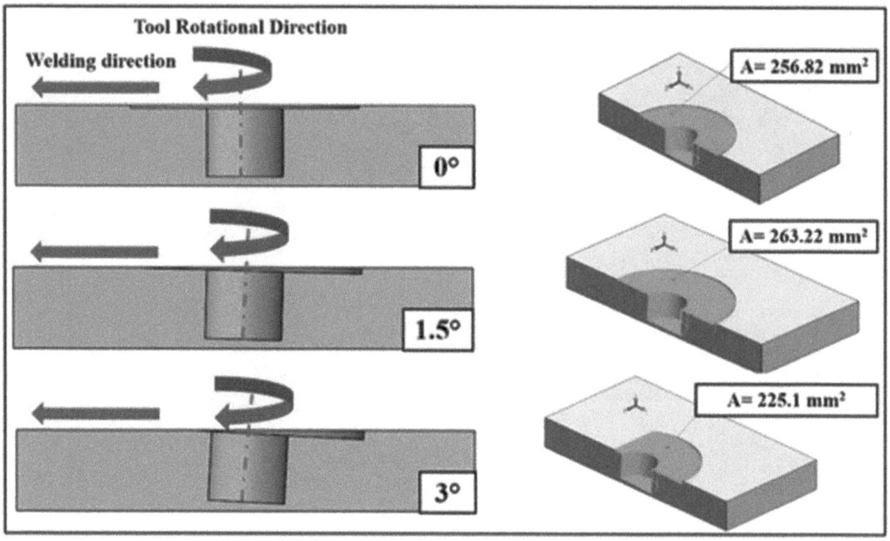

FIGURE 4.12 Illustration of the relationship between the tool insertion depth (0.2 mm) and the current approximate contact area (TTA). ([29] [Open access].)

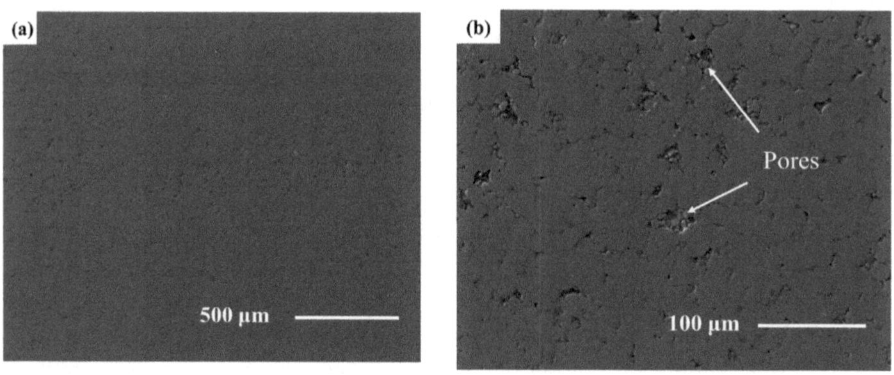

FIGURE 4.13 Uses mold and CNC die cutter (jigging machine) for high-speed hard milling: (a) general view, (b) spindle and drive motor, and (c) concrete structure similar to a monobloc bridge. ([7] [Open access].)

subjected. To machine die and mold parts as accurately as possible, warp control is critical, especially for finishing operations that require high-precision machining and long cutting sessions (several hours) with the same cutter. In this way, a positioning accuracy of ±0.002 mm and repeatability of 0.001 mm can be achieved. The patented self-adjusting spindle with preload CNC router shown in Figure 4.13 can grind material hardened to HRC40, HRC50, or even HRC62 at high speeds up to 20,000 rpm with maximum feed speed 200 m/min. (800 in/min) and a maximum feed speed of 125 m/min (500 in/min) [7].

It is designed to isolate the vibration and runout of spindle drive motor, improve surface smoothness and machining accuracy, and extend the life of cutter when used coaxially with a diaphragm clutch without any backlash (Figure 4.13b) [7].

4.2.2 Materials and Cutting Tools

The materials used in cutting tools have a significant impact on the effectiveness and performance of manufacturing processes. Cutting tools are essential parts of production processes, combination of application, design and material quality determine their effectiveness. HSS, carbide, ceramics, and advanced composites are popular materials for cutting tools. Each of these materials offers specific benefits in terms of durability, hardness, wear resistance, and thermal stability.

While carbide tools are preferred for their high hardness and wear resistance, making them ideal for high-speed applications, HSSs are valued for their toughness and adaptability. Ceramics offer a balance between wear resistance and hardness, while ceramics are used for their outstanding toughness and resistance to high temperatures. New materials such as polycrystalline diamond (PCD) and cubic boron nitride (CBN) are offering superior performance for specific applications.

New developments in materials science such as advanced coatings that increase wear resistance and reduce friction are aimed at improving tool life and performance. Furthermore, real-time adjustment of manufacturing parameters is made possible by integrating smart technologies and monitoring systems, enhancing tool performance. Gaining greater precision in manufactured parts, reducing costs, and streamlining production processes all depend on understanding how materials and cutting tools interact. The future of manufacturing technology will be greatly influenced by ongoing research and development in cutting tool materials as industries continue to seek higher levels of efficiency and quality. CBN provides superior performance for specific applications.

New developments in materials science such as advanced coatings that increase wear resistance and reduce friction, aim to improve tool life and performance. Furthermore, real-time adjustments of manufacturing parameters are made possible by integrating smart technologies and monitoring systems and enhancing tool performance. The success of manufacturing processes depends largely on the interaction between the materials and cutting tools. The creation of advanced cutting tool materials has become essential to increase productivity, accuracy and efficiency as manufacturing needs change. Cutting tool technology has historically relied on traditional materials such as carbide and HSS, but advances in metal ceramics, ceramics, and ultra-hard materials such as PCD and CBN have also played a significant role in increasing the capabilities of cutting tools for specific applications.

Recent developments have further enhanced tool performance by reducing friction, enhancing wear resistance and facilitating real-time monitoring. These developments include better coatings and the use of smart technologies. These developments reduce operating costs, improve machining accuracy and extend tool life. As companies place greater emphasis on efficiency and sustainability, they will need to continue researching cutting tool materials. To meet the challenges of modern production, future developments will likely focus on developing environmentally friendly materials, improving recycling techniques and enhancing performance features.

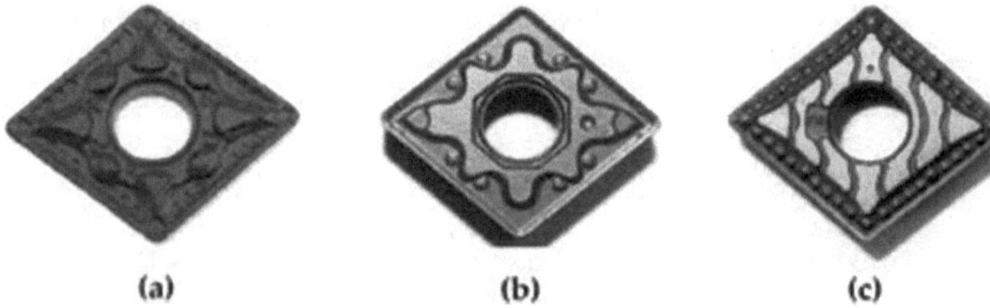

FIGURE 4.14 Indexable inserts utilized in tests: CNMG 120408-MP3 WPP10S, CNMG 120408-PM 4315, and CNMG 120408E-M GRADE T9315. ([7] [Open access].)

Finally, the advancement of manufacturing technology depends on the continuous development of materials and cutting tools, ensuring that manufacturers can meet the increasing expectations placed on them in terms of product quality, productivity, and sustainability. There are three categories of cutting tools:

- Single-point cutting tool
- Double-cutting point tool
- Multi-purpose cutting tool

Ti(C,N) + Al_2O_{3+}TiN chemical vapor deposition (CVD) coating deposited on the HC substrate in CNMG 120408-PM 4315 (Figure 4.14a). The main application of the additive is steel machining, while cast iron is another possible application. TiCN + Al_2O_3 CVD coating deposited on the HC substrate in CNMG 120408E-M GRADE T9315 (Figure 4.14b). The main application of the additive is steel machining, while it can also be used for machining cast iron and stainless steel.

TiCN + Al_2O_3 CVD coating deposited on the HC substrate in CNMG 120408-MP3 WPP10S (Figure 4.14c). The main application of the additive is steel machining, while cast iron is another possible application [7].

Figure 4.14 shows several recently developed CBN inserts. Miniature inserts (NEW PETIT CUT from Mitsubishi Carbide) with a CBN tip soldered directly to the main carbide insert are shown in Figure 4.14. As a result, CBN blank is stronger and can absorb more heat generated.

Based on the above test, the arc inclination angle is changed for testing, and a combination of v_c = 150 m/min, f = 0.2 mm/rev, and a_p = 0.15 mm of main cutting force of the minimum cutting dose is selected to conduct the FE simulation test of the variable arc inclination angle to investigate the best tool for the inclination angle of the biconvex arc. The bending angles are in the following order: 5°, 7°, 10°, 13°, 15°, and 19°. Under the same conditions, a 3D model is built and tested. Figure 4.15 shows the 3D model [3].

Four types of micro-textured tools are designed on the comb-shaped cutting surface of a concave bio-tool to analyze the effect of micro-textured tools on cutting performance of tool (the four texture patterns are shown in Figure 4.16). These types of micro-textured tools are micro-textured tools with a 45° groove, parallel to the main one [3].

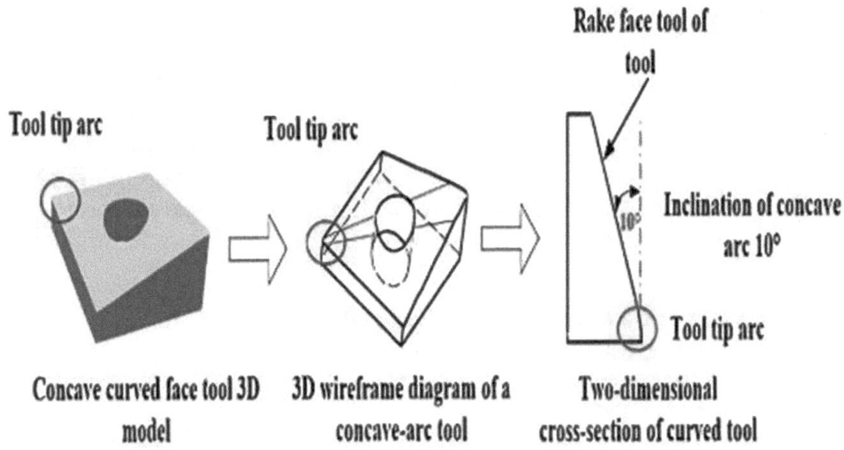

FIGURE 4.15 Concave deflection model in three dimensions. ([3] [Open access].)

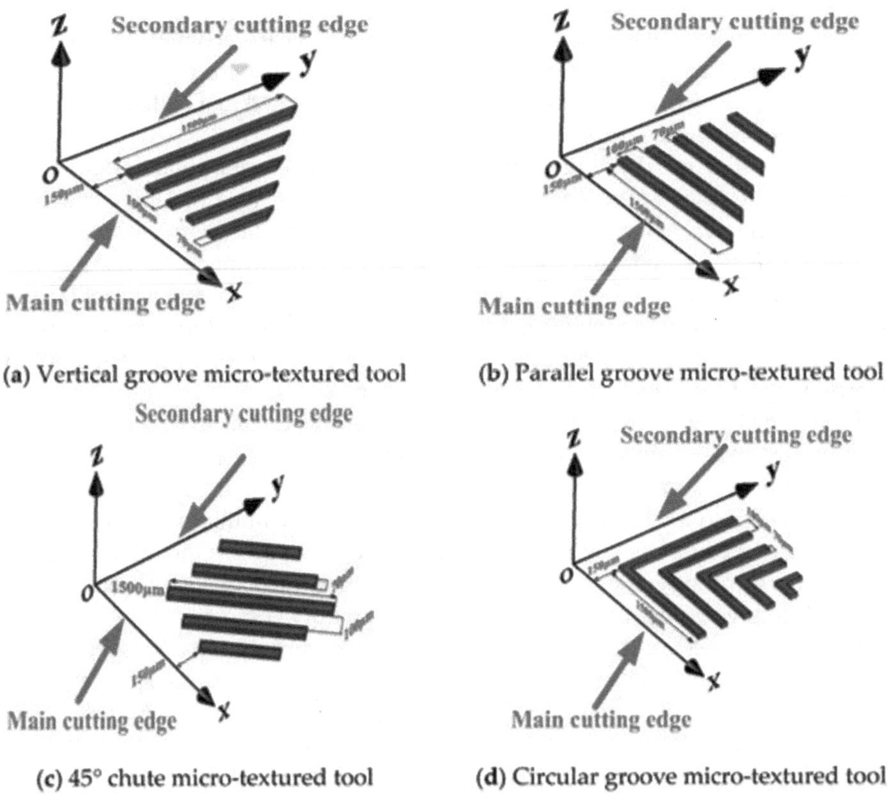

FIGURE 4.16 Four morphologies of micro-textured tools. ([3] [Open access].)

The cutting edge and vertical groove are perpendicular to the main cutting edge, and the circumferential groove is parallel to the arc of the tool cutting edge. To ensure that all four structures have the same size, cutting experiments are conducted with following parameters: $v_c = 150$ m/min, $f = 0.2$ mm/rev, $a_p = 0.15$ mm. The pitch of micro-texture grooves is 100 μm, width is 70 μm, length is 1500 μm, distance to the edge is 150 μm and depth is 70 μm. Since the cutting surface of the ridge is curved within the same dimensional parameter, the depth and width of texture decrease with increasing of concavity of cutting surface of ridge.

4.2.3 Complete Machining Using Hybrid Processes

Hybrid manufacturing combines several manufacturing technologies to improve the accuracy, speed and adaptability of production processes. This strategy blends advanced processes such as additive manufacturing (3D printing) and laser machining with traditional processes such as turning, milling and grinding. Hybrid methods eliminate material waste and manufacturing time while producing complex geometries and high-quality surfaces by combining the advantages of each technology.

Adopting hybrid manufacturing reduces setup changes and enhances workflow productivity by enabling the simultaneous execution of multiple processes. For example, combining subtractive and additive technologies makes it possible to create complex components that would be difficult or impossible to accomplish using traditional manufacturing alone. Furthermore, process control is improved by incorporating technologies such as real-time monitoring and automated tool path development, providing consistent accuracy and quality.

Greater production flexibility, higher material removal rates, and better surface finishes are the results of recent developments in hybrid manufacturing technology. Hybrid manufacturing technologies have become a critical tactic for gaining a competitive advantage as companies seek to optimize manufacturing processes and meet the growing demand for customized goods. In short, the development and application of hybrid manufacturing technologies, which have the potential to transform traditional workflows and meet the demands of contemporary production settings will determine the direction of future manufacturing [6–8].

There are also plans to establish process monitoring systems to evaluate, regulate and improve hybrid processing operations. Numerous parameters and their interactions need to be monitored. Numerical modeling methodologies are needed to optimize and gain a deeper understanding of hybrid machining processes. Hybrid machining procedures can be created and combined with in-situ tooling techniques. A machine metrology system is necessary to achieve and maintain high quality. Most hybrid processing methods have been successfully commercialized for industrial purposes; most are still in the research stage. To improve the cost-effectiveness of hybrid processing methods, more research into their industrial applications is urgently needed. From inception and industrial application to technological maturity, hybrid processes are expanding at an increasingly rapid pace. To meet the industry's demands for extremely complex product specifications made from state-of-the-art materials and, on another hand, to produce parts in a more productive, cost-effective, and energy-efficient manner, further improvements are urgently needed [8–11].

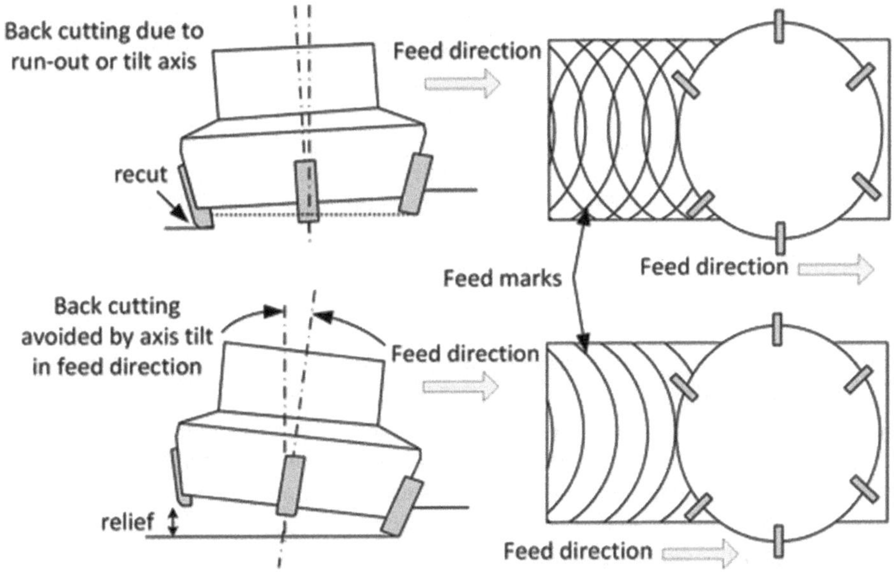

FIGURE 4.17 To reduce the back cut effect, tilt the spindle slightly in feed direction (enlarged image). ([20] [Open access].)

Back-cutting phenomena can also be caused by gradually tilting the spindle toward the back of tool, which lowers the back of cutter relative to top (Figure 4.17). In addition, vibration can be caused by increased cutter engagement and small chips remaining on edges after cutting, which can scratch the surface from behind [20].

4.3 HARD MACHINING PROCESS CHARACTERIZATION

A close examination of methods and settings used to manufacture materials with hardness greater than 45 HRC (Rockwell hardness grade C) is essential for characterizing solid machining processes. This approach to machining has become popular as a competitive alternative to conventional milling techniques, providing increased accuracy and efficiency in producing high-quality components. Cutting tool materials, tool shape, machining parameters (such as feed rate, depth of cut, and cutting speed) and workpiece material properties are important aspects that influence the solid machining process [15].

The goal of characterization studies is to gain an understanding of wear mechanisms that directly affect the life and performance of cutting tools such as adhesion, thermal fatigue, and wear. Furthermore, minimizing workpiece deformation and maintaining dimensional accuracy depends largely on the thermal effects generated during milling. Manufacturers can enhance process stability and optimize machining conditions by using advanced monitoring techniques and real-time data analysis. Recent advances in tool materials, such as creation of ultra-hard-cutting tools and advanced coatings, have further improved the efficiency of solid machining procedures. Manufacturers can increase productivity, shorten cycle times, and improve surface finishes by examining the connections between these different components [22].

Ultimately, the best possible performance and reliability in manufacturing processes depends on characterization of solid manufacturing processes. The future of precision engineering will be greatly influenced by continued research and innovation in this area as industry searches for high-quality and affordable manufacturing solutions.

4.3.1 Cutting Forces

Cutting forces are important variables in manufacturing processes that have a significant impact on workpiece quality, tool performance, and overall process productivity. These forces occur during material removal and are influenced by some variables, including the mechanical properties of workpiece material, depth of cut, feed rate, cutting speed, and tool shape. To improve tool life, reduce energy consumption, and optimize manufacturing conditions, it is essential to understand and accurately measure cutting forces.

Real-time monitoring of cutting forces has been made possible by recent developments in measurement technologies, such as the use of force gauges and sensor technologies, which provide insights into the manufacturing process. Manufacturers can find the best settings to reduce vibrations, increase surface finishes, and improve dimensional accuracy by examining the correlations between cutting forces and various process factors.

In addition, the development of cutting force models and simulations has helped create more efficient cutting tools and manufacturing techniques by predicting force behavior under different conditions. In production situations, this predictive skill facilitates improved planning and resource allocation.

In short, the development of manufacturing technology is critically dependent on research into cutting forces. Through continuous research and development of cutting-pressure monitoring and evaluation technologies, the industry can increase productivity, reduce costs, and improve the overall quality of manufactured goods. Improving manufacturing processes and increasing industrial efficiency require an understanding of cutting forces. Accurate measurement and monitoring of these forces are essential to efficient manufacturing processes because they directly impact tool performance, workpiece quality, and energy consumption.

Our ability to monitor cutting forces in real time has improved recently due to advances in measurement technology and modeling methodologies. This has allowed for faster changes and improved process control. By establishing a correlation between cutting forces and various manufacturing parameters, manufacturers may be able to identify the most favorable conditions that increase tool life, reduce vibration, and lead to improved surface finishes.

Furthermore, the design of cutting tools and manufacturing techniques continues to be influenced by the ongoing study of cutting force dynamics, which helps improve industrial efficiency. Understanding how to save costs and increase efficiency is critical as industries look to forces to be essential in driving innovation and upgrading manufactured parts [4].

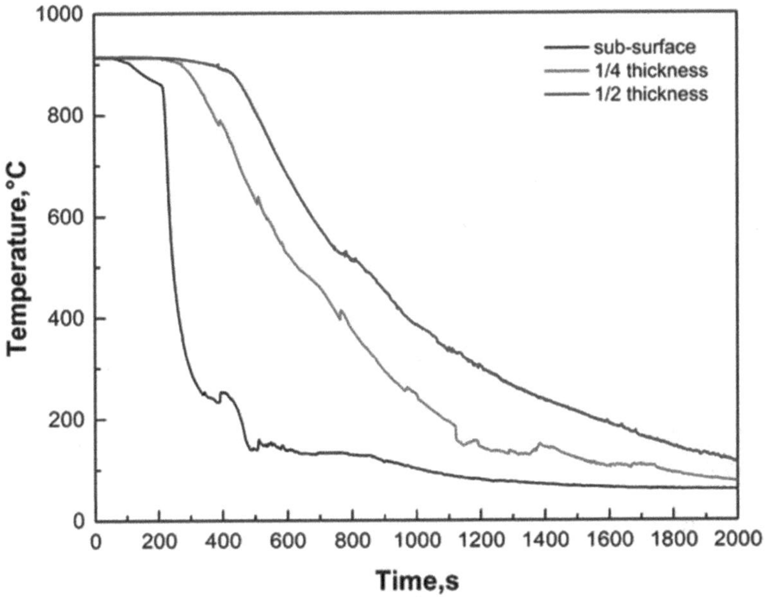

FIGURE 4.18 Measured industrial quenching temperature curves of the 178 mm thick steel plate. ([4] [Open access].)

In short, the advancement of manufacturing technology and ability of manufacturers to meet the increasing demands for precision and efficiency in today's production environments depend on the characterization and analysis of cutting forces. Figure 4.18 displays the sub-surface, *t*/4, and *t*/2 quenching temperature curves of the steel plate under investigation, demonstrating the stark variations in temperature decrease between the locations. The cooling rate at *t*/2 was roughly 0.6°C/s for the 600°C–300°C range, which is significantly less than the sub-surface position's 6.8°C/s. The measured temperature findings will be extremely helpful in examining the variety of transformation behaviors caused by the thickness effect when combined with the continuous cooling transformation (CCT) diagram.

These tests fall into two main categories; those that do not require machining and those that do. This means that since the machinability rating does not have a consistent scale, there is usually no indication of degree of machinability. The shortcomings of existing machinability criteria are also demonstrated in an early study, which emphasized the need for complete machinability [13]. Figure 4.19 illustrates the variation in the quantity of scientific articles mentioning "Machinability". Depending on the percentage of copper in the alloy, even a single alloying element can have a significant and non-linear effect on the machinability of the material. Thus, as shown in Figure 4.19, we can show that even alloys from the same material group can differ greatly in machinability. To do this, the flank wear in helical drilling of three commonly used ferritic-pearlite steel inserts is compared to evaluate their machinability [2].

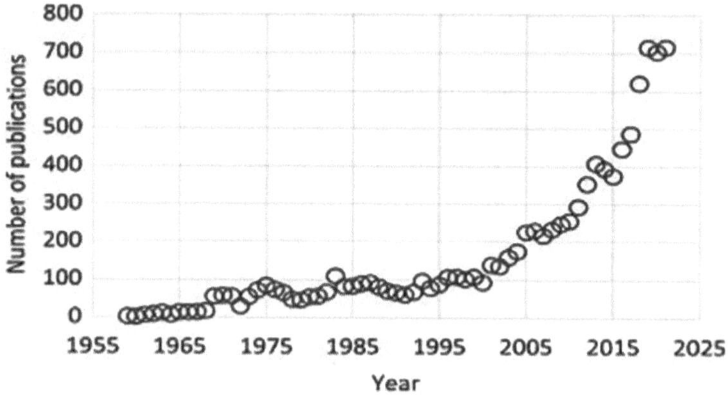

FIGURE 4.19 Change in the number of scientific papers that mention "Machinability". ([2] [Open access].)

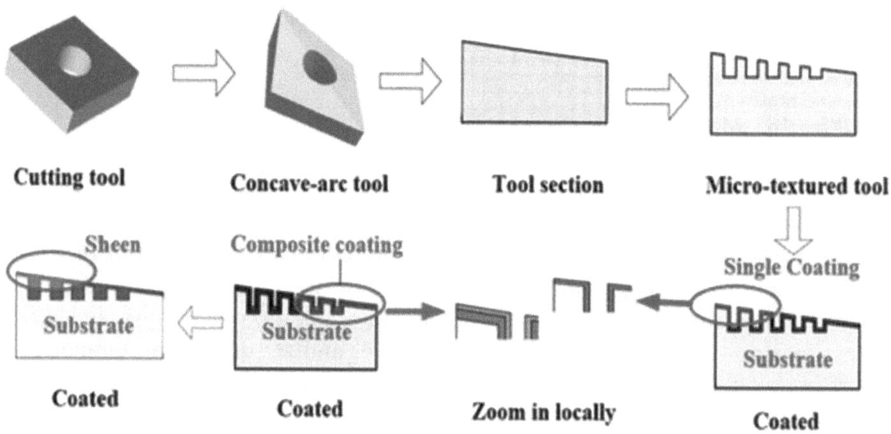

FIGURE 4.20 Scheme of the technological process of applying paint to a tool. ([3] [Open access].)

This is demonstrated by the results of the machinability study of $4Cr_{16}Mo$. When copper is subjected to the same cutting conditions, it can be shown that flank wear is generally lower. However, when the copper concentration increases from 1.4 to 1.9 wt%; this effect is not linear and even inverse [5]. Finally, there may be noticeable differences in machinability due to heat treatment characteristics. The coating process is carried out using a micro-textured bionic tool with 45° flute, which is selected previously. In this study, the TiN material, the highly oxidation-resistant Al_2O_3 material, and composite coating combining the two are selected based on the characteristics of coating materials. The coating process is applied to the carbide surface, and identical cutting parameters are used for the turning test [3]. The effect of coating material on the machinability is investigated. The flow chart of the tool coating procedure is shown in Figure 4.20 [3].

The wear studies in milling are designed to achieve different impact entry durations. The average time before a coated cutting edge deforms most during edge penetration into the workpiece is defined as the parameter t_e in this study. The study examines the relationship between tool diameter (*D*) and radial depth of cut (*a*), in particular ratio *a*/*d*, which affects the primary cutting speed at a fixed cutting speed. The depth of cut is varied from 3 to 51 mm, while *D* remains constant at 60 mm. This increases the ratio *a*/*d* from 5% to 85%. With this configuration, milling tests can be carried out, with t_e values ranging from approximately 0.21 to 4.5 ms. The results, which are summarized in accompanying table and depicted in Figure 4.21a, have shed light on the effects of depth of cut on milling performance and tool deformation.

The machinability or tool life of aerospace alloys is particularly sensitive to cutting speed and resulting tool temperature. Consequently, this component has often considered in machinability evaluation tests. In light of Figure 4.21b, it is used a flank wear to compare two titanium alloys, Ti54M and Ti6Al4V, at different cutting speeds to determine their machinability. At cutting speeds $v_c = 60–80$ m/min, the wear of both materials is approximately the same; however, at $v_c = 90$ m/min,

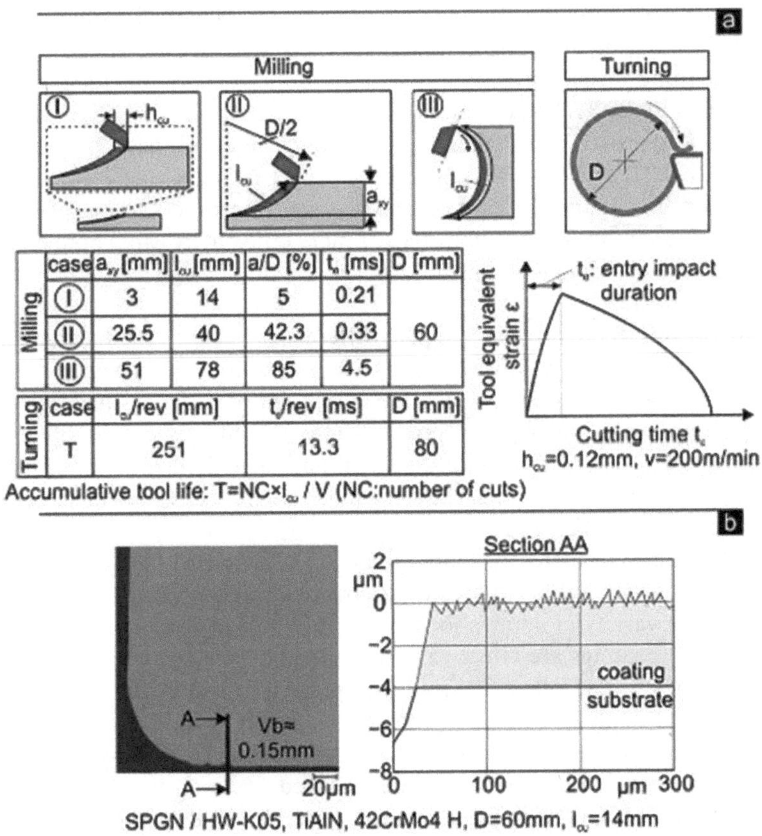

FIGURE 4.21 (a) Cutting parameters used in milling and turning, (b) using confocal microscopy to record the wear condition of the rake face. ([4] [Open access].)

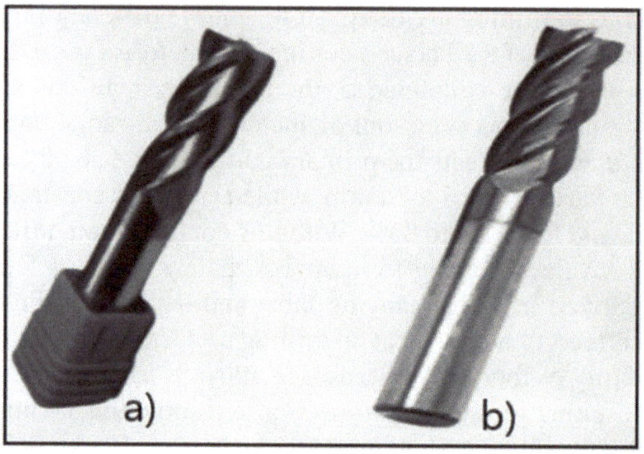

FIGURE 4.22 Uncoated carbide tool (a) and carbide tool with TiAlN coating (b). ([5] [Open access].)

it increases sharply for Ti6Al4V and very slightly for Ti54M [5], when estimating the production cost of Ti-48Al-2Cr-2Nb and Ti-6Al4V, tool life is considered as a function of cutting speed. Only Ti-48Al-2Cr-2Nb could achieve the range of tool life that is possible for Ti6Al4V at cutting speeds in the range of $v_c = 100$ m/min if cutting speed is lower than $v_c = 40$ m/min [5]. The comparison of these two parts can be seen in Figure 4.22. The process is optimized using the Taguchi technique and most important factor affecting the surface roughness is determined to be spindle speed using variance analysis. The method is made more efficient by using both coated and uncoated tools. According to the results, the coated carbide end mill provided excellent surface polishing [5].

Because of this, there are fundamental differences between the machinability of different materials, as well as differences of how it is affected by specific influencing elements. Fiber composites, which are composed of different materials, should be considered when assessing the processability of different types of materials. Two examples of how different reinforcement and matrix alloy qualities affect the machinability of composites under the same machining process boundary conditions are shown in Figure 4.22 [5].

Advanced thin-film materials called diamond-like carbon (DLC) coatings have high wear resistance, low friction, and exceptional hardness. These qualities make them ideal for a variety of applications in the automotive, aerospace, and industrial industries. DLC coatings are effective in extending the life and performance of cutting tools, bearings, and other precision components because they have properties similar to natural diamonds such as strong thermal stability and chemical inertness.

In short, DLC coatings are a major advancement in surface engineering that provides manufacturers with longer tool life, reduced friction, and improved performance in a range of applications. Continued research and development in this area is expected to further expand the capabilities and uses of DLC coatings, leading to more efficient and environmentally friendly production methods. Thus, DLC coatings can prevent early heat generation during machining by reducing friction in the

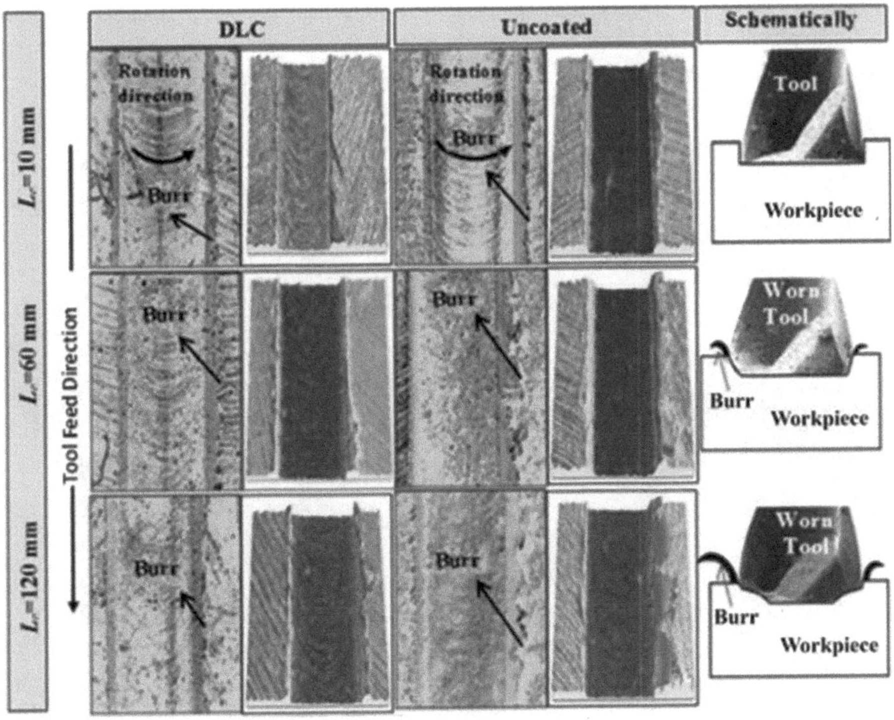

FIGURE 4.23 During the cutting process, burrs are formed (f=5 μm/groove, a_p=0.15 mm). ([5] [Open access].)

contact zone. To make the coating tougher, DLC can be mixed with hard coatings. For example, the properties of three-layer CrN/CrCN/DLC coating are evaluated [7,8]. These coatings are compared with the untreated surface and TiAlSiN coating. Compared with the untreated surface, their wear resistance is 58 times higher.

Figures 4.23 and 4.24 of [8] show the tool performance during machining, tool wear, surface roughness, burr formation and cutting forces, but due to the low friction of the DLC coating, the cutting forces for coated tool test are lower than the uncoated tool test. The longer the cutting time in milling with an untreated tool, the greater the burr formation. There is no significant difference in burr formation between coated and uncoated tools at short cutting lengths. In short, DLC-coated tools provide less friction and wear, resulting in a smoother polished surface [5].

Typically, DLC coatings are deposited using methods such as CVD and physical vapor deposition (PVD), which enable consistent application to a range of substrates. By adjusting the deposition conditions and the sp^2 to sp^3 ratio, these coatings can be made to have specific properties such as improved lubricity or greater durability for sp^3 hybridized carbon.

To further improve performance, recent developments in DLC technology have focused on creating multilayer coatings, enhancing adhesion to substrates and adding nanomaterials. The use of DLC coatings is growing as industries require more and more high-performance components that can withstand harsh conditions.

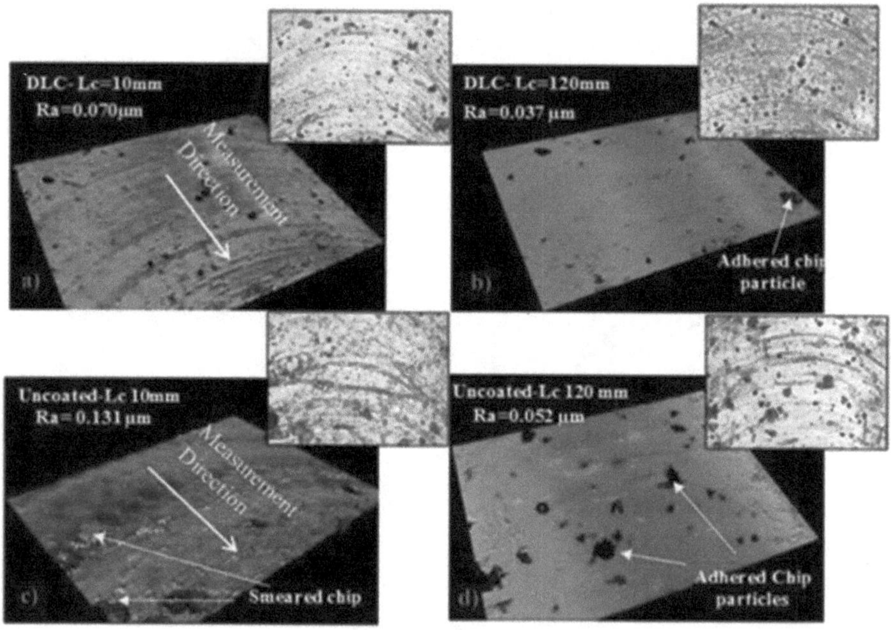

FIGURE 4.24 After 10 mm cutting operation, 3D topographic images of machined surfaces (f=2.5 µm/groove, $a_{p=}$0.15 mm). (Reproduced with permission from Elsevier.) DLC (L_c= 10 mm) in (a). (b) DLC with L_c 120 mm. (c) Uncoated; (10 mm is L_c). (d) Uncoated (120 mm × L_c). ([5] [Open access].)

Additional influencing parameters such as matrix alloy [6], reinforcement material type [6], reinforcement volume fraction [7], and reinforcement particle size [8] should be included in the life-based machinability study of modern materials such as composites. Cutting performance of various high-speed machining (HSM) operations on graphite templates using diamond-coated tools. To evaluate the scratch resistance of the coating, tests are conducted on microcrystalline diamond (MCD), submicrocrystalline diamond (SMCD), nanocrystalline diamond (NCD) and micro/nanocomposite diamond (MCD/NCD) [5]. Figure 4.25 shows that although microcrystalline diamond has the highest adhesion to the substrate, microcrystalline diamond (NCD) is found to have the lowest coefficient of friction. In addition, under these settings, the performance of SMCD coating is lower than MCD and NCD. On the other hand, MCD/NCD composite diamond adhered strongly to WC-Co [5].

With consistent edge tuning superior to standard sandblasting methods and a noticeable improvement in surface roughness, finishing improves broaching productivity. A new milling cutter modifies the contact force between the robot and workpiece to maintain stability throughout the milling process to further improve stability in robotic machining. It summarizes the previous discussed tool parameters that affect the machinability determined by tool life. Heating the workpiece material to a specific temperature is another fundamental process modification method to extend tool life and improve machinability. On the other hand, this method requires

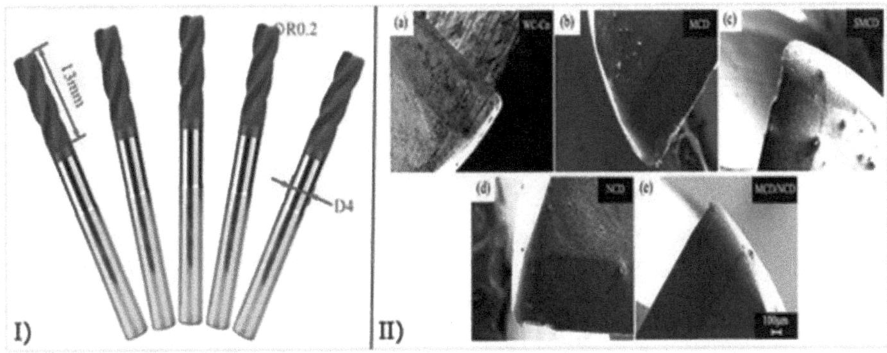

FIGURE 4.25 (I) Diamond-coated end mills with corner radius. (II) Wear profile of different types of grinding tools: (a) WC-Co, (b) MCD, (c) MC, (d) NCD, and (e) MCD/NCD. ([5] [Open access].)

precise calibration of workpiece temperature. Compared to traditional machining without external heating, if they are set too high, the additional heat load on the tool can lead to increased wear [9]. The different impacts of process characteristics on the measured tool life and machinability can be attributed to factors such as cutting speed, feed rate, tool material, and cooling conditions, all of which influence wear patterns, surface finish, and overall efficiency.

4.3.2 Torque and Cutting Force for Assessing Machinability

When assessing the suitability of materials for machining, two key properties are torque and cutting force. These forces directly impact tool efficiency, surface finish, and overall machinability. Manufacturers can improve machining tactics and tool selection by gaining important insights into the behavior of material under different cutting conditions by measuring torque and cutting force.

Cutting geometry, feed rate, cutting speed, and mechanical properties of workpiece are some of variables that affect cutting force, which is mostly related to the rate of material removal. Conversely, torque refers to the amount of energy are required for machining operations and reflects the resistance encountered by the cutting tool. By examining these forces, the best machining parameters that improve surface quality, reduce tool wear and increase yield are determined.

Recent advances in sensor technology and data acquisition systems have made it possible to monitor torque and cutting force in real time, allowing for rapid adjustments during milling operations. This skill allows for a more accurate assessment of machinability, which helps in creating models that predict tool life and performance.

In conclusion, determining machinability requires measuring and analyzing cutting force and torque. By leveraging these insights, manufacturers can improve manufacturing conditions, reduce operating costs and produce higher-quality manufactured components, ultimately leading to improvements in manufacturing effectiveness and efficiency. High process forces resulting from machining operations can limit the amount of material processed. Based on the shear zones of cutting process, the origin of technological forces can be explained (Figures 4.26–4.28). The actual chip separation process occurs in the primary shear zone. Thus, the technological forces develop to exceed the shear strength of materials. The main need

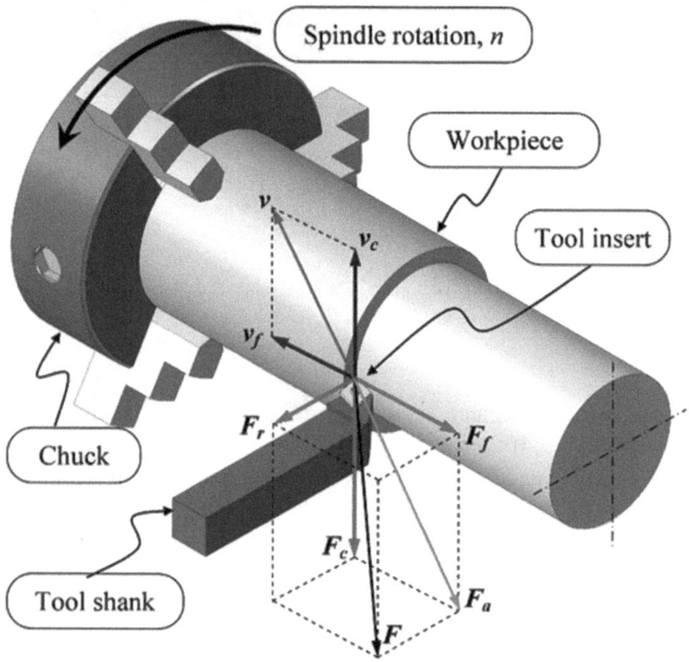

FIGURE 4.26 In the turning process, three axes of cutting force components are involved. ([19] [Open access].)

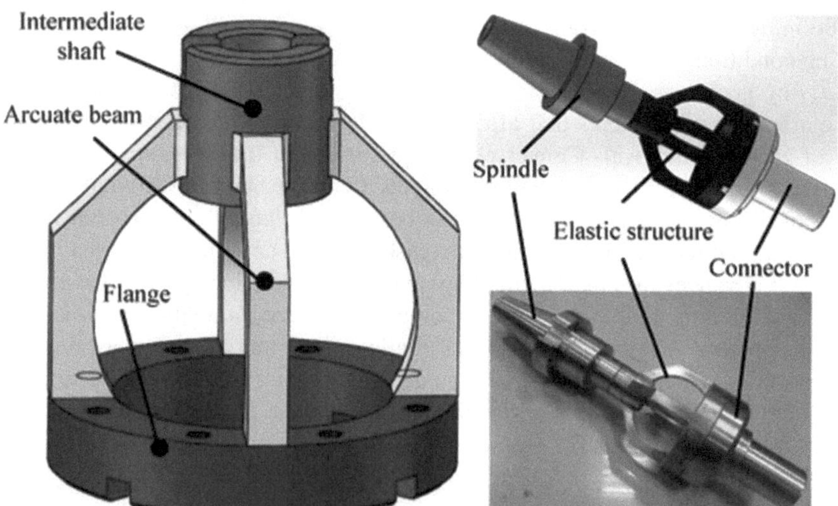

FIGURE 4.27 The measurement structure. ([31] [Open access].)

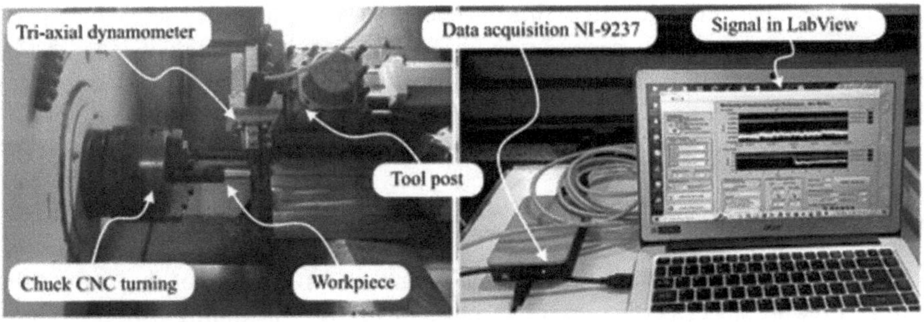

FIGURE 4.28 Heat generation and dissipation. (Modified from [19] [Open access].)

when developing the equipment is that it should be done at the lowest possible cost. Equipment is available to monitor cutting parameters, and it is clear that researchers are interested in developing this technology.

So, many simple and complex devices have been developed to measure cutting torque and feed force. In educational systems, cost factors play a critical role in providing the logistical requirements needed to conduct experiments with undergraduate, graduate and postgraduate students [17–19]. When creating the method and specific equipment, the areas of experimental research in which the local academic environment is involved and wants to develop are considered. The high level of development makes it difficult or blocks the costs associated with the acquisition of measuring instruments. The quality of student experiments will be improved by using the apparatus and methodology proposed in this study.

The workpiece material is fixed in a chuck before the spindle and rotates at a certain rotational speed n (rpm), as shown in Figure 4.26, to illustrate the rotation process. If the insertion tool contacts the workpiece material during chip removal, it will be a barrier. In this case, the material will not be cut by the cutting tool. To compensate for this, force F, known as the machining or resultant force, must be applied. By dividing this force into three separate components, known as triaxial cutting force components, it can be examined [19]. The cutting force F_c and active force F_a combine to produce a feed force in the cutting direction F_f [19]. To measure four-axis loads (i.e., F_x, F_y, F_z and T) for milling or drilling operations, this work aimed to design an integrated rotating dynamometer. Figure 4.1 shows a new elastic structure consisting of two semi-octagonal rings that are mutually perpendicular. The connector, the elastic structure, and the spindle constitute the measuring structure. The connector connects the tools, and the spindle attaches to the machine head. However, the multi-dimensional force device is the intended application for this connector. The flange, the arc beam and the intermediate shaft constitute the elastic framework. The cutting force is converted into deformation by the arc bar. The four arc bars generate stress and deformation when the end of the connector is subjected to four-axis loads [31].

As the cutting speed increases, the yield stress of tool decreases and surface roughness decreases too. Due to increased temperature in the deformation zone and the rate of material removal, surface roughness increases as feed rate and depth of cut increase. The cutting length of the tool decreases as the approach angle increases, resulting in a decreasing in surface roughness [12]. Increasing the rake angle results in a shorter tool-to-chip contact length, which facilitates chip removal and reduces surface roughness. Log *C*, log *t*, log, log *d*, log *f*, log and log *V* are predictors. The suitability of the model is confirmed by a regression coefficient of 78.8% [19].

4.3.3 Cutting Forces

Cutting forces are important to the machining process because of they affect the quality of workpiece, performance of tool and overall machining process. These forces are affected by several factors including cutting speed, feed rate, depth of cut, tool geometry, and properties of workpiece material resulting from the interaction between the cutting tool and workpiece. To increase surface finishes, extend tool life, and improve machining conditions, accurate measurement and analysis of cutting forces are essential.

It is now possible to monitor cutting forces in real time due to advances in measurement technology such as force gauges and advanced sensor systems, which provides a useful information for improving machining tactics. Furthermore, modeling techniques such as analytical and computational methods help predict the behavior of cutting forces under different conditions, which helps improve process planning and tool design.

An understanding of cutting forces encourages advances in tool materials and coatings, further improving performance and helping to optimize current manufacturing processes. A comprehensive understanding of cutting forces will continue to be essential for advancing manufacturing techniques and achieving better production results as industries demand increasing levels of accuracy and efficiency. The creation of the white layer and distribution of residual stresses are determined by penetration of heat into the sub-surface layer, so the thermal consideration of HM operations is critical. Many scholars argue that the hardness of workpiece material affects the cutting temperature in HM in addition to cutting conditions, which are primarily determined by tool nose radius and cutting speed [13]. As shown in Figure 4.5, a static calibration test was conducted to determine the static performance of the dynamometer in four directions: F_x, F_y, F_z, and T, respectively. The spindle rotation experimental platform we designed ourselves is used to conduct the calibration experiments in the X and Y directions. Using a caliper, the arc-shaped beam A was made horizontal before the calibration test in the X direction. Then, the flywheel of the radial force device loads the F_x force. We used the same calibration procedure as in the X direction and rotated the dynamometer 90° for the calibration in the Y direction. A static universal materials testing machine (MTS, Eden Prairie, MN, USA) was used to conduct the calibration test in the Z direction [19] (Figure 4.29).

The relationship between the increasing in hardness of material being processed and temperature of tool edge near the side surface. The area of interaction between the cutting tool and workpiece heats up significantly when cutting metal. Temperature affects the residual stress, hardness and surface roughness of the workpiece, as well as the wear rate of the cutting tool. One of the most important variables

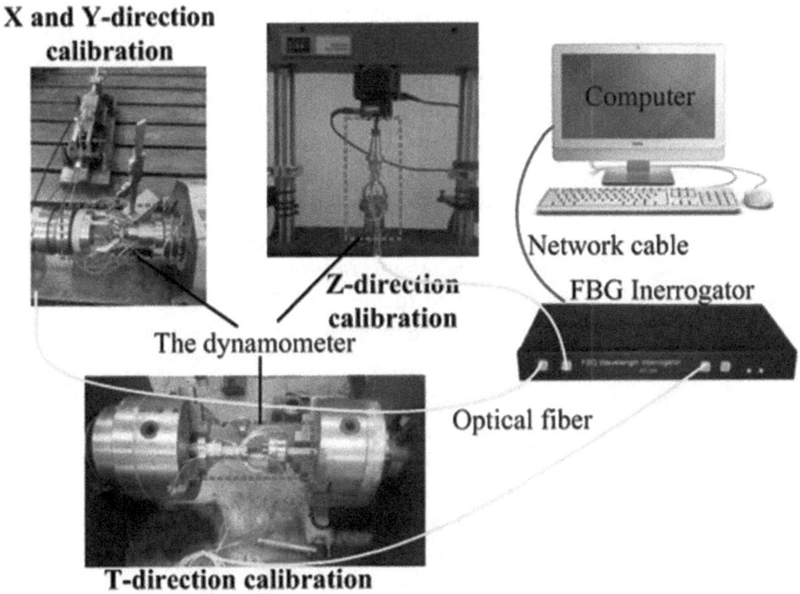

FIGURE 4.29 Static calibration device. ([31] [Open access].)

to pay attention when studying the mechanics of the cutting process is the cutting temperature. Another important element of HSM operations is cutting temperature, which has a significant impact on surface quality, tool life and wear. Therefore, to accurately predict the cutting temperature and then optimize the machining parameters, it is necessary to establish theoretical models. Numerous studies have been conducted to study the temperature and its distribution in cutting zone [14]. A numerical model based on the finite difference method is established to predict the steady-state temperature fields of the tool and chips, as well as the transient temperature changes in continuous and intermittent orthogonal machining processes. Their method can be used to select the best cutting speed, feed, and tool geometry such as tool rake angle and clearance angles to prevent excessive heating.

4.3.4 Temperature and Cutting Power Combined to Assess Machinability

To improve manufacturing processes and ensure superior production results, a machinability assessment is essential. Manufacturers can gain a comprehensive understanding of behavior of workpiece materials during machining and performance of cutting tools by integrating temperature and cutting force measurements. Temperature has a significant impact on surface integrity, material deformation and tool wear. High temperatures have the potential to shorten tool life and degrade the quality of the finished surface by causing the workpiece material to thermally soften. Additionally, cutting force which represents the energy required to complete a machining process provides information about the effectiveness and efficiency of the process. A more comprehensive understanding of the machinability of different materials under different conditions can be gained by analyzing the relationship between temperature and cutting force. Modern sensor technology allows for real-time monitoring of cutting force and temperature, allowing for rapid adjustments to machining parameters to achieve optimal results. By combining these metrics, manufacturers can find optimal cutting conditions, use less energy, and increase productivity. In short, combining cutting force and temperature measurements provides a solid basis for evaluating machinability. This method promotes innovation in the machining sector by helping to improve tool life, surface quality, and production efficiency as well as helping to improve machining processes. Typical tool wear characteristics seen in finished turning on CBN tools with most noticeable wear impact are shown in Figure 4.30. The most important aspect of cutting process is the cutting temperature.

The sawdust then absorbs 90% of heat generated during cutting, and the remaining heat is transferred to the tool and workpiece [19]. This study used artificial thermocouples to measure cutting temperature and examined whether the metal could be processed in terms of cutting temperature.

Artificial thermocouples are drilled through screw holes on the end face and placed just below the cutting tip of insert to study the machinability of the material [19]. Thus, a simple, reliable, accessible and accurate approach to machinability research has been established. The study included the use of thermocouples and infrared cutting temperature measurement methods. The comparative study concluded that increasing the cutting speed did not reduce the cutting temperature; rather, it increases the cutting temperature. This result demonstrates phenomena that contradict those reported by Abdullah et al. [29], who proposed a new approach to predict grinding temperature. An infrared camera is used to record the temperature of workpiece during the milling process. CNC machine shops work with a variety of materials to create parts that meet the needs of their customers. Typical materials that are easy to handle include brass, aluminum and various polymers. However, to get the desired results when machining difficult-to-cut materials such as titanium alloys like Ti-6Al-4V, one will need sophisticated equipment and knowledgeable technicians.

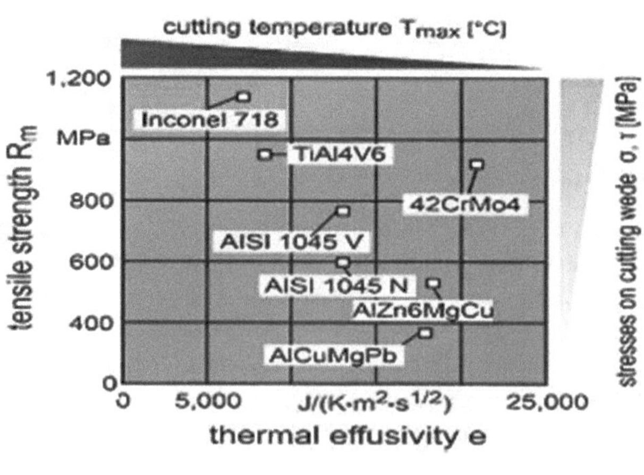

FIGURE 4.30 Classification of materials by machinability. ([19] [Open access].)

Therefore, it is important to understand why one material is easier to produce than another. The definition of machinability, its rating and variables that affect the machinability of material are discussed in this article. Typical examples of materials with excellent machinability and standard methods for improving the machinability of material can be found at the end of the article [16].

4.3.5 The Influence of Chip Formation Mechanisms on Material Machinability

Titanium alloy is widely used in gas turbines and aircraft fuselages due to its exceptional thermal stability, corrosion resistance and high strength-to-weight ratio. However, titanium alloy is one of the most difficult materials to process in the world. One common phenomenon when machining titanium alloys is the formation of sawtooth chips, which creates a cyclic cutting force and reduces the tool-chip contact area. As a result, vibration increases sharply during the cutting process and cutting temperatures rise to extremely high values, which limits the material removal rate and exacerbates tool wear. Therefore, the main purpose is to study the mechanism of sawtooth chip production and how it interacts with cutting tool. Analysis of previous studies shows that the main response is the tool-chip interface. The three processes in machining titanium alloys are abrasion, adhesion, and diffusion [17].

Studies have been conducted on the relationship between the chip morphology and machining conditions, feed rate and cutting speed. In addition to Ti-6Al-4V alloy, non-conventional processing methods and other titanium alloys such as β-titanium alloy have also been studied. Separation of chips from the workpiece is the basis of all machining operations (Figure 4.31a). The material is pierced by a cutting edge, which cuts it along the "primary shear zone". The material begins to flow when the maximum shear stress is reached. The developing chips flow down the front surface of the cutting edge after elastic and plastic deformation; this area of deformation is called the "secondary zone" [18]. Chips can be classified by their

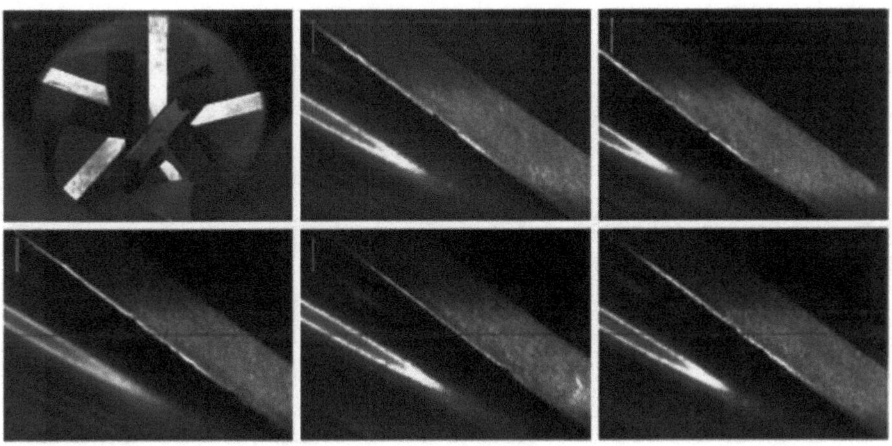

FIGURE 4.31 Tool wear for constant *F*. ([17] [Open access].)

appearance or, more specifically, by shape or microscopic structure associated with the type of chip formation [16–19]. The four types of chip formation are shown in Figure 4.31b: (a) continuous chips, (b) plate chips, (c) segmented chips, and (d) discontinuous chips. Segmented chips can be considered as a special case of plate chips. Figure 4.16 illustrates one possible chip shape classification. Chip control is essential to achieve reliable machining processes with high levels of productivity. This involves efficient chip breaking and removal. Their shape determines how well chips are removed from the cutting zone and working area of the machine. Unfavorable chip formation hinders automation and subsequent digitalization of processes, causing unplanned interruptions in the machining process when the machine operator manually removes chips. Process variables have an influence on the chip shape at macro and micro levels during precision turning of Ti-6Al-4V. It discusses how to control the feed rate, cutting speed, and depth of cut during turning tests to limit chip formation and improve chip evacuation and friability at the macro level. The chips are examined under a scanning electron microscope (SEM) to better understand the micro-scale mechanics underlying chip curling. The surface roughness of machined specimens is evaluated to understand the effect of chip evacuation on the achievable surface quality. Figures 4.31 and 4.32 show the tool wear obtained with constant *F* and with variable *F*, respectively [17].

According to the data obtained, the regulation of chip formation is significantly influenced by how technological factors interact. In particular, at all feed rates, the combination of higher cutting speeds and deeper cuts resulted in chip entanglement in the workpiece. When machining at higher cutting speeds, using relatively higher feed rates at moderate depths of cut produces good chip-breaking results [17].

As the cutting speed increases, as shown in Figure 4.32a, the segmentation becomes more intense until individual chip segments are formed [24]. Transverse flow of material occurs as chip lamellae accumulate (Figure 4.32b) [20]. The type of chip formation exposes the cutting tool to high, variable mechanical and thermal loads. Therefore, the phenomena of fatigue and crack formation occur [24]. The material's tendency to harden, combined with the flake chip formation process, results in a hard, sawtooth chip edge that is sharp and causes abrasive wear on the

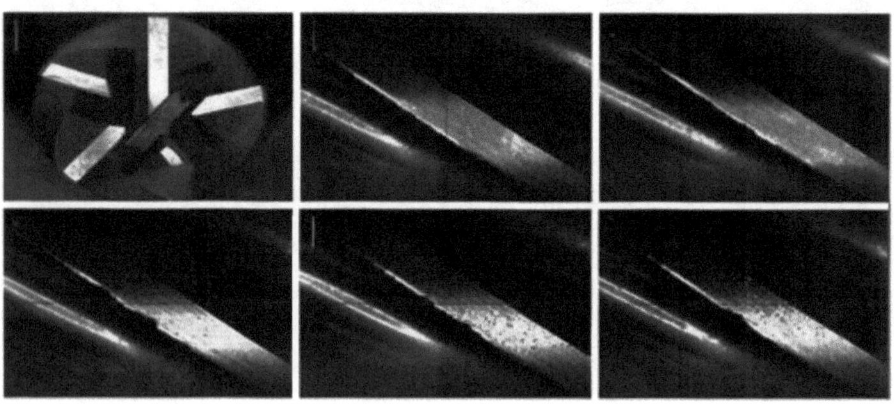

FIGURE 4.32 Tool wear for variable *F*. ([17] [Open access].)

cutting tool as chips flow off [18,21]. The Ra and Rz values obtained using the contact and non-contact procedures are almost identical for specimens 5 and 6; however, the non-contact method gave better results for specimens 8 and 12. The flat waveform of sample roughness can be clearly seen on the 3D surface using false color rendering for R_z. This indicates that linear evaluation indicates a flat waveform. Part 12 in Figure 4.32d provides the best result in spatial contrast representation [18].

The strength, hardness, and toughness are closely related to the processing of this alloy. This nickel alloy also includes hard abrasive carbide particles, which results in hardening of workpiece material and abrasive wear on the cutting tool. In other words, large mechanical and thermal loads always have an impact on the machining of Inconel 718. This study discusses the chip-making method of Inconel 718. The ability of Inconel 718 to produce segmented or jagged chips at various feeds and cutting speeds is one of its main machining properties. Research has shown that chip segmentation or serrations due to shear localization affect the integrity of the machined surface and contribute to chip evacuation and automation of machining processes. Research has concluded that it is desirable to use the serrated or segmented chip phenomenon to reduce the cutting force. Efficient and effective machining of Inconel 718 requires careful analysis of models and approaches to understand the chip formation mechanism. Therefore, it presents a description of chip production models and methodologies during Inconel 718 machining [25].

Others seek to characterize the relationship between chip morphology and changes in cutting process parameters or relate chip morphology to one or more metrics related to machinability assessment (tool wear, cutting forces, surface roughness, etc.). Comprehensive methods that incorporate chip morphology into a multi-criteria evaluation leading to comprehensive machinability characterization are generally lacking in the literature. Here are the exact tools that Dormer Pramet used: Tool holder (left version) PCLNL 2525M 12 and replacement insert (CNGA 120408 S 01020B), consisting of polycrystalline CBN (TB310); it can be used without the cutting fluid [18].

Finally, special tables are provided for classifying previously introduced methods of coolant supply to production. They should be compared according to the main mechanisms [25]. Compared to other concepts, the solid scheme shows the greatest potential for improving the manufacturability of the corresponding main mechanism. Convective heat transfer from the glass surface to the cooling jets is assessed by simulating the static quenching of a square flat glass sample in an industrial setup.

The regularity of flow regime at the jet outlets is first confirmed by examining the flow in the air supply nozzles. The accuracy of numerical simulation of static tempering of a flat glass sample is verified by comparison with surface temperature measurements obtained by pyrometry and residual stress measurements obtained by photoelasticity. Heat transfer coefficients are calculated for four different jet Reynolds numbers. It is found that the average Nusselt number correlates with the relative deviation of local heat transfer from the average heat transfer, which does not depend on Reynolds number [26–28].

The effect of different workpiece materials and cutting speeds on surface roughness. It is obvious that as a result of BUE formation, surface roughness increases as cutting speed decreases. In addition, BUE has a significant effect on tool wear, which is related to surface roughness [7]. It is recommended to use coolants and cutting tools with lower adhesion tendency and higher cutting speeds to minimize BUE formation [26]. At relatively low cutting speeds, BUE roughness control can also be achieved by reducing the undeformed chip thickness, increasing the rake angle and using weaker inserts [7].

PROBLEMS

1. Describe the basic features of HM materials needed to manufacture cutting tools.
2. Mention some grinding processes.
3. Identify problems related to procedures related to technology: steel manufacturing.
4. Describe complete processing using the hybrid processes.
5. Explain several CBN entries that have recently been developed.

REFERENCES

1. Wang, M., Zhu, M., Hu, X., Liu, K., Fan, X., Meng, X., … & Wang, J. (2024). Synergistic effect of elliptic textures and H-DLC coatings for enhancing the tribological performance of $CuAl_{10}Fe_5Ni_5$ valve plate surfaces. *Coatings*, 14(9), 1161.
2. Zhao, S., Liu, S., Xue, Y., Li, N., Xu, K., Qiu, W., … & Shi, C. (2023). Microstructure and properties of monolayer Ta and multilayer Ta/Ti/Zr/Ta coatings deposited on biomedical Ti-6Al-4V alloy by magnetron sputtering. *Coatings*, 13(7), 1234.
3. Li, Q., Ma, C., Wang, C., Wang, B., & Zhang, S. (2023). Analysis of the cutting performance of coated micro-textured bionic tools for dry cutting AISI 52100. *Machines*, 11(9), 886.
4. Wang, Q., Ye, Q., Wang, Z., Kan, L., & Wang, H. (2020). Thickness effect on microstructure, strength, and toughness of a quenched and tempered 178 mm thickness steel plate. *Metals*, 10(5), 572.
5. Dabees, S., Mirzaei, S., & Sobola, D. (2022). Characterization and evaluation of engineered coating techniques for different cutting tools—Review. *Materials*, 15, 5633.
6. Zhao, G., Li, Z., Hu, M., Li, L., He, N., & Jamil, M. (2019). Fabrication and performance of CVD diamond cutting tool in micro milling of oxygen-free copper. *Diam. Relat. Mater*, 100, 107589.
7. Šramhauser, K., Náprstková, N., Sviantek, J., Stančeková, D., Van Tuong, N., & Novotný, J. (2022). Analyses of tool wear and chip type for different coated carbide inserts in turning hardened 1.6582 steel. *Coatings*, 12(7), 974.
8. Mirzaei, S., Alishahi, M., Souček, P., Buršíková, V., Zábranský, L., Gröner, L., … & Vašina, P. (2020). Effect of substrate bias voltage on the composition, microstructure and mechanical properties of W-B-C coatings. *Appl. Surf. Sci.*, 528, 146966.
9. Mirzaei, S., Alishahi, M., Souček, P., Zábranský, L., Buršíková, V., Stupavská, M., … & Vašina, P. (2019). Effect of bonding structure on hardness and fracture resistance of W-B-C coatings with varying B/W ratio. *Surf. Coat. Technol.*, 358, 843–849.
10. Mirzaei, S., Alishahi, M., Souček, P., Ženíšek, J., Holec, D., Koutná, N., … & Vašina, P. (2020). The effect of chemical composition on the structure, chemistry and mechanical properties of magnetron sputtered W-B-C coatings: Modeling and experiments. *Surf. Coat. Technol.*, 383, 125274.
11. Bohley, M., Kieren-Ehses, S., Heberger, L., Kirsch, B., & Aurich, J. C. (2018). Size limitations and wear behavior of TiB2 coated micro end mills (Ø <50 μm) when machining cp-titanium. *Procedia CIRP*, 71, 187–191.

12. Swain, N., Venkatesh, V., Kumar, P., Srinivas, G., Ravishankar, S., & Barshilia, H. C. (2017). An experimental investigation on the machining characteristics of Nimonic 75 using uncoated and TiAlN coated tungsten carbide micro-end mills. *CIRP J. Manuf. Sci. Technol.*, 16, 34–42.
13. Haja Syeddu Masooth, P., Jayakumar, V., & Bharathiraja, G. (2020). Experimental investigation on surface roughness in CNC end milling process by uncoated and TiAlN coated carbide end mill under dry conditions. *Mater. Today Proc.*, 22, 726–736.
14. Martinho, R. P., Silva, F. J. G., Martins, C., & Lopes, H. (2019). Comparative study of PVD and CVD cutting tools performance in milling of duplex stainless steel. *Int. J. Adv. Manuf. Technol.*, 102, 2423–2439.
15. Silva, F. J., Martinho, R. P., Martins, C., Lopes, H., & Gouveia, R. M. (2019). Machining GX2CrNiMoN26-7-4 DSS alloy: Wear analysis of TiAlN and TiCN/Al_2O_3/TiN coated carbide tools behavior in rough end milling operations. *Coatings*, 9, 392.
16. Hosokawa, A., Hoshino, G., Koyano, T., & Ueda, T. (2018). Cutting characteristics of PVD-coated tools deposited by filtered arc deposition (FAD) method. *CIRP Ann.*, 67, 83–86.
17. García-Hernández, C., Garde-Barace, J. J., Valdivia-Sánchez, J. J., Ubieto-Artur, P., Bueno-Pérez, J. A., Cano-Álvarez, B., … & Kyratsis, P. (2021). Trochoidal milling path with variable feed. Application to the machining of a Ti-6Al-4V part. *Mathematics*, 9(21), 2701.
18. Osička, K., Zouhar, J., Sliwková, P., & Chladil, J. (2022). Cutting force when machining hardened steel and the surface roughness achieved. *Applied Sciences*, 12(22), 11526.
19. Rizal, M., Ghani, J. A., & Mubarak, A. Z. (2022). Design and development of a tri-axial turning dynamometer utilizing cross-beam type force transducer for fine-turning cutting force measurement. *Sensors*, 22(22), 8751.
20. Abellán-Nebot, J. V., Vila Pastor, C., & Siller, H. R. (2024). A review of the factors influencing surface roughness in machining and their impact on sustainability. *Sustainability*, 16(5), 1917.
21. Rodríguez-Barrero, S., Fernández-Larrinoa, J., Azkona, I., de Lacalle, L. N. L., & Polvorosa, R. (2016). Enhanced performance of nanostructured coatings for drilling by droplet elimination. *Mater. Manuf. Processes*, 31, 593–602.
22. Sousa, V. F. C. & Silva, F. J. G. (2020). Recent advances on coated milling tool technology—A comprehensive review. *Coatings*, 10, 235.
23. Sousa, V. F. C. & Silva, F. J. G. (2020). Recent advances in turning processes using coated tools—A comprehensive review. *Metals*, 10, 235.
24. Wake, William Charles. "Adhesion and the Formulation of Adhesives." *(No Title)* (1982).
25. Martha, P. K., Kishore, K., & Ramadevudu, G. J. (2019). Lifetime and performance improvement of HSS drill bit by Titania (TiO_2) nano-coatings and cryogenic treatment. *Int. J. Eng. Adv. Technol.*, 8, 1389–1393.
26. Zafar, M. Q., Uddin, G. M., Asim, M., Khan, A. A., Tahir, Z.-U.-R., Hayat, N., … & Jawad, M. (2020). Comparative analysis of low-temperature PVD-based TiN nano-thin-film-coated and -uncoated TNMG inserts in dry machining. *J. Chin. Inst. Eng.*, 43, 143–152.
27. Ji, W., Zou, B., Zhang, S., Xing, H., Yun, H., & Wang, Y. (2018). Design and fabrication of gradient cermet composite cutting tool, and its cutting performance. *J. Alloy. Compd.*, 732, 25–31.
28. Lee, Y., Resiga, A., Yi, S., & Wern, C. (2020). The optimization of machining parameters for milling operations by using the Nelder–Mead simplex method. *J. Manuf. Mater. Process.*, 4(3), 66.
29. Abdullah, M. E., Rohim, M. N., Mohammed, M. M., & Derazkola, H. A. (2023). Effects of partial-contact tool tilt angle on friction stir welded AA1050 aluminum joint properties. *Materials*, 16(11), 4091.

30. Wang, M., Zhu, M., Hu, X., Liu, K., Fan, X., Meng, X., … & Wang, J. (2024). Synergistic effect of elliptic textures and H-DLC coatings for enhancing the tribological performance of CuAl10Fe5Ni5 valve plate surfaces. *Coatings*, 14(9), 1161.
31. Liu, M., Bing, J., Xiao, L., Yun, K., & Wan, L. (2018). Development and testing of an integrated rotating dynamometer based on fiber Bragg grating for four-component cutting force measurement. *Sensors*, 18(4), 1254.
32. Manea, H., Lu, H., Liu, Q., Xiao, J., & Yang, K. (2024). Micro-grinding parameter control of hard and brittle materials based on kinematic analysis of material removal. *Mathematics*, 12(10), 1589.

5 Advantages of Tool-Holding Systems

5.1 INTRODUCTION

A wide variety of machine types are used in production. All of these machines are unique in their way. Although many other machining processes are possible, turning, milling, drilling, and boring are some of most commonly performed. On some machines, the cutting tool is fixed in place, while the workpiece moves. In other cases, the work element remains stationary while the tool is moved. Some machines, such as various types of grinders, have moving parts, including the cutting tool and workpiece. In addition to the metal-cutting (or "chip-making") procedures, some specialized machines are used for a variety of other processes such as riveting, welding, flame cutting, tube bending, stamping and even inspection [1]. The improvement in the production efficiency and miniaturization of the product size are ongoing trends. In addition, with the increasing demand for microstructure processing, the machining accuracy of ultra-precision machining (UPM) has been steadily improved. As shown in Figure 5.1, a five-axis machine consists of several critical components that work together to provide precise and flexible operation. The base

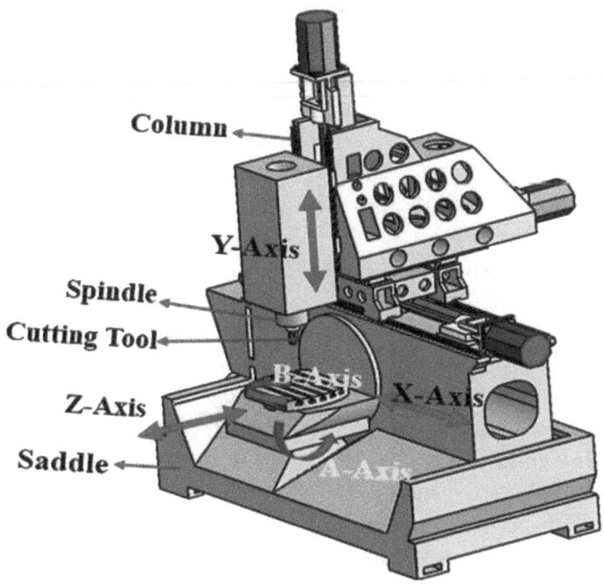

FIGURE 5.1 The physical structure of the 5-axis machine used for analysis. ([1] [Open access].)

DOI: 10.1201/9781003604648-5

and column, typically made of cast iron or welded steel, provide vibration absorption and basic support for stable operation. The spindle, made of materials such as high-speed steel (HSS), carbide, or ceramic, rotates the cutting tool at high speeds for precise operation. Rotary axes made of high-strength steel or aluminum (A and B) and linear axes (*X*, *Y*, and *Z*) provide the degrees of freedom needed for complex geometries. Ball screws and linear guides made of hardened steel ensure smooth and precise movements [1,2].

Over the past century, machine tools have become increasingly sophisticated as our understanding of materials and machining processes has expanded. As far back as the 1600s, lathes were used to work metal and wood. As the knowledge of materials, especially steels, increased, precision machines have been developed to perform cutting and forming operations on a wide range of materials. During and after the First and Second World Wars, machines that could cut metal quickly and accurately were in demand [3,4]. The numerical automatic use of paper tape with holes punched like a piano was introduced [1]. These computer-controlled machines, called computer numerical control (CNC) machines, were eventually replaced by computer numerical control (CNC) machines, devices in which the machine's functions are controlled by an electronic computer.. Anisotropic wettability can be achieved by a focused ion beam (FIB); a multi-step FIB fabrication method is developed to achieve precise and reliable fabrication of non-conductive materials such as single-crystal diamond tools, and a new method is proposed to prepare anisotropic wetted surfaces with high yield and good transparency [1]. Injection molding, molding and lithography are combined into the German Lithographie Galvanformung and Abformug (LIGA) technology [5,6]. It primarily performs micromechanical processing using deep X-ray irradiation, microelectroforming, microplastic molding and other technologies. LIGA technology is suitable for complex micromechanical structures because it can create structures with huge aspect ratios, widths down to submicrons, and depths of hundreds of micrometers or even millimeters [7,8]. It can be applied to various materials such as glass, ceramics, metal, plastic, and polymer materials. It can also be mixed with other materials. In addition, the microstructure obtained by the LIGA process has precise tolerances, straight and smooth sidewalls, and well-defined shapes and dimensions. After creating metal molds using deep X-ray irradiation and microelectroforming, plastic molding can be used to create highly accurate polymer structures suitable for large-scale manufacturing. It applies to the complex microstructure of multiple R&D and product manufacturing units [9]. However, fabricating microstructures on curved substrates is challenging, and LIGA technology requires expensive processing equipment [10]. Ultraviolet (UV)-LIGA deposition methods such as mask fabrication, substrate processing and UV-LIGA deposition technology were used to create multilayer metal microstructures with exceptional quality and accuracy. Figure 5.2 shows the electroplating photographs of their multilayer metal microstructure [1,11].

Considering that the main factors determining machinability are the workpiece material, tools, and cutting parameters such as feed, speed, and depth of cut, cutting steel and sintered materials can be considered difficult to machine. Porosity, microstructure, chemical composition, and advanced technology can be important factors in this situation. Drilling is the most common processing method for sintered parts.

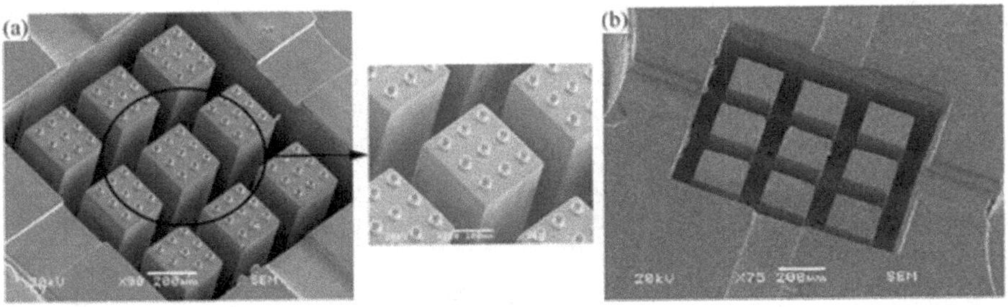

FIGURE 5.2 (a) SEM images of the multilayer metal microstructure was fabricated using UV-LIGA overlay technology, and (b) SEM image of the first layer microstructures. ([1] [Open access].)

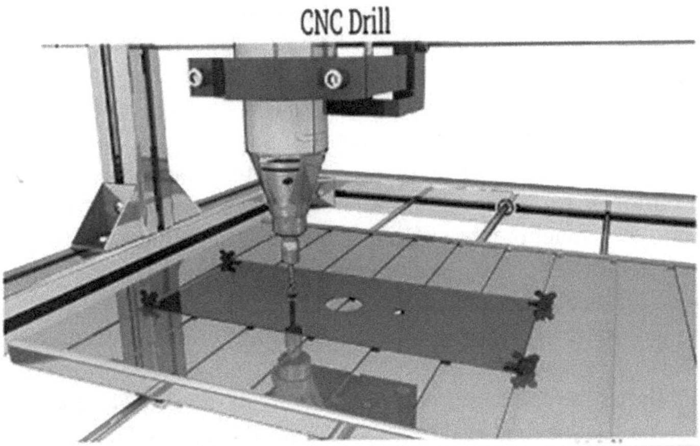

FIGURE 5.3 Computer is used to control the manufacturing process in the modern manufacturing center. To ensure operator safety during grinding, the doors are locked in a closed position. ([5] [Open access].)

Numerous variables, including optimization of tool geometric characteristics and their relationship to tool wear progression, require the measurement of cutting forces. Monitoring of tool wear is possible based on force changes. These initial CNC machines have evolved into machining centers where many processes are fully automated. Modern machining centers can automatically load one or more workpieces, perform complex series of machining operations, change cutting tools and remove the finished product from the machine. All these tasks can be performed without help of a human operator. In most cases, the operator's only task is to set up the automated processor and turn on machine. Writing a program that instructs a machine what tasks to perform and in what order is one of responsibilities of a modern machine. An example of the modern machining center is shown in Figure 5.3. Machining centers are now so sophisticated that a part designed using computer-aided drafting (CAD) software can be sent electronically to a CNC machine for machining [5].

5.2 LATHES

To create precision components, the combination of CAD software and computer-aided manufacturing (CAM) procedures is needed. The input instructions that make up the CAD program can be defined using a variety of coding techniques. Although 3D printing is a popular method for creating images for CNC machines, it uses G-code and requires more machining steps before the final commands are executed. After CAD, 3D-printed components may require additional finishing unlike CAD, which creates parts ready for manufacturing. Furthermore, the FIB technology is emphasized as a means of improving and preparing diamond tools with specific properties. FIB can produce diamond tools with edge radii of a few nanometers, making them suitable for UPM [1]. The tool edge radius schematic, illustrated in Figure 5.4, highlights improvements in tool design to increase manufacturing accuracy. The minimum ratio of chip thickness to tool edge radius in nano cutting is 0.3–0.4 [1,11]. To produce microscopic surfaces, a new cutting technology based on force modulation for multi-edge diamond tools is used. The machine tool for turning or facing operations is a lathe. These processes involve rotating the workpiece while removing metal from the inside or outside diameter using a cutting tool. Parts that are symmetrical about axis of rotation are made on lathes [1]. A CNC machine is an automated computer numerical control device that shapes and removes access materials to create a precise product that is unique to the user. The CNC machine downloads 3D image programmed using CAD software to its computer. The CNC machine is used to machine a part made of metal, plastic, wood, ceramic, or composite materials using coded programming instructions. A manual operator is not needed for anything other than the initial setup of the gadget [11].

Figure 5.5 shows the schematic cutting structure of this technology. Using the FIB technology, a multi-edge diamond tool with a periodic sinusoidal microstructure was created [1].

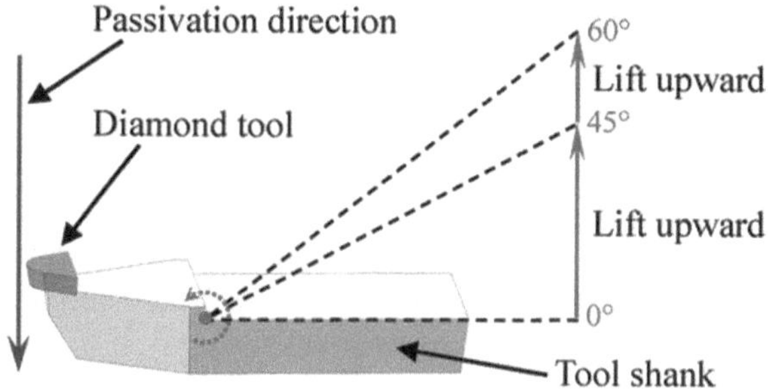

FIGURE 5.4 The tool edge radius is measured by a field emission scanning electron microscope. The cutting edge, measuring setup and diamond tool result obtained by FIB with an edge radius less than 20.95 nm, as well as rotation of X and Y axes of the platform on which the tool is manufactured are three main possible sources of inaccuracy. ([1] [Open access].)

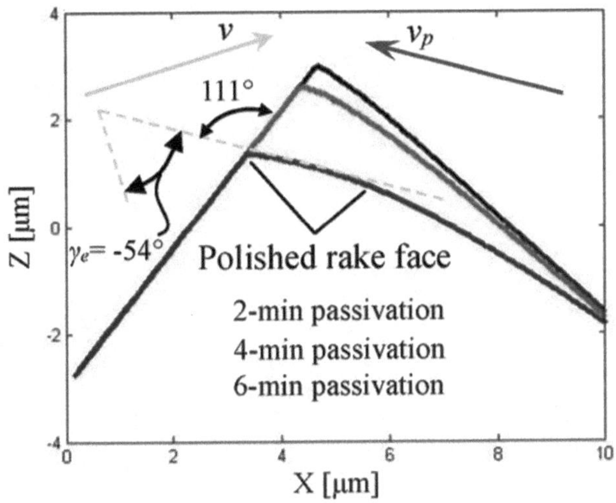

FIGURE 5.5 Schematic diagram of the cutting process using multiple cutting edges of the diamond tool. ([1] [Open access].)

5.2.1 Types of Operation of Lathes

The operating principle of a lathe depends on the specific operation and desired cut. The simulation results obtained using Surface Evolver (SE), a computer software created by Kenneth Brake, are shown in Figure 5.6. The three-dimensional (3D) structures are treated by Surface Evolver (SE) as surfaces that are subject to energy optimization standards only. Figure 5.6a illustrates how the simulation run time reduced the surface energy and surface area of one contact hole and nine contact holes after the resistance flow procedure (RFP). The remelted images of one contact hole are plotted against the simulation run time for surface areas of 38.3685 and 32.9430, respectively, in Figure 5.6c and d. The remelted images of nine contact holes for surface areas of 155.0468 and 144.5697, respectively, are shown in Figure 5.6f and g. However, the Surface Evolver (SE)-based model did not require detailed knowledge of the material-specific properties and assumed how the simulation run time would affect the shape evolution [12].

Some of the most commonly performed turning operations are as follows:

- Get your bearings
- Cone rotation
- Parallel turning, parting and chamfering
- Thread cutting boring drilling
- Extension
- Grabbing

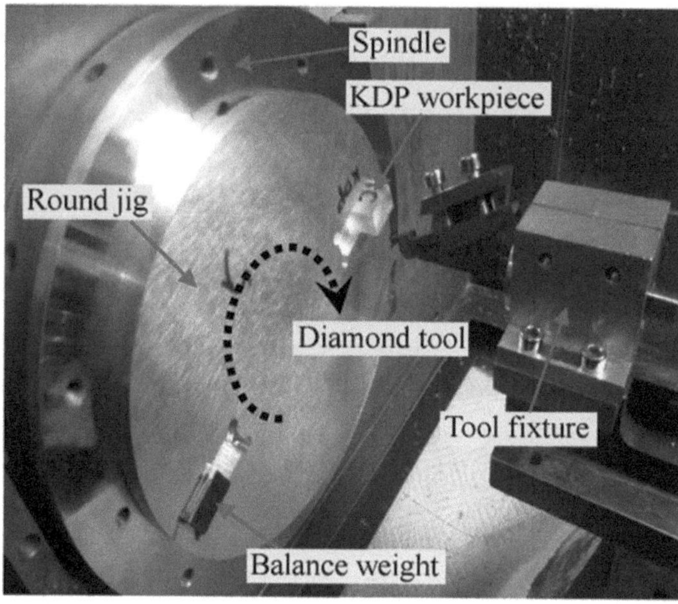

FIGURE 5.6 In terms of simulation times, Surface Evolver's simulation results show that (a) surface energy and surface area for one contact hole pattern and nine contact holes were (b, e) at surface areas of (c) 38.3685, (d) 32.9430, (f) 155.0468, and (g) 144.5697, respectively, prior to the resist flow process (RFP). ([12] [Open access].)

1. Facing: Turning the workpiece is the first step in most turning operations. To establish an exact right angle with the axis and eliminate surface defects, this operation involves cutting the metal from the end.
2. Change: Turning is the most common operation performed on a lathe, among the most frequently performed. Turning is a machining operation in which the outer cylindrical or conical surface of workpiece is machined to remove excess material and reduce the diameter of object. As a result of this operation, the metal surface acquires excellent polishing.
3. Rotate the cone: Using a compound caliper, the metal is formed into a nearly conical shape by tapering. By modifying the compound caliper, it provides flexibility in adjusting the angles between parallel rotation and facing.
4. Composite support method: This method involves rotating a compound support at an angle corresponding to the desired taper angle of part. This method allows for a variety of taper angles to be created, allowing for precise turning of internal and external tapers.
5. Tailstock method: This method is used for making outer tapers only. This method tilts the workpiece relative to lathe axis and determines the feed direction by moving the tailstock from its center position to one side of bed. As a result, when the tool moves and engages the workpiece at an angle relative to axis, a cone is formed.

6. Fixture for the taper turning method: To apply the taper turning method, tilt the slide paths by the same amount as the component's taper angle. This tilt of seat occurs automatically, as it moves along the guides, forming a conical section.
7. Shape tool technique: This technique is used to create outer cones. This method uses a tool that reflects the intended shape of the component being manufactured. The main application is chamfering processes.
8. Parallel rotation: Reducing the diameter of metal while maintaining its alignment with the axis is the main goal of parallel turning. The purpose of this procedure is to make the diameter of the metal uniform.

Splitting is the process of separating a portion of a piece to produce a distinct finish. For cutting, the parting tool must be advanced carefully. The cutting tool is removed and set aside to deepen the cut without breaking the tool. The process of obtaining a beveled edge on the ends of a workpiece is known as chamfering [2,5,13].

5.2.2 Applications of Chamfering

Improving the aesthetic appearance of the workpiece allows the nuts to slide more easily over the threaded parts of workpiece. Sharp edges and burrs are removed. The workpiece end is protected from possible damage. Knurling is the process of applying rough texture to the surface of the workpiece using a knurling tool to create a textured diamond-shaped pattern. The knurling process is performed at slower speeds and requires a lot of lubrication. The purpose of knurling is to improve grip on the workpiece, which is especially useful for reducing slippage when handling the workpiece by hand. Drilling is a machining technique used to remove material from a workpiece. This is achieved by making a hole in the workpiece using a stationary drill bit mounted in the tailstock. Drills are usually made of carbon and high-speed steel. An internal turning process called boring is used to enlarge existing holes. It belongs to one of two subtypes.

To increase the size of entry, an internal turning technique called boring is used. Conical expansion of hole entrance is a necessary step in countersinking. This process requires a drill bit with a diameter larger than size of original hole. After drilling, the reaming is a machining procedure that produces accurate inside diameters of holes [14]. The reaming process removes very little material from pre-drilled holes, ensuring very accurate final hole dimensions. When using finger cutter tools to process helical surfaces, the kinematic principle is illustrated in Figure 5.7 [5].

Some people believe that little has changed in terms of turnover over the previous century. However, those who hold this opinion are unaware of recent metal-cutting incidents. Lately, manufacturers of turning tools and equipment have been announcing significant advances. According to these manufacturers, employees who implement these innovations will experience numerous benefits such as reduced cycle times. The kinematic basis of producing in the slotting tool processing of helical surfaces is depicted in Figure 5.8 [5].

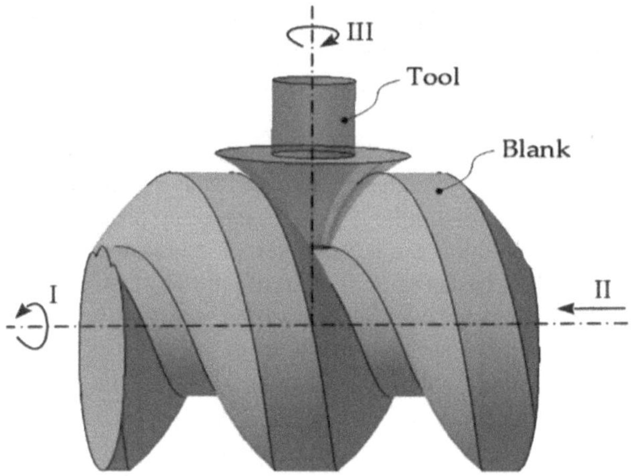

FIGURE 5.7 Kinematic principle. ([5] [Open access].)

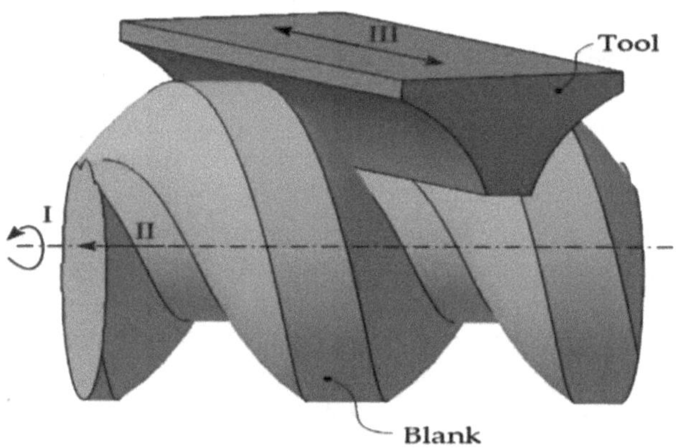

FIGURE 5.8 The kinematic principle of generating: I—represents the rotational movement of the blank, II—the translational movement of the blank along its own axis, and III—the cutting movement performed by the tool. ([5] [Open access].)

It can be argued that one of these events is taking a different turn. Feed is directed along *Z*-axis toward the chuck during normal rotation. However, Prime turning technology makes multi-directional turning possible with just one tool. Prime Turning users can even pass the tool back and forth as long as it stays within the cut. The result is faster machining, especially in multi-pass roughing [6]. This study presents the proposed intelligent tool holder using the Tongtai TS-85 bedroom CNC lathe. The intelligent tool holder is adapted to the shown turret of the CNC lathe (Figure 5.9b) and is manufactured for the conventional turning tool model SVJCR-2525M16 (Figure 5.9a). The wedge design enables the turning tool to be mounted on a turret (Figure 5.9c). The technical specifications of turning tool model SVJCR-2525M16 are provided [2].

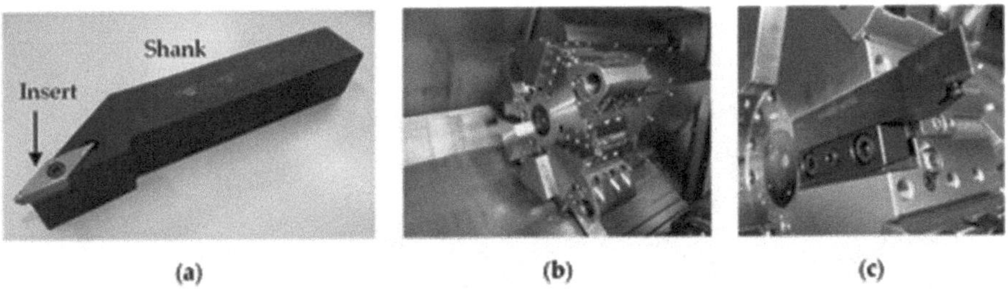

FIGURE 5.9 The proposed intelligent tool holder can be used with the following three devices: (a) standard turning tool (model: SVJCR-2525M16), (b) turret of Tongtai TS-85 CNC lathe, and (c) turning tool fixed on the turret. ([2] [Open access].)

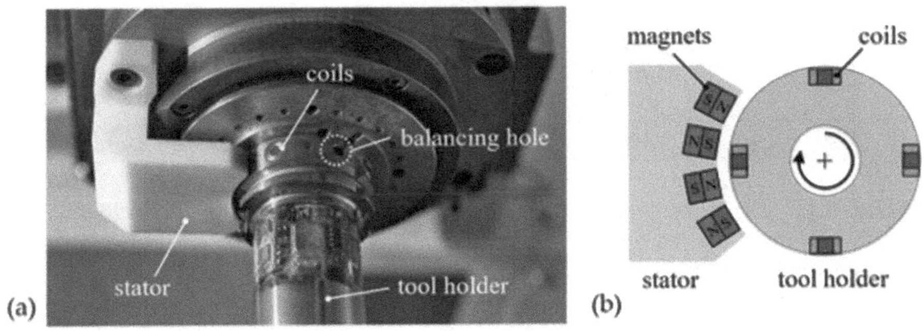

FIGURE 5.10 The five components that make up the smart tool holder structure include (a) parts 1 and 2, which are held together by ten M6 hexagon-head screws; (b) parts 3–5, which form a wedge; and (c) four M6 adjusting screws, which adjust the tightening of the conversion tool. ([2] [Open access].)

Figure 5.10 illustrates type a for roughing and type b for light roughing, finishing and profiling. There are two variations available. The narrow lift angle relative to the feed direction that is designed to speed up cutting and improve surface quality when using both machining designs. The RollFeed process requires a unique tool body, unique inserts and a unique B-axis motion unit. The process requires special programming because the method uses an additional axis [2]. The RollFeed software is developed by Emag in collaboration with Open Mind Technologies AG, Wessling, Germany, a CAD/CAM software company, using the HyperMill CAM solution from Open Mind Technologies [2].

The roll feed rotary insert method has an exceptionally wide cutting-edge radius for turning and is designed to improve surface quality. Rakwid noted that the large cutting-edge radius, however, speeds up the system in applications where good surface quality is not required. The five parts of smart tool holder are shown in Figure 5.11a. Parts 1 and 2 are the front and rear tool covers, respectively, which are used to hold the tool. Parts 3–5 form a wedge structure used to hold the smart tool holder on lathe turret. Two M6 hexagon-head screws hold the wedge structure together, and M6 lock screw is used to change the height of wedge to ensure that

FIGURE 5.11 Experimental setup including the static load test of intelligent tool holder. It starts with installing the conversion tool and intelligent tool holder on optical table and simulating their installation on the tower. ([2] [Open access].)

smart tool holder is securely fixed to the lathe turret. Parts 1 and 2 are assembled using 10 M6 hexagon-head lock screws (Figure 5.11b). Figure 5.11c shows how four M6 lock screws can be used to change the clamping force of the lathe tool.

The cutting force consists of three parts (Figure 5.11a): feed force (F_f), negative force (F_p), and initial cutting force (F_c). The authors build a device for holding the turning tool and assemble the intelligent tool holder on optical table to reproduce conditions on turret (Figure 5.11b). The authors transmit the initial cutting force to cutting edge of turning tool by hanging standard weights, where the intelligent tool holder is used to measure initial cutting force (Figure 5.11b). The standard weights include 100, 200, 500, 1000, and 2000 g. The data is transmitted by the wireless data collector of the intelligent tool holder to a personal computer via Bluetooth [2].

A torque wrench is used to tighten the turning tool and intelligent tool holder assembly with a constant torque, ensuring that the condition is the same for each test. In addition, the pre-stress of film strain gauge is the same for each test by tightening the four adjustment screws (Figure 5.12) with a constant torque. The output signal of film strain gauge at different fixed cutting forces is shown as a voltage in Figure 5.12. At a constant force, the output signal should theoretically remain constant; unfortunately, this is not the case and will change over time. In force sensors, this is known as thermal drift phenomenon [15]. In equipment that always requires accuracy, such a change is unacceptable. The two main sources of heat transfer of sensor are thermal drift of sensing element, which is related to the temperature compensation method of the system [2].

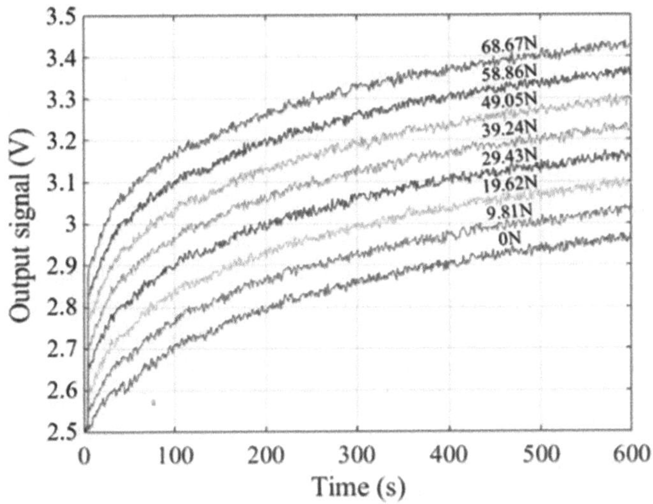

FIGURE 5.12 Output signal of the film tension force sensor at different cutting force values. ([2] [Open access].)

5.2.3 Drilling Machines

If you have a beautiful framed picture that you can hang on the wall, but there is nowhere to put it, you can use a drill press. This equipment is integral to every industrial sector and has no limitations. A mechanical engineer needs to understand the workings of this machine [16]. In this post, we will tell you about the working principle, types of drilling machines, various tools and their vocabulary, drilling operations, advantages, disadvantages, and applications of drilling machines. Technical information about drilling machines is given below:

- Power source: The main power source of drilling equipment is electricity, which guarantees reliable and stable operation. To increase power and precision for certain applications, some modern versions additionally include hydraulic systems. Portable drill presses can be used in remote areas as they are battery-operated, allowing for greater ease and flexibility [17].
- Engine power: With a variety of motor powers, drilling machines can handle a wide range of workloads and applications. There are solutions to suit a variety of requirements, from rugged machines designed for heavy-duty applications to miniature models ideal for smaller jobs. Engine power gives a clear indication of machine's power output and is usually measured in horsepower (hp) or kilowatts (kW).
- The maximum drill diameter is determined by the size of drill chuck that machine can hold. Commonly used chuck sizes are 1/2, 3/4 and 1 inch to suit different drilling needs. Due to their flexibility, drill presses can handle a wide range of projects with different-sized holes.

- Spindle speed: It is usually expressed in revolutions per minute. Spindle speed is the speed at which drill rotates. Different drilling operations require different spindle speeds to achieve the best results. A common feature of drill presses is their ability to have speed settings that can be changed depending on materials and hole sizes.
  - The maximum depth to which the machine can penetrate the material is determined by drilling depth. Selecting a drilling machine with the correct depth is essential to ensure accurate and productive drilling. This specification ensures that the equipment can meet the specific needs of different types of drilling projects. As shown in Figure 5.13, the vibration-assisted drilling process involves using a mechanical or piezoelectric oscillator to impose a sinusoidal axial oscillatory motion of the drill in the axial feed direction during the drilling process [18].
- Feed rate: During drilling, the feed rate is the rate at which the drill penetrates the material. Operators can adjust the drilling speed depending on material being drilled and desired result thanks to variable feed rate. This feature provides greater accuracy, efficiency, and flexibility in a variety of drilling settings.
- Drill bit compatibility: Drilling operations can be performed differently due to the compatibility of different sizes and types of drill bits with drilling equipment. Commonly used types of drills are step, annular, spade, and twist. Drills and drill presses are compatible as long as the machine chuck size and configuration match, ensuring seamless integration [7].

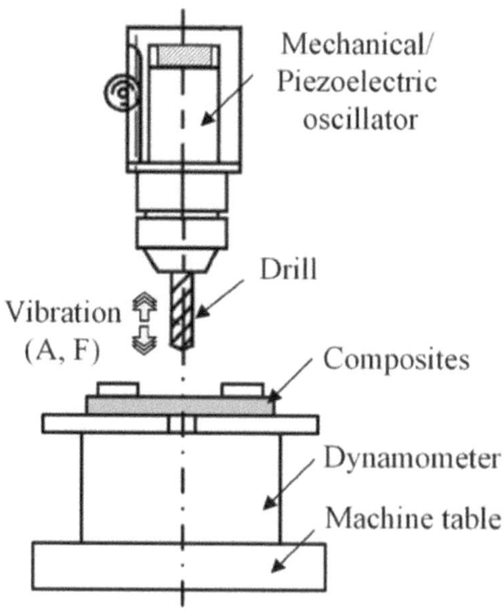

FIGURE 5.13 Illustration of vibration-assisted drilling. ([18] [Open access].)

5.3 MILLING MACHINES

Drilling and milling machines are similar, in that the workpiece is placed on a table that can move along the *x*, *y*, or *z* axes, and the cutting tool is often mounted along a vertical or horizontal axis. One of the most interesting technologies of the past is the milling machine, which is used to create grooves, as well as flat and curved surfaces. It handles all tasks, including machining flat surfaces and cutting teeth and slots. These devices are sometimes called multi-tasking machines (MTMs) that are versatile tools designed to perform multiple functions, often combining several processes into one unit. We will talk about the operation of these devices based on parts in this article. In addition, we will look at the many types of milling machines and their advantages, disadvantages, and applications. This subject is so famous that test takers have made it their favorite theoretical topic like GATE ME and SSC JE ME. Now we will describe in detail the horizontal milling machine. Its orientation is horizontal, as the name suggests, with the spindle parallel to the work table [9]. As mentioned earlier, the plastic flow of the material creates the entrance hole burr, while the conformation of the material caused by high compression velocities at the center of the hole causes the exit hole burr. Figure 5.14 shows the mechanism by which exit hole burrs are formed during the drilling operation [18].

A horizontal milling machine, like everything in life, has its advantages and disadvantages. On the other side, its design allows for multiple cuts at the same time, increasing efficiency and speed. The downside can be difficult to set up and may not be suitable for all complex applications [5]. The splitting between two adjacent layers is a sign of drilling-induced damage when the thrust force is greater than the material's interlaminar bond strength [13]. Both the machined-hole entry

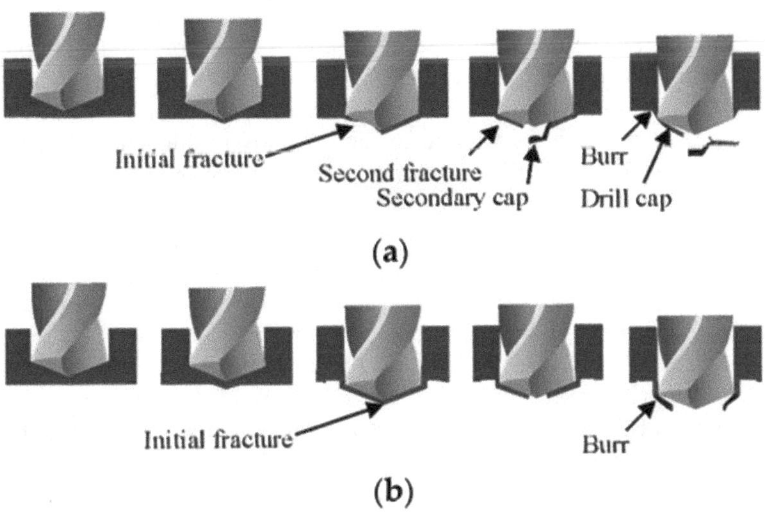

FIGURE 5.14 Mechanism of burr formation in the drilling process: (a) uniform burr and (b) crown burr. ([18] [Open access].)

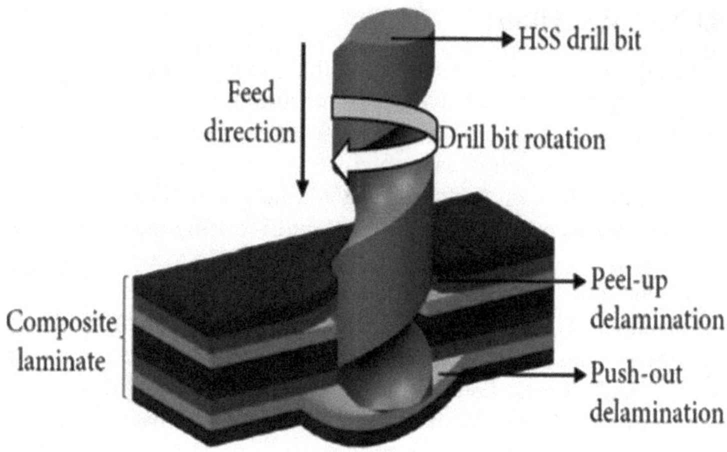

FIGURE 5.15 Mechanism of delamination in composite laminate drilling. ([18] [Open access].)

(peel-up delamination) and departure (push-out delamination) exhibit this phenomenon, as Figure 5.15 shows [18]. Vertical milling machines are often preferred for tasks such as slotting and grooving due to their design and functionality. In short, the design features of a vertical milling machine, combined with its precision and versatility, make it well suited for tasks such as slotting and grooving.

- Precision and control: The cutting tool can be precisely controlled with vertical milling machines. This precision is critical for creating precise slots and grooves.
- Versatility: The vertical axis positioning allows for a large number of movements and adjustments. This adaptability is very useful when creating complex cuts and fine work.
- Ease of setup: Since the workpiece can be easily positioned and clamped, setting up a vertical milling machine for slots and grooves can be easier.
- Tool options: Vertical milling machines are versatile enough to accommodate a range of slotting and groove needs, as they can hold a wide range of cutting tools and accessories.

Kawasegi's research group has conducted a large number of studies on the preparation of diamond tools using FIB technology. A method for preparing surface-textured diamond tools is presented. To improve the efficiency of diamond tool processing, Figure 5.16 shows the production of a pattern with depth 43 nm and width 1.8 μm using FIB on the tool torch surface with subsequent heat treatment. This technique provides superior cutting performance of diamond tools compared with direct FIB sputtering and femtosecond laser [1]. Removing the non-diamond phase generated by FIB can prevent the negative impact on the cutting performance due to the high resolution of FIB radiation [19].

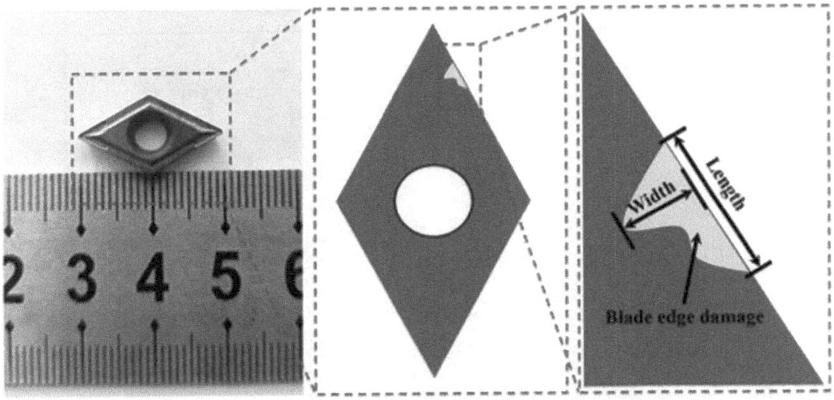

FIGURE 5.16 Textured diamond cutting tool produced using heat treatment and FIB radiation. (a) Magnified image of texture determined by AFM; (b) SEM image of rake face of textured diamond cutting tool. A coherent scanning interferometer is used to determine the topography of rake face with texture in the perpendicular (c) and parallel (d) directions. (Reprinted with permission from [1] [Open access].)

A method to improve diamond cutting and forming tools at nano- and micrometer scales using FIB micromachining is reported. The surface treatment of NiP using FIB and treated tools is shown in Figure 5.17a and b [1]. Although the FIB method is well suited for creating tool shapes at nano- and micrometer scales, ion irradiation may lead to alloy formation and defects and reduce the tool's performance. FIB treatment on a single-crystal diamond tool without compromising the tool performance by removing Ga ions produced by ion irradiation using heat treatment at 500°C with Al deposition is reported. NiP and Al alloys are treated to evaluate the method. The surface morphology of NiP is shown in Figure 5.17c–e [1].

The improvement of tool alignment accuracy to less than 1 μm, fast data synchronization and conversion, and integration of machine surface measurements and surface feature functions such as Alvarez lens and lens surface microgroove measurements are demonstrated in Figure 5.18 [1]. By modifying and synchronizing the data flow, the real-time surface measurement system is seamlessly integrated into the ultra-precision fast tool servo (FTS) system. This integration significantly improves the efficiency of real-time surface measurement and the accuracy of the machining process, providing precise feedback and control during operation [9]. To support this advancement, a custom controller was developed to implement the internal data processing algorithms required for real-time decision-making and system optimization [1].

Closed-loop process control for free-form surface manufacturing and measurement can be achieved by synchronizing and controlling the entire integrated sensor systems and CNC machine tools. Let us change our focus and focus on a vertical milling machine. Named for the vertical position of spindle, this machine is the opposite of its horizontal cousin. However, what makes it different? What makes the vertical milling machine the preferred option? Like their horizontal counterparts, vertical

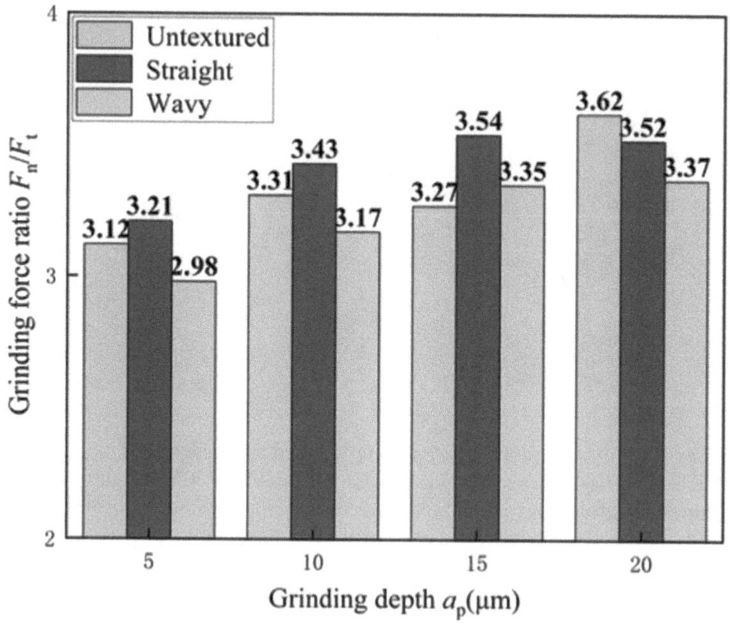

FIGURE 5.17 Cross-sectional images of NiP surface after machining with a cutting distance of 19,040 m using (a) FIB tool and (b) heat-treated tool. White light interferometry is used to measure the surface topography of NiP after machining with (c) non-FIB tools, (d) FIB tools, and (e) heat-treated tools. ([1] [Open access].)

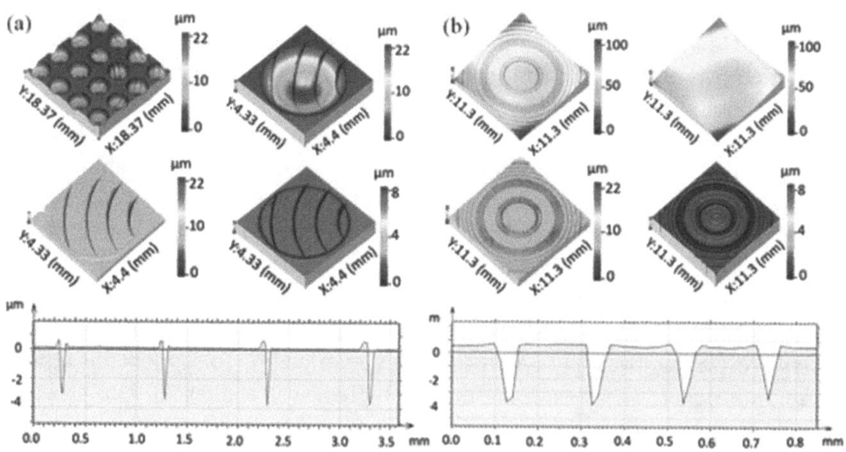

FIGURE 5.18 (a) Surface characterization of a 4×4 microscope lens array produced on the machine tool. It includes an overview of microscope lens array, one microscope lens, microgrooves extracted after demolding, lens shape and measurement anomalies removed and microgrooves shown in cross section. (b) Surface characterization of Alvarez lens with microgrooves: raw data measured on the machine tool, Alvarez lens shape calculated using a robust second-order regression filter, surface microgrooves extracted after removing optical measurement anomalies, and microgrooves shown in cross section. ([1] [Open access].)

milling machines have their advantages and disadvantages. Their main selling points are adaptability and small size. A vertical milling machine can perform tasks such as drilling, plunge cuts and complex contouring operations. Conventional milling can be compared with up-milling. However, their disadvantages are associated with working with large or heavy workpieces [20]. They are usually less effective for this type of work due to their design. A vertical milling machine is a tall machine that protrudes from the frame, with a spindle positioned perpendicular to the work table.

5.3.1 Tool Holders

Tool holders are vital parts of milling machines as they keep cutting tools safe and ensure the precision and efficiency of machining operations. There are many types of milling machine holders, each with its special qualities and benefits. In this detailed guide, we will look at many types of tool holders for milling machines, with a particular focus on shrink-fitting and hydraulic tool holders [19], at their features, benefits, and uses, to help you determine which one best suits your processing needs. What are tool holders? Before delving into the different varieties, it is very important to understand the general function and importance of tool holders in routing. A tool holder is a device designed to safely store cutting tools including drills, end mills and inserts. It performs several important functions:

- Tool safety: During machining, tool holders hold cutting tools securely in place and prevent them from slipping or being thrown away.
- Tool positioning: They provide precise and repeatable machining operations through precise tool positioning.
- Vibration damping: To minimize tool wear and improve surface smoothness, many tool holders are designed to dampen vibrations.
- Quick tool change: Machine productivity can be increased by using certain tool holder systems that allow quick tool changes.
- Types of tool holders: We understand the importance of tool holders in general, let us look at few different types.

5.3.2 Hydraulic Tool Holders

Hydraulic tool holders are often used in precision milling processes. They use hydraulic pressure to provide exceptional rigidity and precision, holding the cutting tool securely in place. Here are some important features and benefits of hydraulic tool holders:

- Extreme precision: High-precision machining results depend on exceptional tool concentricity of hydraulic tool holders.
- Vibration dampening: To improve surface smoothness and increase tool life, vibration is absorbed by hydraulic system during cutting.
- Easy tool replacement: These holders are known for their easy and quick tool changes, which reduce downtime.

- Versatility: A wide range of cutting tools can be used with hydraulic tool holders, making them an adaptable option for a wide range of machining tasks.
- Applications: They are widely used in sectors such as aerospace and medical device manufacturing where precise measurements and superior finishing are critical.

5.3.3 Smaller Tool Holders

Shrink tool holders are another type of tool holders that is becoming increasingly common in modern machining. The tool is fixed by these holders using theory of thermal expansion. Here is why so many people like them:

- Outstanding runout accuracy: Shrink-fit holders guarantee precise machining results, providing exceptional runout accuracy.
- Extended tool life: By minimizing tool deflection and reducing tool wear, powerful clamping force extends tool life.
- Balanced tool: For high-speed machining operations, balancing these holders is essential.
- Low maintenance: Because shrink-fit holders have no moving parts, they require less maintenance.
- Applications: They are often used in the production of dies and molds, in the automotive industry, as well as in other industries where high-speed processing is necessary.

Although the shrink-fit and hydraulic tool holders provide advantages in terms of precision and efficiency, the decision between the two is based on the requirements and preferences of particular application [21]. The efficiency of the drilling process of multi-component stacks, especially CFRP/Al stacks and fiber metal laminates (FML) composites, as well as the quality of the drilled holes and the production costs, can be significantly affected by tool wear. To identify the main wear processes that occur during drilling operations and are responsible for the reduction in tool life, which can lead to catastrophic tool failure, this element must be carefully monitored, according to Figure 5.19 [18].

Conversely, shrink-fit holders perform exceptionally well in high-speed machining where tool balance and runout are critical. When choosing between them, think about the materials you will be milling, tolerances required, and machine's capabilities. The availability and price of these tool holders may also influence your choice [22].

5.3.4 Tools for Cutting in Turning and Milling Processes

Lathe-cutting tools are wooden, hand, or CNC tools that are mounted on a lathe and used to make turning products. The final shape of workpiece is determined by its trajectory along axis of lathe. The cutting operations supported on lathe and actual application are determined by the distinctive features and designs of different types of lathes, both CNC and non-CNC. For this reason, choosing the right cutting tool

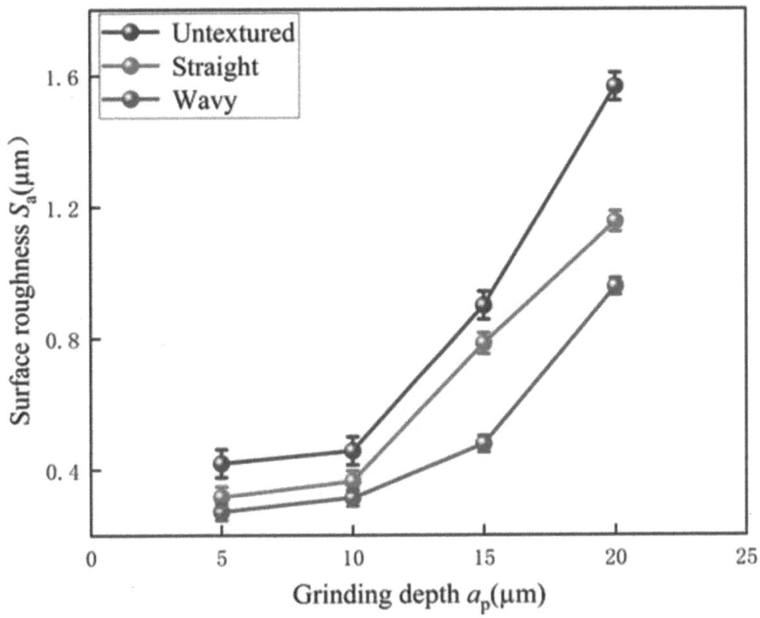

FIGURE 5.19 Types of wear on the drill. ([18] [Open access].)

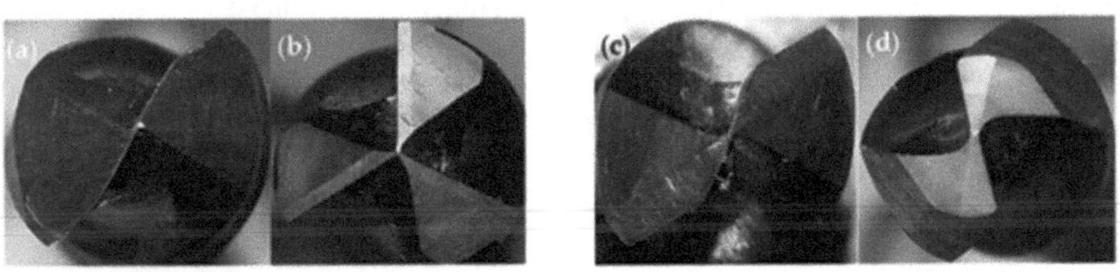

FIGURE 5.20 Drill point geometry: (a) 2-fluted, (b) 3-fluted, (c) 4-faceted, and (d) 8-faceted. ([18] [Open access].)

requires careful handling. This study discusses popular lathes, their features, design and use. Now let us get started on single-point and multi-point tools. What tools you need for this task will depend on the desired shape of material. Single-point and multi-point tools are the two main categories of cutting tools. To conduct the drilling experiments on the GLARE 5 and GLARE 6 specimens, uncoated 6.35 mm diameter carbide twist drills with a 30° helix angle, 120° point angle, and four different drill point geometries (i.e., two-, three-, four-, and eight-facet drills) were used, as shown in Figure 5.20 [18]. Unfortunately, this can also come at a cost, adequately maintaining the equipment and tools [23].

A wide range of materials are available for making instruments. The most widely used are high-speed steel and carbide. For general-purpose milling, HSS can be used. However, for harder and stronger tool steels, carbide should be used. When machining, factors such as depth of cut, feed rate and cutting speed must be taken into account. These characteristics will be influenced by the dimensions, tool material and workpiece material. Cutting speed is the speed at which the cutting tool penetrates the workpiece material. Surface feet per minute is a unit of measurement [24]. The speed of workpiece along its axis, as it approaches the cutting tool is called cutting feed. It is measured in inches per minute. During turning, the workpiece rotates and cutting tool moves linearly. The result is a cylindrical shape. The tool of choice for all turning work is a lathe. Like most machining tasks, turning can be done manually or automatically. The disadvantage of hand turning is that it requires a constant supervision to propose a new design of flexible parallel constraint-based motion mechanism rotating along three translation directions (*X*, *Y* and *Z*) to obtain high rigidity and high decoupling motion [7]. A unique three-degree-of-freedom piezoelectric FTS motors are small machines that convert electrical energy into mechanical motion using piezoelectric materials. These motors usually work on the principle of piezoelectricity, which states that when an electric field is applied to some material, it changes its shape with high natural frequency and decoupling characteristics. The tool path and cutting pattern are shown in Figure 5.21 [1].

During milling, the material is removed from the workpiece using multi-point rotating cutters. Face milling and peripheral milling are the two main categories of milling processes. Face milling creates flat surfaces and flat-bottomed cavities in workpiece. Both vertical and horizontal feeding are possible. Gear teeth, threads, and deep grooves are cut during peripheral milling. There are two methods for feeding the workpiece into the cutting tool. In traditional milling, the workpiece is fed against the rotation of the cutter. This is how manual milling machines should be used. In contrast, in climb milling, the workpiece is fed in the same direction as the rotation of the cutter. A unique variety of tool steel that has exceptional strength, hardness, and heat resistance is called high-speed steel (HSS). The unique elemental composition of

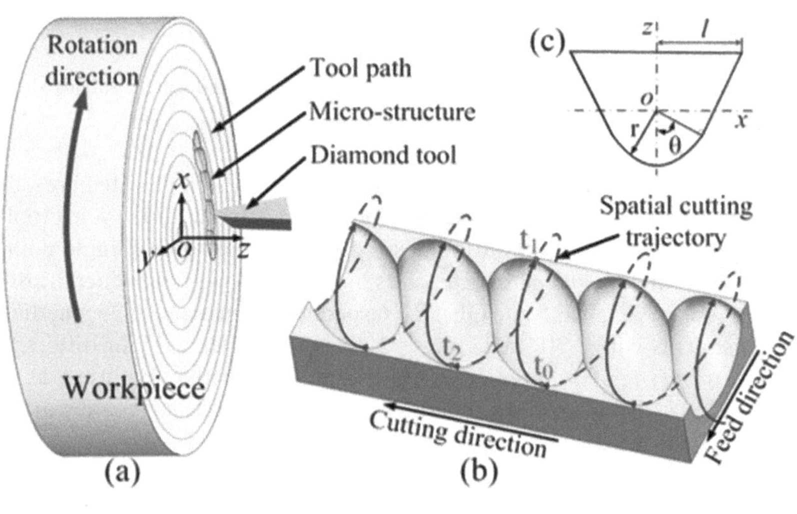

FIGURE 5.21 (a) Tool tip shape, (b) cutting path, and (c) expected cutting path plot for 3-degree-of-freedom FTS. ([1] [Open access].)

HSSs—iron, carbon, tungsten, chromium, molybdenum, and other elements—allows their tools to withstand high mechanical loads and maintain a cutting edge at elevated temperatures. The composition, characteristics, and applications of HSS are described in more detail on this page. A cutting tool consisting of metal injection molding (MIM) powder or extruded carbide has been sintered [25]. A tool path generator is established to produce wave structure patterns on roller surfaces. Various microstructures, wave patterns, and grooves are obtained through modeling and simulation. Based on the original experimental work of tool path generator, a four-axis ultra-precision machine tool is described. Figure 5.22 [1] shows the generation of individual microwave rollers in the microstructure mode. To perform cutting operations efficiently, single-step prismatic machining with a rotating tool typically uses a technique in which the tool rotates. The reference to STS (perhaps a "spindle tool system" or a related idea) indicates that improving the accuracy and productivity of machining operations is the main goal.

Cemented carbide powder, mainly tungsten and cobalt carbide, is formed into inserts by extrusion or MIM and then sintering. This process allows for higher cutting

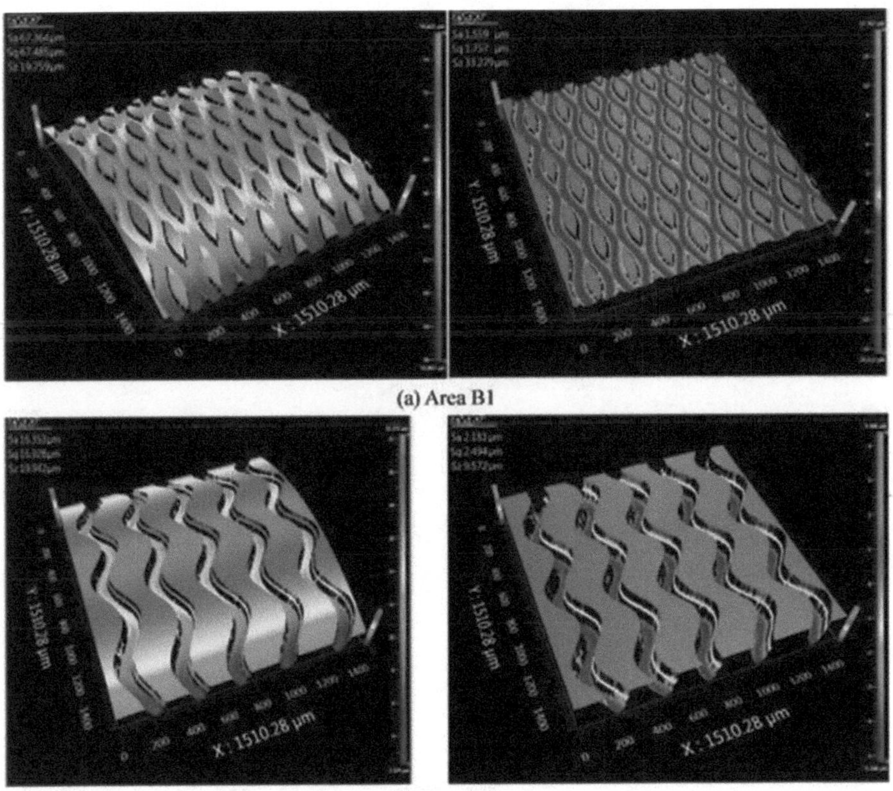

FIGURE 5.22 The three-dimensional topography on a roller surface after removal of the cylindrical shape (5.5X objective, 1X magnification). ([1] [Open access].)

speeds because the inserts can withstand higher temperatures better than HSS. These inserts, known by another name, indexable inserts, are placed on a cutting tool holder and can be moved to accommodate different cutting edges before being discarded. Cemented carbide inserts are typically coated with one or more surface layers to improve their wear resistance. Tool path planning, tool geometry selection and tool radius compensation were included in a systematic approach to tool path generation, which is investigated [1]. A model was developed for dynamic surface generation by analyzing the servo dynamics error components in servo motion such as dynamic deformation, resonant vibration, and path tracking error, using a micro-lens array (MLA) as an example. The surface generated by rotation is shown in Figure 5.23 [1].

Typically, chemical vapor deposition (CVD) or physical vapor deposition (PVD) is used to apply surface coatings. There are six different ISO classifications for cutting workpiece materials to help the user select the appropriate cutting tool. They can be divided into:

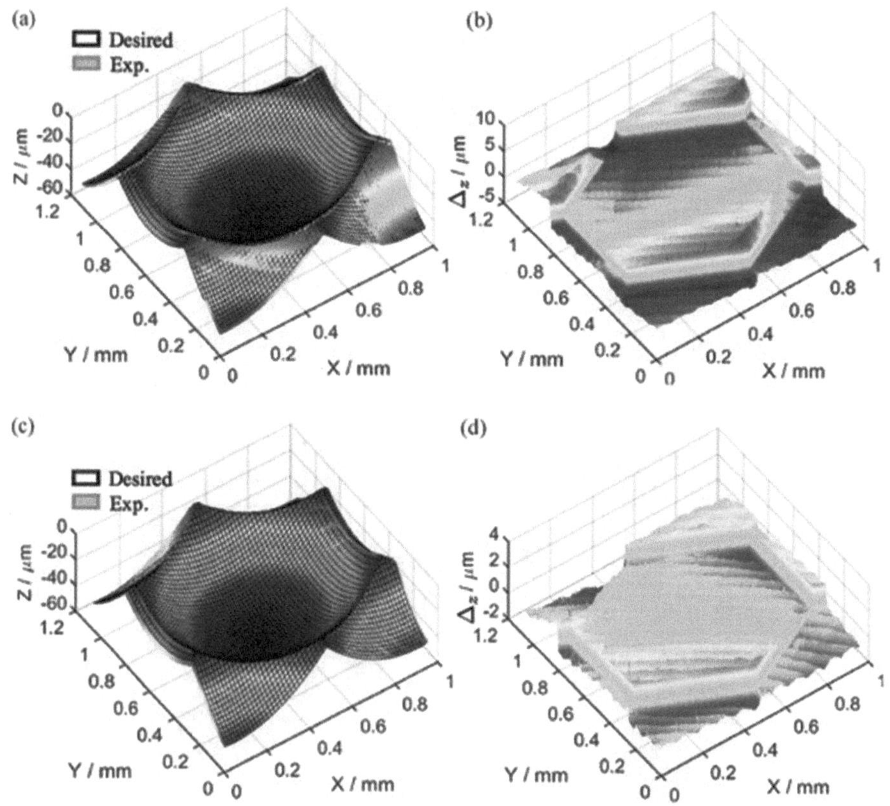

FIGURE 5.23 Surfaces obtained by practical turning have the following characteristics: (a) three-dimensional shape, (b) surface shape error obtained by STS turning, (c) three-dimensional shape, and (d) surface shape error obtained by servo cooperative turning. ([1] [Open access].)

- P: Steel
- M: Carbon steel
- K: Iron
- N: Non-metallic substances
- Super and titanium alloys (S)
- H: Hardened steel.

The insert shape, exit angle, tolerance class, chip breaker, holder model, size, cutting direction, and other data are additionally indicated by other symbols [15]. The surface defect machining (SDM) method, as shown in Figure 5.24a, has the combined advantages of both porosity machining [11] and pulsed laser pre-treatment [6]. The SDM method involves first machining the workpiece by creating surface defects at depth smaller than unmachined chip thickness by mechanical and/or thermal means and then performing the standard conventional machining operation. By cutting the material at lower input energy, SDM easily facilitates deformation [6,14]. In addition, a reduction in associated residual stresses on the machined surface is made possible by a significant stress concentration in the cutting zone compared to the subsurface [13].

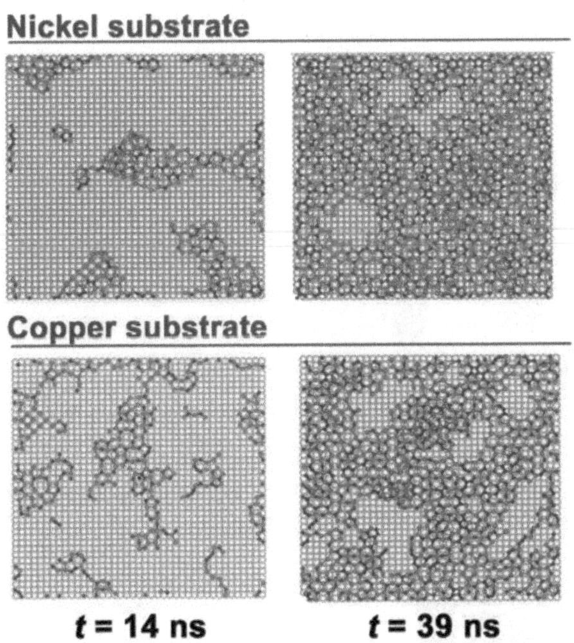

FIGURE 5.24 Finite element analysis (FEA) simulations of solid steel and molecular dynamics (MD) simulations of silicon carbide show differences in deformation modes between conventional and surface defect machining (SDM) are shown in a schematic diagram (a); effect of adding nanogrooves to tool (b): (a) surface defect machining and (b) nanogrooving of tool. ([13] [Open access].)

5.3.5 Indexable Inserts

Compared to other cutting tools, indexable inserts have a major advantage; their cutting edges can simply be returned to an unused state, eliminating the need for re-sharpening. This ensures that continuous production is stopped momentarily and eliminates the need for time-consuming tool setup. The type of production, material being processed and its degree of hardness will determine the cutting insert used. Choosing the right replacement inserts can be difficult, as there are so many different types, shapes and materials available [26]. For specific production applications, a large variety of indexable insert forms are available (see Figure 5.25) [13]. The larger size and edges between the cutting edges make it stronger. There are several ways to install indexable inserts on the shank of tool holder (*B*), courtesy of Sandvik Coromant. The shape accuracy of a free surface is measured using a high-precision 3D sensor with a user-defined function [1]. A deterministic machining method has been introduced for low-speed servo machining in the production of optical components with free-form surfaces [1].

Before manufacturing, the procedure carefully investigated the root causes of errors. Then, using a feedforward approach for modeling and compensation, the shape error of the workpiece after compensation could be accurately predicted. An experiment was conducted on cutting 2D sinusoidal mesh on single-crystal silicon to demonstrate the effectiveness of the proposed approach (Figure 5.26) [1].

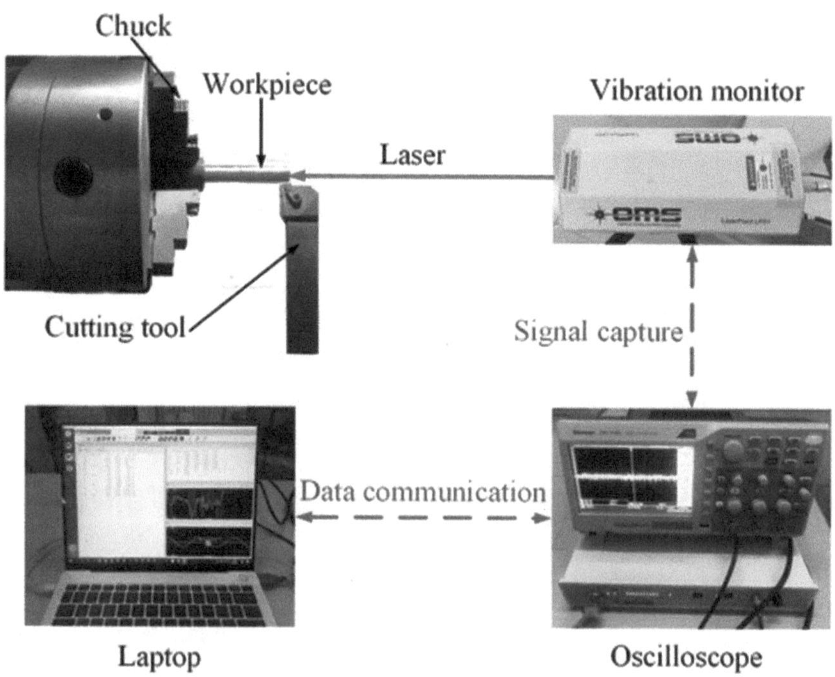

FIGURE 5.25 A wide range of indexable insert shapes are available for specialized manufacturing applications. ([13] [Open access].)

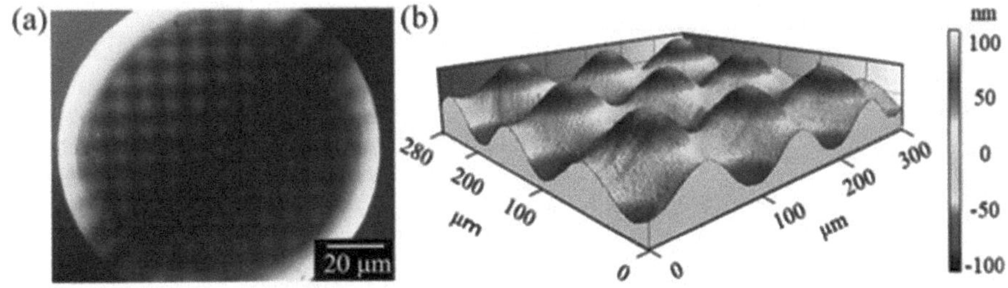

FIGURE 5.26 (a) 3D images of nano pocket networks on single-crystal silicon are obtained and (b) the surface morphology of 2D pocket networks on the workpiece is obtained. ([1] [Open access].)

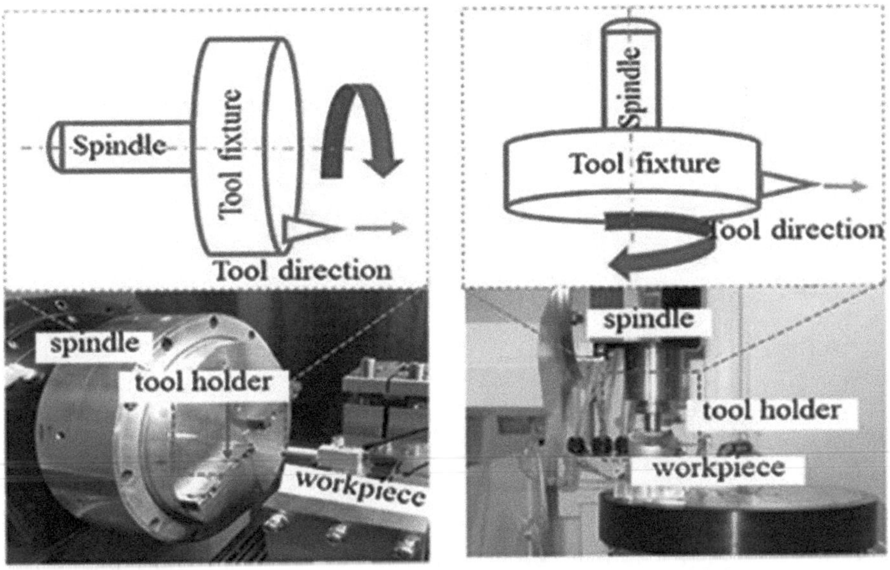

FIGURE 5.27 Fly-cutting methods: two varieties. ([1] [Open access].)

The unified power flow controller (UPFC) technology can be classified into two types depending on tool installation direction: fly cutting, which occurs when the tool installation direction is parallel to the spindle axis [19,20], and ultra-precision raster milling, which occurs when the tool installation direction is set along the radial direction of the spindle [21,22]. Figure 5.27 shows its schematic diagram. Figure 5.28 demonstrates some microstructures and SEM images of machined structures manufactured using fly cutting [1].

Complex microstructures with complex shapes are created by *Z*-axis servo drive such as FTS or STS acting on the workpiece. Due to its outstanding advantages in production, the fly-finishing cutting is suitable for specialized complex micro/nanostructures, including complex multilayer microstructures, pyramid arrays, and triangular pyramid arrays. Figure 5.29 shows the configuration diagram of the fly-finishing cutting system [1].

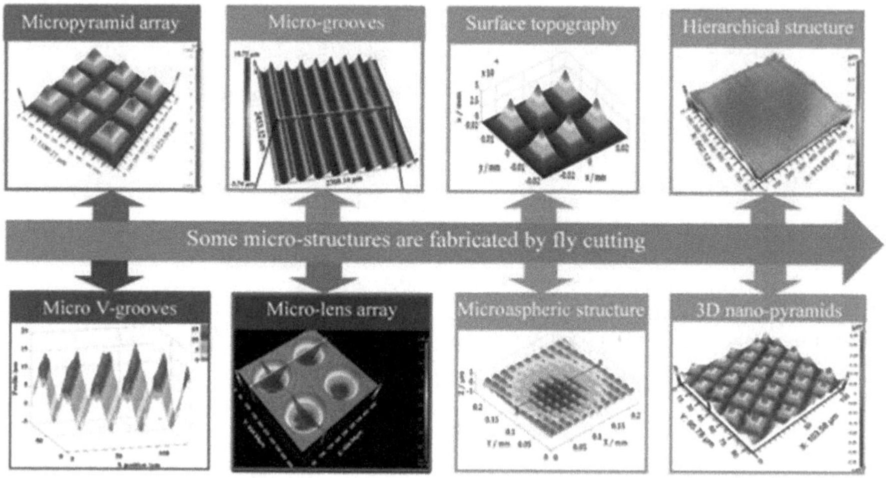

FIGURE 5.28 Fly cutting is used to create specific microstructures. ([1] [Open access].)

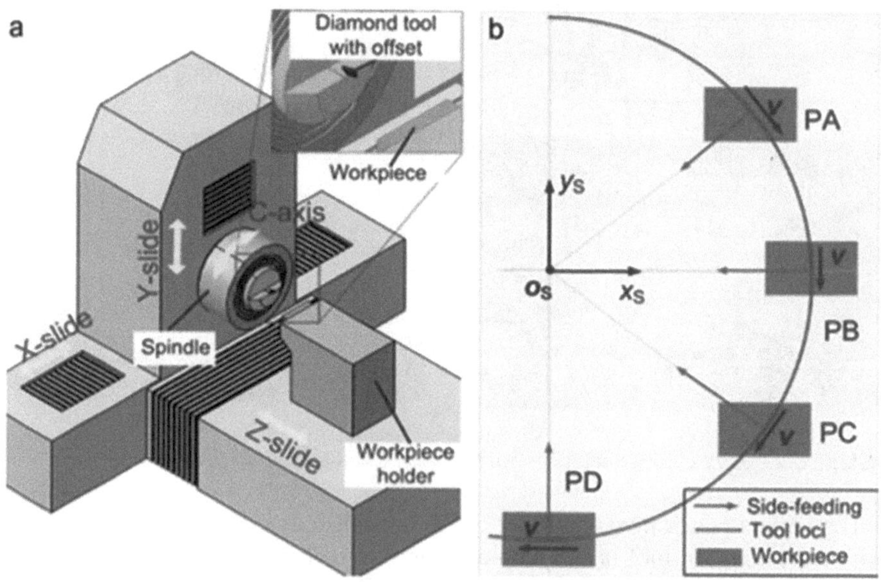

FIGURE 5.29 The configuration diagram of final flying cutting system is shown in (a); induced cutting conditions are shown in (b), where v represents the relative cutting speed and os-xsys refers to the coordinate system fixed on the spindle axis. ([1] [Open access].)

5.3.6 Ceramics

Aluminum oxide, silicon nitride, yttrium, and other oxides and nitrides can be used to make ceramic inserts. Nitrogen combined with an electropositive element such as silicon or boron is called a nitride. Oxide is a chemical compound consisting of an element and oxygen, for example, silicon dioxide (SiO_2). Due to their strong bonding, oxides are incredibly stable. Many metal oxides are much stronger than pure or alloyed metals. Aluminum is usually a fairly soft metal and oxygen from aluminum oxide (Al_2O_3). It is made of ceramic that has a hardness of at least RC50. Unlike most engineering materials, ceramics generally have relatively high chemical stability [27]. They are naturally resistant to corrosion and chemical reactions. In addition, ceramics are generally inert and do not react with bases and acids. However, chemical characteristics may vary depending on the specific composition and type of ceramic. While some types of ceramic can be extremely chemical resistant, others may be more prone to corrosion or breakage in certain situations. In addition, certain ceramics have special chemical characteristics, such as the ability to conduct electricity or act as a catalyst [28]. The general chemical behavior of ceramics is briefly described from the chemical point of view, which is largely impermeable to liquids, and gases and even melts at high temperatures. Generally, inert and non-reactive toward bases and acids. Workpiece surface finish is influenced by the radius and feed rate (Figure 5.30) [13].

A stronger cutting edge and superior surface quality are achieved with a larger nose radius. It is also possible to achieve higher feed rates at larger radii. However, large radii can cause vibration and chatter. Manufacturers usually recommend that the cutting depth be greater than the half diameter of the nose. If you know the other two numbers in Figure 5.30, they can be used to determine nose radius, feed rate or surface polishing.

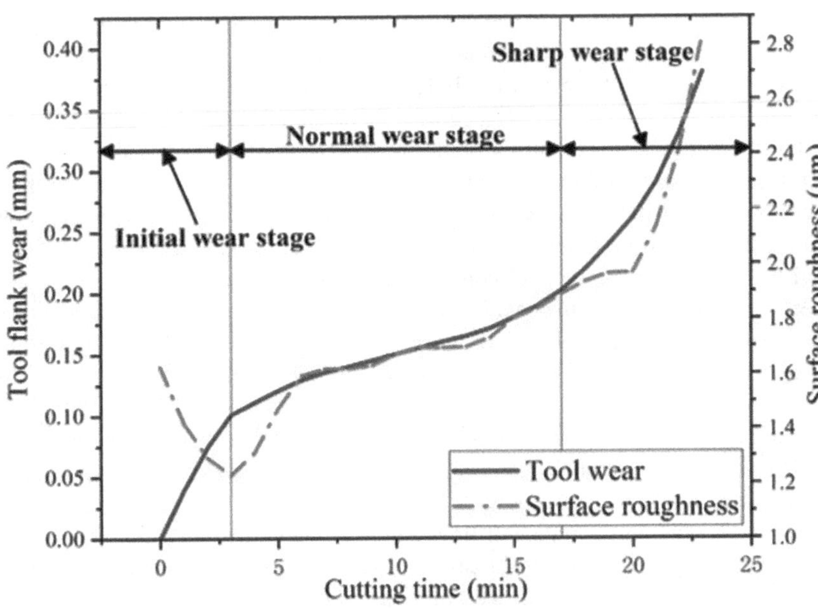

FIGURE 5.30 Workpiece surface finish is determined by the feed rate and radius. ([13] [Open access].)

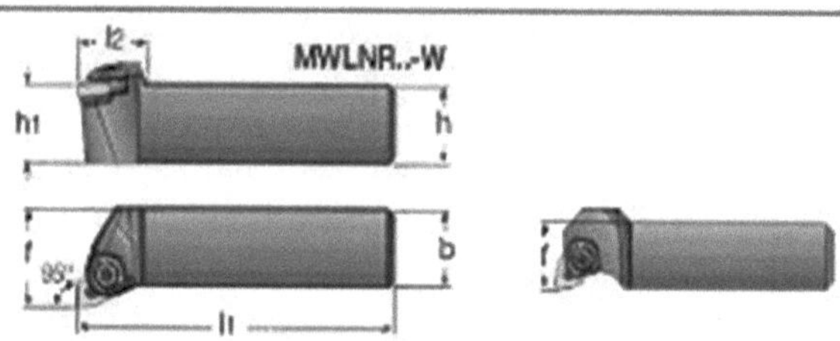

	h	b	h1	l1	l2	f
MWLNR 2525M-08W	25.0	25.0	25.0	150.00	30.0	32.0

FIGURE 5.31 Cross sections and dimensions of the tool holder (mm). ([24] [Open access].)

5.3.7 Insert Shape and Thickness

When designing a holder, the shape of insert is taken into account. Figure 5.31 shows the range of shapes. Only certain shapes can be stored in the tool holder pocket. Cutting inserts composed of carbide, ceramic, coated carbide, coated ceramic, or other hard materials are examples of replaceable cutting inserts for cutting tools [29]. The cutting tool used is ISCAR WNMA 080,408 IC5005 CVD-coated cutting inserts with tip radius, 0.8 mm. The cutting tool is used as a tool holder. Figure 5.31 illustrates the dimensions of a tool holder [24].

Cutting inserts usually consist of a single block with one or more cutting edges located at different corners or around the periphery of the insert. Although the replacement cutting inserts are mechanically attached to a tool holder, they can be removed and adjusted relative to the holder. Replaceable cutting inserts are easily moved (indexed) to give the workpiece a new cutting edge, or they can be replaced in the tool holder, say, if the cutting blades become dull or break. Essentially, indexable cutting tools are modular cutting tool assemblies that come with a tool holder and at least one cutting insert. Combining the insert geometry with the tool holder allows for the formation of several advanced combinations [24]. In design, the shape of the liner is usually important because it determines how well it fits the bearing and performs its intended purpose. Depending on the intended use, liner shapes can be square, rectangular, cylindrical, or more complex geometries. The design of these liner shapes typically takes into account the following factors:

- Compatibility: To ensure a tight fit, the liner shape must match the bearing shape. This may require certain measurements, tolerances or specifications to ensure a proper connection.
- Function: The performance of liner can be affected by its shape. For example, certain geometries may be present in bearings used in manufacturing or cutting to improve the efficiency or strength of the part.
- Manufacturability: The ease of manufacturing the bearing and liner can influence design decisions. Some shapes may be easier to manufacture or assemble than others.

- Corrosion: The durability and maintenance requirements of bearing and internals can be affected by how the shape affects wear distribution over time.
- Cost: More complex designs may require more precise manufacturing or increased manufacturing costs.

Therefore, this work uses three-dimensional surfaces in graphs to investigate the influence of group parameters on vibrations. This strategy, which is almost an optimization method, aims to achieve intermediate values for parameter ranges. Since the workpiece material is hardened steel, the rapid acceleration of tool wear can cause forced vibrations in the cutting mechanism. The highest feed rate and depth of cut seem to generate the highest vibration. Under these conditions, the amount of the material removed from the surface reaches its maximum value, which exerts a huge force on cutting tool and causes a vibration mechanism. In addition, intermediate cutting speed values are the cause of high vibration levels.

5.3.8 Holders of Tools for CNC Applications

The spindle of CNC machine controls the precise process and drives the cutting tool. However, intervention is necessary to ensure the safety of spindle and cutting tool. The space required between these two parts is provided by the tool holder. The cutting tool is designed to be held securely in a precise location by CNC tool holders. The cutting tool can float in some tool holders to prevent tool breakage, while others provide unprecedented precision and secure grip. All holders must be professionally designed to ensure the cutting accuracy and durability required. Purchasing the right tool holder and monitoring wear indicators is critical, as a broken tool holder can be a huge inconvenience and costly to replace. A CNC machine can continue to produce excellent workpieces of all kinds with the help of a precision tool holder. Tool holders can help you maximize the productivity of a CNC machine. Universal tools and accessories fit universal equipment [30].

5.3.9 Types of Tool Holders

The ideal type of tool holder for your project depends on the spindle, tool and CNC machining process. For help finding the perfect tool holders for CNC machine, turn to a leading manufacturer. Here are some typical alternatives to choose from:

- Quick change tap holder
- Floating collet chuck radially
- Jacobs taper tool
- Equipment with Morse taper
- Personalized tool holder
- Quick change tap holders are ideal for use with a variety of tools on specialized CNC machines. Reducing the time required to change tools can shorten production cycles and increase productivity in a manufacturing plant.

Precise cuts and longer tool life are ensured by conical tooling. CNC machine can handle unique projects with special tools thanks to a dedicated tool holder. To create the perfect container for your project, partner with a leading tooling provider [31].

5.3.10 Proper Care of Tool Holders

With proper care, the tool holder can withstand continuous high-performance CNC machining. However, with frequent use, the tool holder experiences enormous pressure. This important part will have to be repaired or replaced over time. The most important maintenance procedure is thorough cleaning and inspection. To avoid contamination, corrosion, and other problems, disassemble and clean tool holders regularly. Maintaining the accuracy and durability of your tools and equipment depends on proper care of your tool holders.

By following these guidelines, you can help your tool holders last longer and continue to perform well regularly. The exact timing may vary depending on holder and how much you use it [13]. If collars are used continuously throughout the day, they should be changed every two to three months. A worn collet will cause your tool to become misaligned, which will affect the quality of product and life of your cutting tool [3]. Here is a detailed guide to help you to keep tool holders in a tip-top condition:

- Vacuum frequent cleaning: To remove chips, dirt, and dust, clean your tool holders after each use. For larger particles, use compressed air or a brush.
- Deep cleaning: To remove any buildup of resin or dirt, periodically disassemble and clean your tool holders with a suitable solvent or cleaner. Be sure to remove all debris before reassembling.
- Visual inspection: Regularly inspect your tool holders for signs of deterioration, wear, or deformation. Look for any obvious problems such as scratches or cracks.
- Check accuracy: Use a micrometer or gauge to check the accuracy and alignment of your tool holder. Any deviation in size may indicate wear or damage.
- Storage dry and clean environment: To prevent rust and corrosion, keep tool holders dry and clean. Consider using rust inhibitors or protective coatings if necessary.
- Proper handling: To avoid physical damage, do not drop or handle tool holders improperly. Use protective covers when transporting or storing.
- Maintenance lubrication: To ensure smooth operation, lubricate moving parts such as clamping mechanisms or clamping systems with a small amount of grease.
- Tightening: To maintain the integrity of the tool holder, regularly check and tighten any loose screws or bolts.
- Adjustments frequent calibration: Ensure that tool holders are always the correct size and alignment by calibrating them regularly. Follow the manufacturer's instructions and recommendations for calibration intervals.
- Check for compatibility tool holder fit: Make sure the tool holders are compatible with the tools and equipment. Incorrect fit can cause malfunctions and potential damage.
- Documentation records: Keep track of all maintenance, calibration and inspection procedures. This makes it easier to track the condition of your tool holders and schedule any necessary maintenance.

It should be noted that the range of applied lubricating and cooling compounds is from 1 to 4.5 μm. The feed rate has the greatest influence on the average surface roughness. This has already been repeatedly proven by the authors using theoretical and experimental methods [3,4]. This is logical, in essence, since the waviness of the workpiece is determined by the horizontal movement of the cutting tool relative to the feed rate. However, it is important to remember that roughness is affected by changes in the cutting depth and cutting speed [24]. Such changes affect the surface integrity of the workpiece since the wear index of cutting tool and definition of chip shape are crucial to understanding the relationship between the strain rates. The effect of parameter sets on the acoustic emission during the machining of hardened steel is the main focus of this study. At this stage, it is important to note that the cutting tool is subjected to strong material hardness effects that cause plastic deformations. These effects are depicted using paired sets of depth of cut, feed rate and speed, respectively, in three-dimensional plots. At low feed rates, increasing the cutting speed seems to reduce the acoustic emission. The depth of the cut shows a tendency to increase to a certain value before decreasing to its high level. To a large extent, the feed rate and depth of cut do not affect the noise emission. At low feed rates, the depth of the cut in this graph has a similar effect on acoustic signals [24]. An important conclusion from this detailed study is that determination of acoustic emission is very complex since it depends on the interaction between the complexity of machining processes and cutting parameters [29–31].

In addition, it helps to create perfectly formed edges or cutting surfaces that are clean, smooth and free of chips and rounding. However, due to internal peel stresses, it is very difficult to apply thick layers (2 μm or more); in addition, the adhesive film has limitations in the choice of the substrate. In addition, the hardness of these coatings will decrease further at temperatures > 300°C due to graphitization [13]. Conversely, with use of the tool, wear and friction rates should decrease, and the outer diamond-like carbon (DLC) layer will turn into graphite [13]. As a result, when the workpiece material is known, the friction and load are correlated according to the equation. As a result, changing the contact area between the tool and workpiece can reduce the generation of cutting forces [32].

The stress profiles of four different tool types along with a comparison with the non-microtextured tools are given. In the case of bionic microtextured tools, the stress slowly penetrated the microtextured sections near the tooltip and mainly accumulated in microtextured areas away from the tool location. Such a cutting progression is observed with the four tool types, especially the non-microtextured ones, starting from the tooltip and gradually increasing to the level of internal cutting stress. As for Z tool, the internal stress distribution is faster than other two tools [32].

The surface roughness of the workpiece is evaluated by a 3D profilometer after machining, and the 3D surface structure of the workpiece after using four tools at the same rotation speed. The results show that the Z3 tool completes the machining operation, and clear tool marks can be seen on the smooth surface of the tool. The reason is that the microtextured structure of the tool generates less cutting force than the non-microtextured structure, which leads to a microtextured structure throughout the friction process. This in turn reduces friction and improves its smoothness, enhancing the machining stability of the machine tool [32]. Surfaces machined at the same feed rate but with one or four inserts, respectively, have the measured roughness. Under these conditions, the theoretical roughness is the same as if there are no insert drift. However, since drift-free multi-point milling tools do not exist in reality, they must be taken into account in the simulation stage [33].

Tool holders for milling machines are essential for precise and efficient machining operations. The precision, performance, and versatility of shrink-fit and hydraulic tool holders set them apart from other varieties. Both options should be selected based on your unique processing requirements and industry standards. Birla Precision Technologies Limited is a trusted brand offering precision engineering solutions and high-quality tool holders. They have a long history of producing and distributing innovative cutting tools, making them industry pioneers. Birla Precision is the company to turn if you are looking for shrink-fit tool holders, hydraulic tool holders or any other precision engineering solutions. Check out their extensive product line by visiting their website and contacting them. For any questions regarding tool holders, please contact us.

With their commitment to providing technical innovation and excellence, Birla Precision Technologies Limited is a trusted partner for companies looking for superior tool holders and engineered solutions. Make a difference in the precision engineering industry by contacting Birla Precision today. To summarize, tool holders are vital to the functionality of milling machines and choosing the right tool holder is critical to achieving the best machining results. Your choice should meet your unique machining requirements, whether it is shrink-fit tool holders designed for high-speed operations or hydraulic tool holders that provide precision and ease of use. Never forget that efficient machining processes depend on precision engineering, and Birla Precision Technologies Limited is a brand you can trust in this regard.

PROBLEMS

1. Explain the types of holders for making cutting tools.
2. Mention proper care of tool holders.
3. Determine the shape and thickness of insert to suit your technique.
4. Describe the operation of hydraulic tool holders.
5. Explanation of cutting tools for turning and milling operations.

REFERENCES

1. Behera, R., Chan, T. C., & Yang, J. S. (2024). Innovative structural optimization and dynamic performance enhancement of high-precision five-axis machine tools. *J. Manuf. Mater. Process.*, 8(4), 181.
2. Tseng, L. W., Hu, T. S., & Hu, Y. C. (2021). A smart tool holder calibrated by machine learning for measuring cutting force in fine turning and its application to the specific cutting force of low carbon steel S15C. *Machines*, 9(9), 190.
3. Schönemann, L. & Riemer, O. (2019). Thermo-mechanical tool setting mechanism for ultra-precision milling with multiple cutting edges. *Precis. Eng.*, 55, 171–178.
4. Huang, X.-F., Tang, P., Yang, S., Feng, J.-Y., & Wan, Z.-P. (2022). Investigation of AlN ceramic anisotropic deformation behavior during scratching. *J. Eur. Ceram. Soc.*, 42, 2678–2690.
5. Moroşanu, G. A., Baroiu, N., Teodor, V. G., Păunoiu, V., & Oancea, N. (2022). Review on study methods for reciprocally enwrapping surfaces. *Inventions*, 7(1), 10.

6. Sukaylo, V. A. & Kaldos, A. (2004). Development and verification of a computer model for thermal distortions in hard turning. *J. Mater. Process. Technol.*, 155–156, 1821–1827.
7. Liu, K. & Jiang, L. (2011). Bio-inspired design of multiscale structures for function integration. *Nano Today*, 6, 155–175.
8. Zhang, Z., He, H., Fu, W., Ji, D., & Ramakrishna, S. (2020). Electro-hydrodynamic direct-writing technology toward patterned ultra-thin fibers: Advances, materials and applications. *Nano Today*, 35, 100942.
9. Brinksmeier, E., Karpuschewski, B., Yan, J., & Schönemann, L. (2020). Manufacturing of multiscale structured surfaces. *CIRP Ann.*, 69, 717–739.
10. Fu, Y. F., Yuan, C. Q., & Bai, X. Q. (2017). Marine drag reduction of shark skin inspired riblet surfaces. *Biosurf. Biotribol.*, 3, 11–24.
11. Guo, Y. B. & Sahni, J. (2004). A comparative study of hard turned and cylindrically ground white layers. *Int. J. Mach. Tools Manuf.*, 44(2/3), 135–145.
12. Felhő, C. & Kundrák, J. (2018). Effects of setting errors (insert run-outs) on surface roughness in face milling when using circular inserts. *Machines*, 6(2), 14.
13. Rizzo, A., Goel, S., Luisa Grilli, M., Iglesias, R., Jaworska, L., Lapkovskis, V., … & Valerini, D. (2020). The critical raw materials in cutting tools for machining applications: A review. *Materials*, 13(6), 1377.
14. Chou, Y. K. & Song, H. (2004). Tool nose radius effects on finish hard turning. *J. Mater. Process. Technol.*, 148(2), 259–268.
15. Geng, R., Yang, X., Xie, Q., Zhang, W., Kang, J., Liang, Y., & Li, R. (2021). Ultra-precision diamond turning of ZnSe ceramics: Surface integrity and ductile regime machining mechanism. *Infrared Phys. Technol.*, 115, 103706.
16. Kong, L. B. & Cheung, C. F. (2012). Modeling and characterization of surface generation in fast tool servo machining of microlens arrays. *Comput. Ind. Eng.*, 63, 957–970.
17. Lee, L.P., Berger, S.A., Liepmann, D., & Pruitt, L. (1998). High aspect ratio polymer microstructures and cantilevers for bioMEMS using low energy ion beam and photolithography. *Sens. Actuators A*, 71, 144–149.
18. Franz, G., Vantomme, P., & Hassan, M. H. (2022). A review on drilling of multilayer fiber-reinforced polymer composites and aluminum stacks: Ooptimization of strategies for improving the drilling performance of aerospace assemblies. *Fibers*, 10(9), 78.
19. Natarajan, E., Inácio Freitas, L., Rui Chang, G., Abdulaziz Majeed Al-Talib, A., Hassan, C. S., & Ramesh, S. (2021). The hydrodynamic behaviour of biologically inspired bristled shark skin vortex generator in submarine. *Mater. Today Proc.*, 46, 3945–3950.
20. Barry, J. & Byrne, G. (2002). TEM study on the surface white layer in two turned hardened steels. *Mater. Sci. Eng. A*, 325, 356–364.
21. Han, Z., Fu, J., Wang, Z., Wang, Y., Li, B., Mu, Z., … & Niu, S. (2017). Long-term durability of superhydrophobic properties of butterfly wing scales after continuous contact with water. *Colloids Surf. A*, 518, 139–144.
22. Ekinovic, S., Doninsek, S., & Jawahir, I.-S. (2004). Some observations of the chip formation process and the white layer formation in high speed milling of hardened steel. *Mach. Sci. Technol.*, 8(2), 327–340.
23. Luo, G., Wu, D., Zhou, Y., Hu, Y., & Yao, Z. (2022). Laser printing of large-scale metal micro/nanoparticle array: Deposition behavior and microstructure. *Int. J. Mach. Tools Manuf.*, 173, 103845.
24. Asiltürk, İ., Kuntoğlu, M., Binali, R., Akkuş, H., & Salur, E. (2023). A comprehensive analysis of surface roughness, vibration, and acoustic emissions based on machine learning during hard turning of AISI 4140 steel. *Metals*, 13(2), 437.
25. Biesuz, M., Saunders, T., Ke, D., Reece, M. J., Hu, C., & Grasso, S. (2021). A review of electromagnetic processing of materials (EPM): Heating, sintering, joining and forming. *J. Mater. Sci. Technol.*, 69, 239–272.

26. Wang, S., Zhang, Q., Zhao, Q., & Guo, B. (2021). Surface generation and materials removal mechanism in ultra-precision grinding of biconical optics based on slow tool servo with diamond grinding wheels. *J. Manuf. Process.*, 72, 1–14.
27. Guo, J., Goh, M. H., Wang, P., Huang, R., Lee, X., Wang, B., … & Wei, J. (2021). Investigation on surface integrity of electron beam melted Ti-6Al-4V by precision grinding and electropolishing. *Chin. J. Aeronaut.*, 34, 28–38.
28. Arnold, T. & Pietag, F. (2015). Ion beam figuring machine for ultra-precision silicon spheres correction. *Precis. Eng.*, 41, 119–125.
29. Zhang, S., Zhou, Y., Zhang, H., Xiong, Z., & To, S. (2019). Advances in ultra-precision machining of micro-structured functional surfaces and their typical applications. *Int. J. Mach. Tools Manuf.*, 142, 16–41.
30. Zhang, S. J., To, S., Wang, S. J., & Zhu, Z. W. (2015). A review of surface roughness generation in ultra-precision machining. *Int. J. Mach. Tools Manuf.*, 91, 76–95.
31. Buhmann, M., Roth, R., Liebrich, T., Frick, K., Carelli, E., & Marxer, M. (2020). New positioning procedure for optical probes integrated on ultra-precision diamond turning machines. *CIRP Ann.*, 69, 473–476.
32. Wang, S., Zhao, Q., & Yang, X. (2022). Surface and subsurface microscopic characteristics in sapphire ultra-precision grinding. *Tribol. Int.*, 174, 107710.
33. Xu, T., Ma, C., Shi, H., Xiao, K., Liu, J., & Li, Q. (2023). Effect of composite bionic micro-texture on cutting performance of tools. *Lubricants*, 12(1), 4.

6 Cermet
Composites Used in Cutting Tools and Technical Developments

6.1 INTRODUCTION

Today's machining industries are continually developing wear-resistant coatings and tooling for the machining, drilling, and mining industries to improve equipment performance and durability while reducing maintenance costs. Producers of technologies can be applied to improve products, maintain competitiveness and ultimately increase revenue. Some essential properties such as excellent wear and corrosion resistance, good toughness and hardness, environmental friendliness, high performance at high temperatures, and fairly low production costs are ideal materials for achieving the above-mentioned objectives [1]. To withstand long-term heavy wear, a component must have a high level of wear resistance and corrosion resistance. If the part can be successfully used in such situations regularly, this is not the case. This reduces maintenance and replacement costs while increasing safety [2].

Equally crucial as having a material with excellent corrosion resistance is having a material with robust mechanical properties. A wide range of metal-ceramic systems are now focusing on extensive study. Among them are titanium carbide (TiC) and titanium carbonitride Ti(C,N) ceramic phases. Conventional "carbide" compounds, which are made of tungsten carbide (WC) embedded in a cobalt link, serve the same purpose as cermets; however, most applications cannot employ them because of their high density and relatively low resistance to oxidation and corrosion. Their applications involve high temperatures in structures. Their primary applications are in cutting and machining due to excellent wear resistance, provided by their high strength, toughness, and hardness combination [3–8]. Historically, metal carbides and nitrides have been made with various techniques, the most cost-effective and efficient being solid-phase reaction techniques [9]. Depending on Ti(C,N), the material's and cermet's hardness can be changed. Since hardness and toughness rise with an increasing TiC content and TiN ratio, respectively, there is a significant amount of TiC and TiN. Nickel, cobalt, stainless steel, and nickel aluminide (Ni_3Al) are typical metal binder materials used in these cermets. Due to its high heat resistance and good strength retention at high temperatures, Ni_3Al is of particular interest [10–12].

However, TiB_2's low fracture toughness and low self-diffusion coefficient make it difficult to densify, limiting its use [4]. Increased impact resistance is one of advantages of combining TiB_2 and AlN due to their different strengthening mechanisms [13]. These mechanisms are necessary for the curing of ceramic materials.

DOI: 10.1201/9781003604648-6

CBides and Cermets are a class of very hard, refractory, and wear-resistant alloys produced by powder metallurgy [14]. These substances are widely used as inserts for drills, cutting tools, wear parts in mining and industrial applications, and high-pressure components [15]. The life of components is largely influenced by properties such as toughness, microstructure, hardness, strength, corrosion resistance, and wear resistance. In carbide alloys and cermets, the hard carbide phase predominates, while in metal matrix composites (MMCs), the dominant component is a binder (or matrix). The advanced plan is divided into two parts. First, performance can be improved due to characteristics, microstructure, composition, manufacturing conditions, carbides, and cermets. Second, new carbides, ceramics, and hard coatings are being produced using primary resources from South Africa to generate new knowledge and expand possible markets for these materials [1]. Under the Registration, Evaluation, Authorisation, and Restriction of Chemicals (REACH), the European Union's chemical registration, evaluation, authorization, and restriction system, two WC-Co wear products, Co_3O_4 and WO_3, are registered [2–5]. As shown by fracture toughness (KIC) tests [2], the low fracture toughness of ceramic materials is the main obstacle to their use in thermal and mechanical applications. Objects of high hardness, such as those made of hardened steel, require the use of ceramics as cutting tools for this purpose.

6.2 STRUCTURAL FORMATION

Structural design is one of the most important factors affecting the effectiveness and performance of cutting tools in manufacturing operations. Cutting tools with increasing strength, durability, and accuracy have been developed as a result of advances in manufacturing processes and materials. By thoroughly understanding the relationship between structural design and cutting performance, manufacturers can build tools that not only meet stringent industry requirements but also evolve to meet the ever-changing needs of contemporary manufacturing applications. Technological advances will drive innovation and enhance production across a range of industries by further improving the effectiveness of cutting tools. To be competitive in the manufacturing sector, structural forming research and development must be given top priority.

6.2.1 Nitride Titanium (TiN)

A refractory material with excellent microhardness and chemical and thermal stability is titanium nitride. TiN has many uses, including a component of specific refractories and cermets, a material for crucibles used in oxygen-free metal casting and precursor to beautiful and wear-resistant gold-like coatings. Experiments on burning pressed samples of titanium powder in nitrogen have shown that one of the main factors influencing the combustion process is the rate of nitrogen filtration inside titanium. Reducing the amount of titanium dioxide in the rutile form will result in TiN. The existence of large homogeneous mononitride and nitride forms of various stoichiometries such as sodium nitride (Na_3N) has been discovered. For example, an oxygen or carbon atom in the interstitial region can easily replace a nitrogen atom in

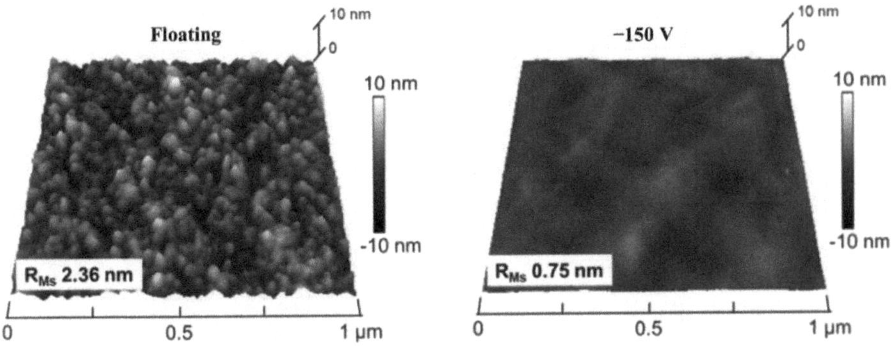

FIGURE 6.1 ATM images of TiN thin films formed under floating substrate bias and −150 V bias voltage. ([1] [Open access].)

titanium nitride (TiN) in the $TiN_{0.1}$ and $TiN_{0.5}$ phases. Figure 6.1 shows the morphology of Ti-N thin films formed on Si under floating substrate bias and −150 V. These thin films had a surface roughness with a root mean square (RMS) value of less than 2.5 nm. When Ti-N thin films are prepared under −150 V substrate bias, their surface roughness is smaller than when they are deposited under floating bias. This is due to high-energy ions bombarded the surface of thin film, smoothing it. The short service life of cutting tools used in various machining processes is due to progressively decreasing hardness, which leads to increased processing costs [4–7].

High-resolution transmission electron microscopy (HRTEM) images of Ti-N thin films formed under floating substrate bias and −150 V on Si and SS are shown in Figure 6.2. It is clear that under floating substrate bias and −150 V, the formed Ti-N thin films have grain sizes 30 and 20 nm, respectively. These results compare favorably with those obtained from X-ray diffraction (XRD) data using the Scherrer's formula [1].

For example, the curve took on a parabolic shape when the temperature increases to 500°C [16]. Under normal conditions, refractory carbides and nitrides such as titanium nitride and carbide are rare. They usually require a lot of time and money to install. The excellent heat resistance and inertness of these materials result in powerful covalent bonds. Thus, this cannot be achieved using traditional methods. Its widespread use is stopped by this rescue [17].

6.2.2 Carbonitride of Titanium (TiCN)

Powders, coatings, fibers, and filaments are typical forms of refractory carbides and nitrides. Nevertheless, powdered TiC, TiN, and Ti(C,N) materials that are currently made by various methods will be taken into consideration to study. TiCN is well known as an intriguing material for cutting tools because of its wear resistance and capacity to be utilized in production of intricate ceramic composites or refractory materials [18–20]. The inclusion of carbon or nitrogen atoms in the interstices may be the reason for the titanium carbonitride ceramic particles' broad dispersion. The octahedral lattice site has random contributions of nitrogen and carbon atoms.

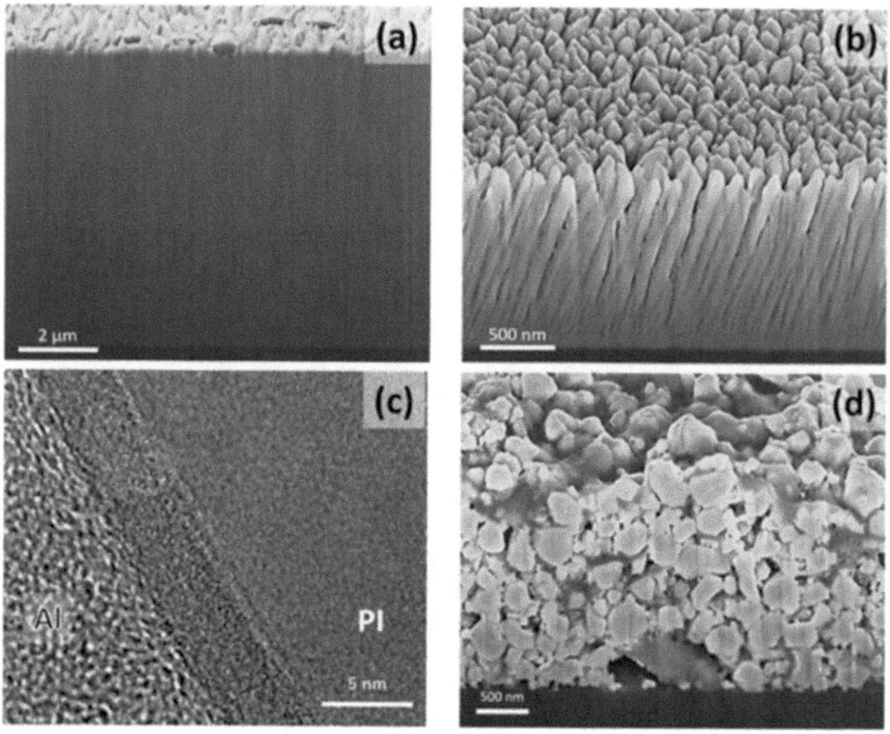

FIGURE 6.2 TiN thin films deposited at floating voltage and substrate bias, −150 V with their HRTEM morphology. ([1] [Open access].)

In contrast to TiN and TiC, TiCN is still face-centered cubic in terms of cubic structural classification [20–22]. The components are carbonized immediately or nitrided to yield Ti(C,N), TiC, and TiN. This process uses a lot of energy since it needs energy. Temperatures vary from 1900°C to 2900°C during 5 to 20 hours [23]. Lastly, Finarhyd requires finer grinding because it is an extremely clumpy powder. Because of the small particle size, further chemical washing is needed to eliminate impurities.

As a result, the procedure is very expensive due to the cost of pure titanium, and it uses a lot of energy [24], which leads to an ordered arrangement of titanium atoms and a bold formation in the presence of N and C atoms.

6.3 COATING OF TICN AND TIN

Researchers are currently studying the production of titanium nitride and its widespread use in engineering services. Solid coatings remain at the forefront of research due to their versatility in many research areas [1–3]. There has been much interest in the mechanical compatibility of coating materials such as TiN and titanium carbonitride (TiCN) with various metal substrates. However, the production of such ceramics on metal substrates is difficult due to differences in mechanical characteristics, especially in tribological and mechanical aspects [1–6]. In this regard, a large amount of research has been carried out to improve the properties of these coatings and expand the scope of their application. Cutting tool development is critical to improve mechanical properties, especially during long-term, high-speed machining. However, there are several problems with the design of cutting tools. Before it can be used, the material must meet the required mechanical properties. Second, the cost of the final product reflects the limitations of traditional sintering methods such as the amount of material and time required to complete the sintering cycle. This creates serious difficulties for the production of such devices [8]. In addition, the materials used to make cutting tools must be resistant to abrasive cutting conditions such as elevated temperatures and friction between the workpiece and cutting tool surface. For cutting tool materials to meet these requirements, they must have several qualities [9].

TiCN coatings provide strong mechanical and wear-resistant qualities as well as great thermal stability. They have demonstrated minimal internal stresses and great resistance to corrosion [4,6,7]. Additionally, they are still utilized today to safeguard mechanical components and materials used in cutting tools [25–28] due to their resistance to damage during transportation [5,8–9,29].

6.4 CHALLENGES AND RESTRICTIONS OF TIN AND TICN

Despite this, the service life of H13 tools is impacted by a number of failure modes due to harsh operating circumstances, including high load, temperature, and stability time. Thermal fatigue and abrasive and adhesive tool surface degradation are the principal effects of such working circumstances. As a result, working tools must be replaced after a given amount of time, which has an impact on the total cost of production operations [11,12]. The stoichiometric arrangement of these components determines the melting point of these composites [30–33]. Refractory linings for furnaces, cutting tools, and hypersonic vehicles, as well as molten metal crucibles, are reported to be manufactured with these defining characteristics from TiN/TiCN. TiN/TiCN has disadvantages in all applications, including low fracture toughness and poor oxidation resistance [34,35]. In addition, important elements such as a long sintering period can limit grain formation and increase particle size. As a result, the performance and overall efficiency of sintered body may be reduced indirectly [36–41]. In any case, it is a critical to improve the synthesis of TiN and TiCN composites and properties such as tensile strength, elastic modulus, fracture toughness, and oxidation resistance required an urgent research [42,43].

6.5 TREATMENT TECHNIQUES

Morphology of materials and components or microstructural integrity influences the processing and performance of composites [44].

Since both materials are classified as ultra-high temperature ceramics, specialized processing and sintering techniques are required for their fabrication [45,46]. These techniques include spark plasma sintering (SPS), pseudothermal hot isostatic pressing (PHIP), quasi-isostatic pressing (QIP), and reaction spark pressure methods.. For some modern applications such as semi-finished manufacturing and final manufacturing, TiCN-based cermet is a promising material [47,48]. However, its low wear resistance does not allow it to be used in the future, for example, in high-speed cutting industries. Due to increased hardness and absence of a ductile metal binder phase contrasting with the entire cermet, the surface layer containing a large amount is called wear, as it can lead to improved cutting operations when the carbide phase is formed in the nitride environment layers of repetition [49–52]. Carbide-containing cermets are manufactured mostly from titanium carbides, tungsten carbides, and possibly cermets or solid materials based on titanium and tungsten carbides in small quantities [52,53]. Therefore, the harder the material, which reduces the wear rate and increases wear resistance, the greater the number of binder phases [53].

6.6 FACTORS THAT IMPROVE SINTERING KINETICS

- In comparison to other conventional sintering techniques, SPS's diffusion kinetics are primarily improved. It is widely known that using pressure to produce plastic can increase the compaction of any composite specimen [54]. The high pulsed energy source causes the SPS to generate [55,56]. According to studies, applying a continuous pulsed current is essential for achieving significant densification of any sintered product in IPS [57].
- High heating rates are necessary to produce optimal densification when producing Al_2O_3 nanocomposites from the resulting powders. [58–63].
- Additionally, studies have revealed that DC pulsing increases the titanium carbonitride's degree of condensation.

6.7 BINDERLESS

The toughness of unbound tungsten carbide (WC) compared to titanium carbonate (TiCN) and alternative materials. Superior mechanical properties, including higher hardness and toughness, are demonstrated by unbound WC, and these properties help explain its remarkable corrosion resistance in demanding applications. Removal of the binder improves the integrity and purity of the material, enhancing performance under high-stress conditions. Comparative studies show that unbound WC is superior to TiCN and other alternatives, highlighting its potential use in wear-resistant coatings and cutting tools. The present study highlights the need for further investigation into the processing methods and applications of unbound WC to optimize its advantages in industrial settings. Binder-free WC, in particular, exhibits exceptional wear resistance compared to titanium carbonitride and some other materials. Binderless carbide systems are beneficial and important in many manufacturing industries, such as high-speed cutting tools, due to their exceptional wear resistance [64,65]. Bondless carbides and cermets are gradually replacing traditional cemented carbides in industry due to deterioration of binder phase in some harsh environments. Bondless WC-MoxC carbides are the most productive and have good combination properties [66].

It has two intertwined fcc lattices, one of which has a comparable crystal structure and composed of carbon and metal. Two-dimensional (2D) nanomaterials have garnered a lot of interest as a novel kind of nanomaterial. There are ultrathin 2D

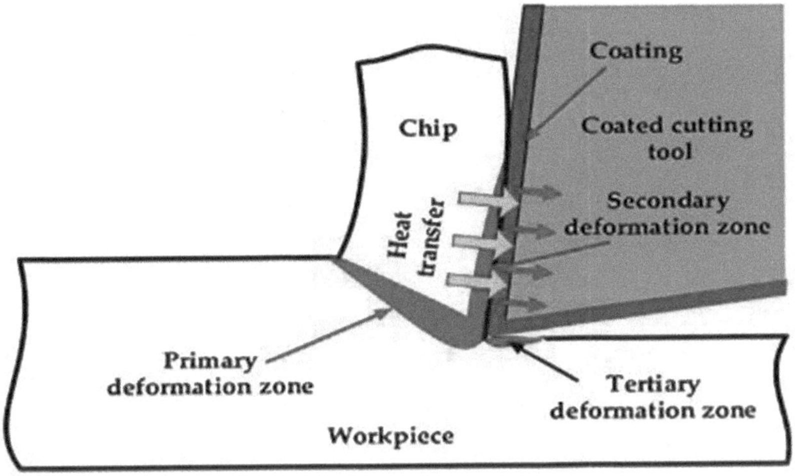

FIGURE 6.3 Critical analysis of science of high-temperature tribology and cutting performance of ceramics and ceramic tool materials. ([66] [Open access].)

nanomaterials that have a special planar structure, in which the side measurements are larger than sub-microns. At the same time, their longitudinal dimensions are limited to the atomic or nanoscale level. Typically, a bond is created by combining d orbitals with 2s and 2p orbitals.

The structure of sp orbitals changes with increasing concentration of electron clouds. The bulk modulus and lattice constants of carbides are theoretically analyzed and correlated using the Murnaghan equation. Based on individual two-dimensional transition metal atoms, we explore 66 transition metal carbides due to their high content; more than 800 of them have different computer characteristics of MAX phases. In addition, fracture toughness is reduced compared to previous sintering methods. Figure 6.3 illustrates the grain distribution of metal and ceramic materials used at high temperatures, either with or without binders [66].

The effect of hardness on different metals Ti(C, N) is determined. It has been shown that when using metal binders with a thickness of 4–5 and 2–3 μm, respectively, the crack resistance of TiCN approaches 20% and 25%, and with an increase in the metal/binder ratio, these values begin to decrease [52].

6.8 PRIOR STUDIES ON TITANIUM CARBONATE NITRIDE AND TIN WITHOUT SINTERING TECHNIQUES USING BINDERS

Understanding the properties and use of TiCN and TiN without binders has advanced greatly from previous investigations using sintering processes. These materials are ideal for cutting tools and other high-performance applications due to their exceptional hardness, corrosion resistance, and temperature stability. The improved performance results from the absence of a binder during the sintering process, which improves the mechanical properties of materials and reduces contamination [67,68].

FIGURE 6.4 FESEM microstructures with different grinding times on TiCN without a binder, including (a) as received, (b) 1 hour, (c) 4 hours, and (d) 10 hours [70].

Superior tolerance control is possible with titanium carbonitride cermet enhanced surface quality, feed rate, and any complex machine part [64,69]. Hard coatings made of Ti(C,N), TiN, and TiC are applied to the surface of hard alloys. These coating materials have recently undergone modifications to better suit particular applications in various sectors [61]. The various structural distributions of hard alloys are shown in Figure 6.4. The majority of materials used in friction and wire drawing applications develop these microstructures.

To further improve the functionality of TiCN and TiN, the door is opened to innovative applications in a variety of industries, and future research should focus on optimizing sintering conditions and investigating new formulations. Unbonded carbides are developed to replace the traditional carbides and supplement the old typical carbide cermet composition [66]. Interestingly, using outdated conventional sintering techniques, it is quite challenging to fuse any materials including tungsten carbides with titanium carbides [71,72]. Due to its exceptional properties, such as high refractoriness and excellent thermal stability, this material is particularly suitable for applications in extreme environments, including aerospace, nuclear reactors, and high-temperature structural components [73].

Different grain is produced by various reinforcing chemical buildings, as depicted in Figure 6.5. However, it has been demonstrated that TiC inclusions can enhance the processing capabilities of high-speed cutting tools [65]. Zhang et al. [64] have investigated the mechanism of tungsten carbide that is created during the initial alloying with 0.5% Cr_3C_2-0.3% VC. According to Kim et al. [69], binder-free ($TiC_{0.5}N_{0.5}$) TiCN nitride, which are ball at different times, have been sintered at 1170°C using high-frequency induction heating methods to show good results.

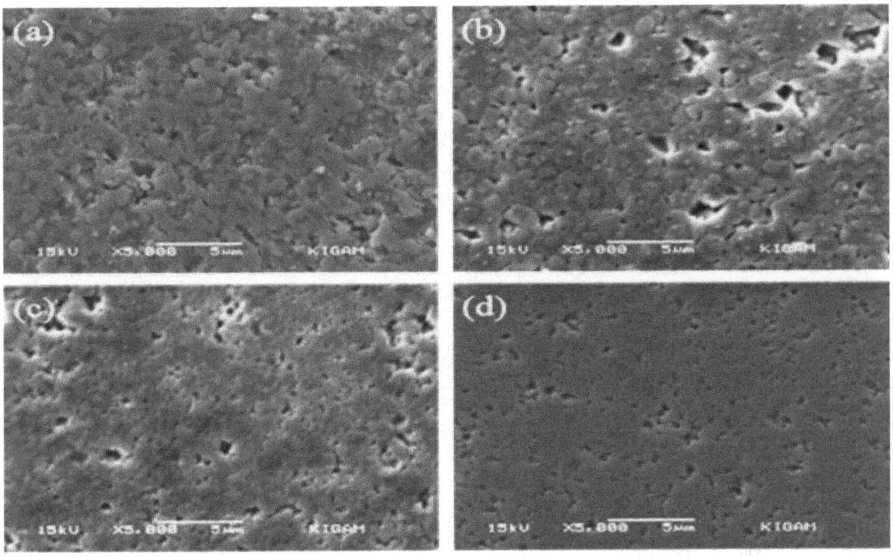

FIGURE 6.5 FESEM microstructures on sintered TiCN without a binder after milling for (a) 1 hour, (b) 1 hour, (c) 4 hours, and (d) 10 hours [70].

Continued investigation into optimization techniques is important to enhance the efficiency and scope of using TiCN and TiN in advanced manufacturing processes [74]. The compaction process has been significantly improved by developing the crystalline structure of ball-milled powders. Shredding is reported to be decreased. After 10 hours of grinding, the powder crystallite size decreases from 241 to 9 nm, as illustrated in Figure 6.6.

After sintering, the rapid sintering procedure preserved the fine grain microstructure. The relationship between variable crystallites, relative density and fracture toughness are insignificant. It is estimated that the crystallite size decreases to a range of 231 to 239 nm after 10 hours of treatment. This process was carried out using high-frequency induction heating sintering (HFIHS), also known as fast combined high-frequency induction thermal sintering [65,75]. After sintering, the process retains a fine microstructure with acceptable mechanical properties, as shown in Figure 6.7. According to the results, 99% compactions are achieved after 4 minutes of exposure.

Relative density and microhardness are in a direct inverse relationship, since the grinding time reached 40 hours. Using the high-energy ball milling technology, the microstructure of binder-free titanium nitride particles has been thoroughly improved, as shown in Figure 6.8. The X-ray diffraction (XRD) patterns show grinding and broadening due to deformation of crystallites after 40 hours and decrease in the crystallite size, which is noticeable after 1 hour of grinding. The ball mills used have granules with a powder in the grinding bottle; there is a tendency for unwanted particles or contaminants to enter. In Figure 6.8, XRD of grinding is clearly shown. The peak broadening of TiN decreases as a result of grain development during the sintering process.

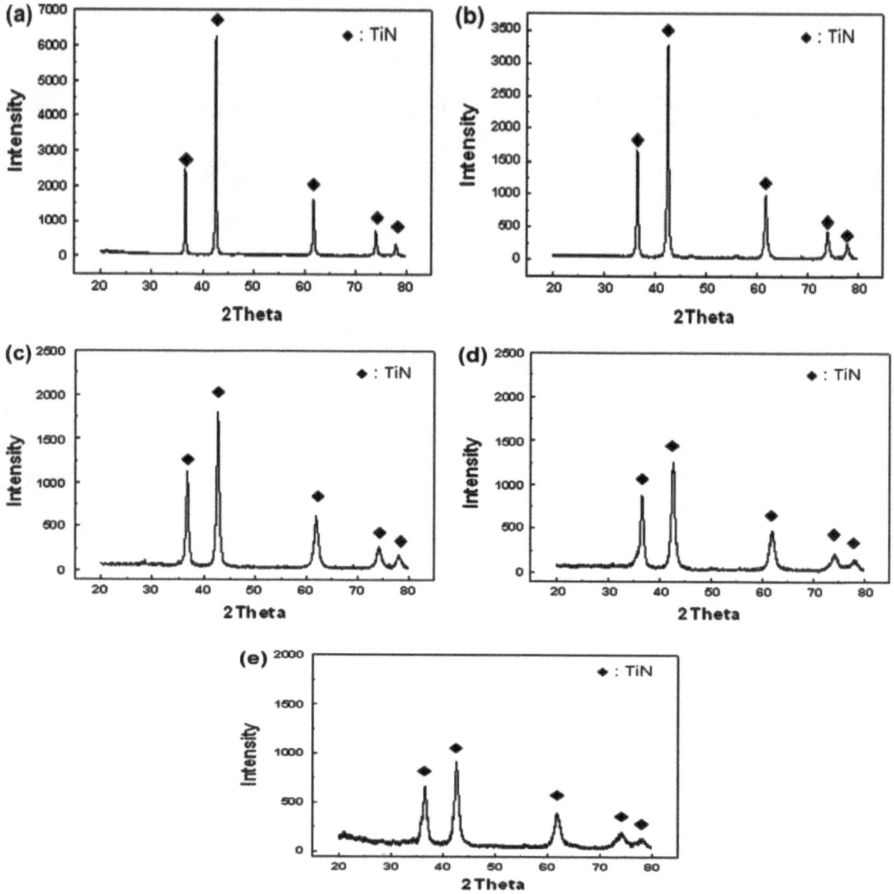

FIGURE 6.6 XRD patterns below illustrate titanium nitride to milling durations without using binders: Cp-TiN, TiN-1 hour, TiN-4 hours, TiN-10 hours, TiN-40 hours, and TiN-1 hour [70].

This chapter reports on sintered TiCN and TiN, which are produced without binders. The exceptional mechanical properties of these materials such as their high hardness, excellent corrosion resistance, and thermal stability have drawn an attention. Because there are no binders used during the sintering process, TiCN and TiN exhibit superior purity and performance, making them particularly suitable for high-stress applications such as protective coatings and cutting tools. The microstructure and properties of these materials are significantly affected by many sintering parameters, such as temperature and environment.

The effect of adding nanometric or micrometric TiN on the mechanical properties and microstructure of TiC-based cermet has been the subject of recent research [6]. These results demonstrated that the addition of nano TiN could stop the melting of TiC grains, and the mechanical performance was improved when TiN was added at a weight percentage of less than 6 wt%. It would be interesting to investigate how the

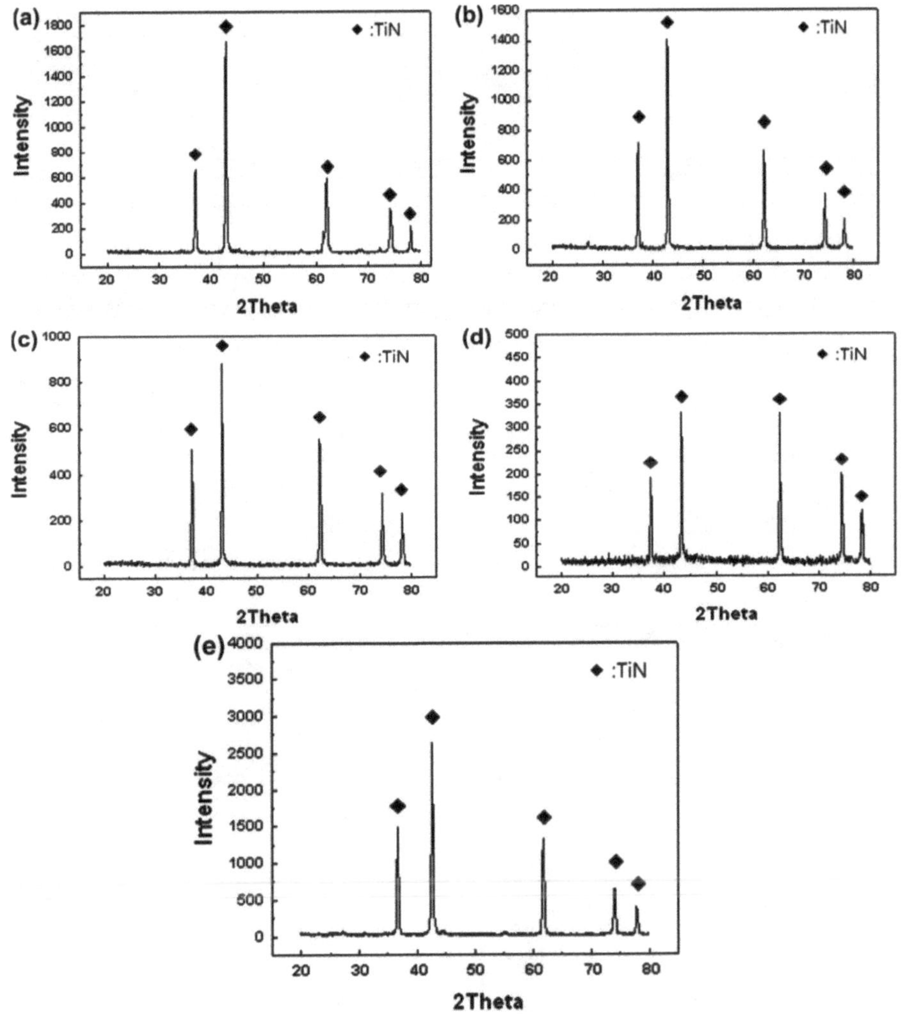

FIGURE 6.7 Results through XRD patterns of titanium nitride to milling periods (without binders): 40 hours, 40 hours, 10 hours, 1 hour, and Cp-TiN [70].

TiN particles generated by self-propagating high-temperature synthesis (SHS) affect the final mechanical properties of bulk cermet in this case to compare its frictional behavior with other solid materials. Nitrogen addition to TiC-Ni-Mo ceramic alloys has been the subject of numerous studies aimed at optimizing the alloys' microstructure, particularly the solids' grain size. These days, titanium carbide (TiC)-based modified cermets are frequently utilized as cutting tools due to their superior hardness and wear resistance compared to $TiC–Mo_2C–Ni$ cermets [3–5].

In addition to better performance in terms of wear resistance and good surface quality of mechanical workpieces, it can replace traditional carbide alloys. Grinding time progress is shown in Figure 6.9 [70].

FIGURE 6.8 TiN powder SEM pictures for (a) Cp-TiN, (b) 1 hour, (c) 4 hours, (d) 10 hours, and (e) 40 hours, respectively, were obtained [70].

Due to their well-known hardness and corrosion resistance, TiCN and TiN are useful materials for cutting tools and wear-resistant coatings. Their formulation has been studied using binder-free sintering procedures in the past, which has improved their mechanical properties and purity. Based on the body of research, binder-free TiCN and TiN made through controlled sintering processes offer significant benefits in mechanical properties and future applications. To further expand their industrial applications, future research should focus on developing new formulations and further simplifying sintering procedures.

The formula developed determines the final size of each particle. After long milling segments of 1, 4, 10, and 40 hours, Norton determined them to be 56, 21, 15,

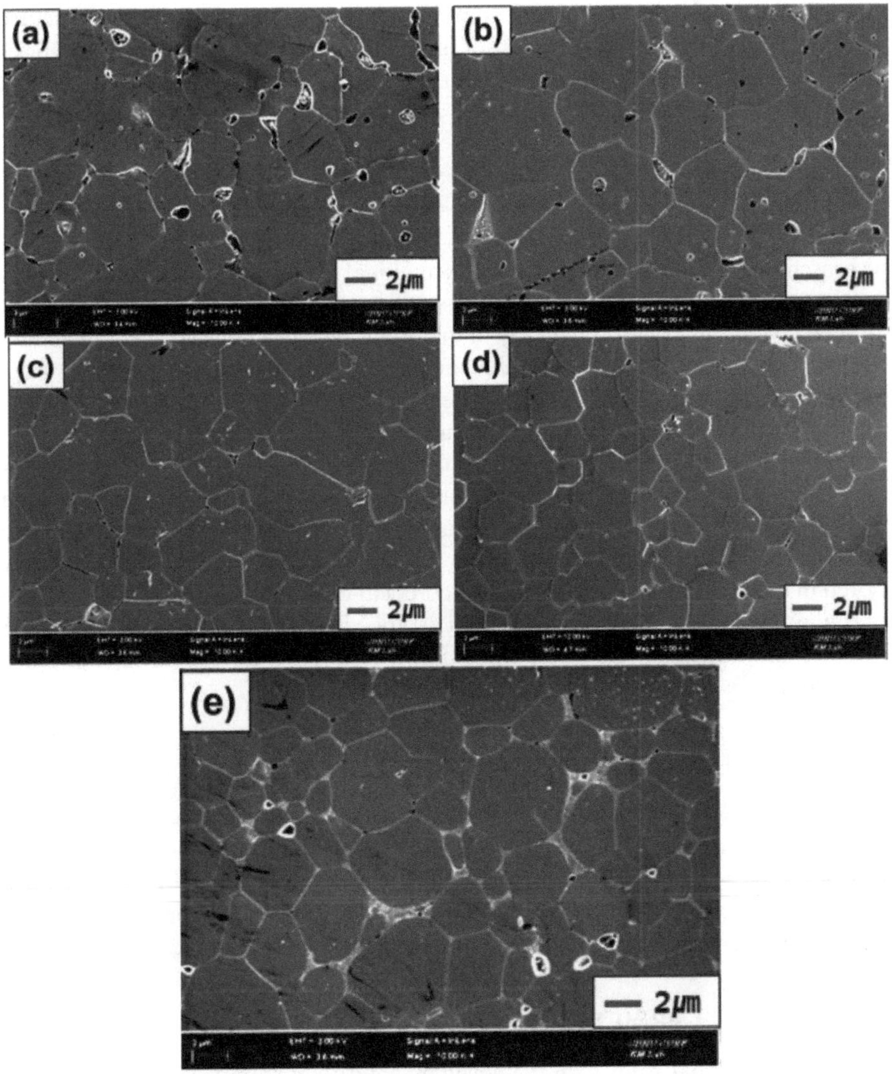

FIGURE 6.9 SEM of titanium nitride taken at milling times of Cp-TiN: (a) 1 hour, (b) 4 hours, (d) 10 hours, and (e) 40 hours [70].

and 10 nm. When TiN is added, the concentration of Mo in bond phase increases, which greatly increases the corrosion resistance of cermet. The grain size and edge thickness of solid particles generally decrease with increasing nitrogen content in cermet. In this regard, TiCN-based cermet shows interesting properties, including low metal sliding friction, excellent chemical stability, high hardness at high temperatures, and good wear resistance. As the TiC content increases in the material, the overall hardness correspondingly improves. To ensure that the metal binder adequately wets the ceramic phase and forms a mixed carbide (Ti,Mo,C) ring around the TiC grains, molybdenum is often supplied as Mo_2C. Harder surfaces are attributed to nickel. The parent combination forms a complex microstructure after sintering, with the development of distinctive fundamental edge structure of the grains (Ti,Mo) (C,N). The edge is enriched with the heaviest element Mo and shares the same crystal structure as the nucleus, whereas the center is virtually unresolved TiC or TiCN.

6.9 NIOBIUM CARBIDE

The search for new materials possessing exceptional qualities such as chemical inertness, hardness, high incompressibility, and thermal stability has regained momentum recently. In this context, a variety of materials have been investigated, including trimetallic carbides, diamonds, nitrides, and boron derivatives. Studying, creating, and improving materials for strength-related applications are the main goals of these endeavors. In the past, indentation tests such as Brinell and Rockwell tests are the main methods used to characterize these materials [16–18,24,76]. By optimizing various properties such as hardness, heat resistance, fracture toughness, strength, and wear resistance, it is a critical to ensure good performance and long component life for all these applications [19]. While other parameters, like using tools, are initially ignored in theory, they are crucial; most theoretical theories rely solely on the overall material unit that is extracted from the radioisotope material. These materials are the conventional glass materials like Almaz and C-BN; however, even these materials have certain uses and needs [19].

6.10 NBC CERMETS

NbCs are fragile without a binder; therefore, cermets and MMCs combine the benefits of, say, the ceramic's hardness and binder's fracture toughness. Numerous parameter studies have been carried out to look at different metal-bonded NbC ceramics, including binders based on cobalt, nickel, and iron. Either a percentage by weight (wt%) or a percentage by volume (vol%) is used to represent the binder content.

6.11 NBC-CO CERMETS

The first fundamental studies on the development of NbC grains in a cobalt matrix during sintering were conducted in 1968, 2001, and 2004, respectively. Considering that cobalt is commonly used as a binder, especially in WC-6Co composites for the production of hardmetals, cobalt-bonded NbC-based cermets were developed, including a binder content of 8 or 12 wt% Co. [8,27].

6.12 NBC-NI CERMETS

Co is being replaced by the Ni binder because Co is associated with health problems, and transition from WC to NbC results in rather poor binder wettability [77,78]. Compared to NbC-Co cermets, NbC-Ni cermets are stronger but less hard [4]. MMCs, called cemented carbides, consist of a ductile metal binder enclosing a hard carbide phase, typically a tungsten carbide (WC)monolith.

The hard carbide phase predominates in hard alloys and ceramics, while the binder (or matrix) is the main phase in MMC. The word "cermet", a combination of the words "metal" and "ceramic", refers to a subgroup of cemented carbides, where TiC or NbC, rather than WC, is the predominant hard carbide phase and embedded in a carbide matrix, Co, Ni, or Fe bonds. Until now, 70% of cutting tools produced are made from tungsten carbide (WC), the material of choice for cemented carbide.

6.13 SECONDARY CARBIDES' IMPACT

In the hydropower, chemical, oil, and petroleum refining industries, corrosion is a serious problem that occurs frequently [4–6]. It is a leading cause of failure of flow components, including pumps and valves used in the transport of slurries or marine oil. Ceramic-metal composites provide an exceptional combination of high strength and high hardness to effectively combat aggressive media in such harsh environments. Numerous fundamental studies have been carried out on corrosion behavior of metal and ceramic composites, and some important results have already been obtained [30,31].

6.14 MECHANICAL PROPERTIES

The materials are easily destroyed regardless of joining method and due to porosity, showing poor mechanical strength and toughness. Carbides become ductile when there is a large stress near the crack source [36]. They say that the only way to increase the strength of carbides is to stop fractures from occurring. Surface porosity, microstructure grain size, and grain boundary phases influence the hardness [37]. Much attention is paid to the development and study of NbC ceramics. NbC-Co metal ceramics have been widely researched between 2000 and 2014. Similar to WC cemented carbide alloys in terms of hardness, NbC-Co is less durable. The hardness of NbC0.88-12Co has been increased by replacing NbC1.0. More recently, NbC cermets with iron bonds have been studied. In contrast to NbC-Ni cermets, the hardness-to-fracture toughness ratio is often worse. The durability and endurance of cutting tools, particularly those used for heavy lifting and heavy-duty applications, are now subjected to stronger limitations due to recent technological advancements [1]. Enhancing the mechanical and physical characteristics of diamond composites, which are utilized to create cutting tools for a variety of technical applications, is a crucial in this regard. High physical, mechanical, and performance characteristics including wear resistance, heat and friction resistance, hardness, elastic modulus, fracture toughness, thermal conductivity, ductility, and high diamond holding capacity must all be present in these materials [2–4]. It is commonly known that the wear resistance of diamond composites is influenced by the elemental composition and matrix properties, whether they are metallic or carbide [5–7].

In this regard, much research has focused on Ni-bonded NbC ceramics with varying concentrations of secondary carbides. Metal ceramics are produced using numerous manufacturing techniques. The focus is on the fabrication of metallic powders and their subsequent consolidation into advanced materials. This includes various powder production and processing techniques such as mechanical milling, which involves repeatedly breaking and welding powder particles to produce fine, homogeneous structures; colloidal powder processing, which improves dispersion and homogeneity; and hot pressing, which simultaneously applies heat and pressure to compact the material. Further compaction is achieved by sintering, as well as more advanced techniques such as spark plasma sintering (SPS) and liquid phase sintering (LPS), which improve grain bonding and microstructural control. In addition, gas pressure sintering (GPS) and hot isostatic pressing (HIP) are used to further reduce porosity and improve mechanical properties. In some cases, casting of Fe_3Al–NbC metal matrix composites (MMC) has also proven successful, offering an alternative route to producing high-performance materials with tailored microstructures. Several key properties of ceramic nanocomposites, such as increased mechanical strength, improved fracture toughness, superior wear resistance, and thermal stability, have been widely reported in the literature [7,9]. The study of wear, lubrication, and friction is known as tribology. Materials can wear either through contact between two surfaces or through contact between an exposed surface and abrasive particles, liquids, chemicals, or heat [6,7].

All in all, there are two basic categories of wear mechanisms: mechanical and all other cases that include an additional degree of active chemistry, like oxidation or corrosion. Four categories can be used to separate the different types of mechanical wear: surface fatigue, adhesion, abrasion, and erosion. Additionally, there is also a fretting mechanism operating under similar wear circumstances, which combines adhesive, abrasive, and occasionally corrosive wear. From an industrial perspective, all of the aforementioned modes result in loss of dimensional tolerances, higher operating costs, and supplementary costs like part replacement [6,8]. In addition, according to some studies, the main wear mechanisms include brittle pullout or separation of carbon nitride grains as well as extrusion and spreading of binder phase. From a mechanical point of view, "water" and "dry" corrosions have certain similarities. However, in solid and liquid corrosive environments, the fluid effect and more complex particle trajectories lead to more complex material failure mechanisms [11].

6.15 CHARACTERISTICS OF CUTTING TOOLS MADE OF A GRADED CERMET COMPOSITE

Gradient metal-ceramic composite tool materials have been developed and produced based on hot pressing sintering after dry pressing to meet the performance requirements of tool materials for high-speed machining of complex materials. The surface layer, subsurface layer, and substrate are graded metal-ceramic composite tool materials, which have graded mechanical and microstructural characteristics. The substrate had high flexural strength, high subsurface adhesive strength, and high surface hardness. Three gradient composite cutting inserts are created on different tool faces. All these microgrooves are aligned with rake edge. As shown in Figure 6.10a and b, the cutting tools under consideration are used in orthogonal cutting, known as tube turning, studies. In orthogonal cutting, the cutting edge and the microgrooves are perpendicular to the chip movement on the rake face. This is done to make the microcutting phenomenon associated with granular cutting tools

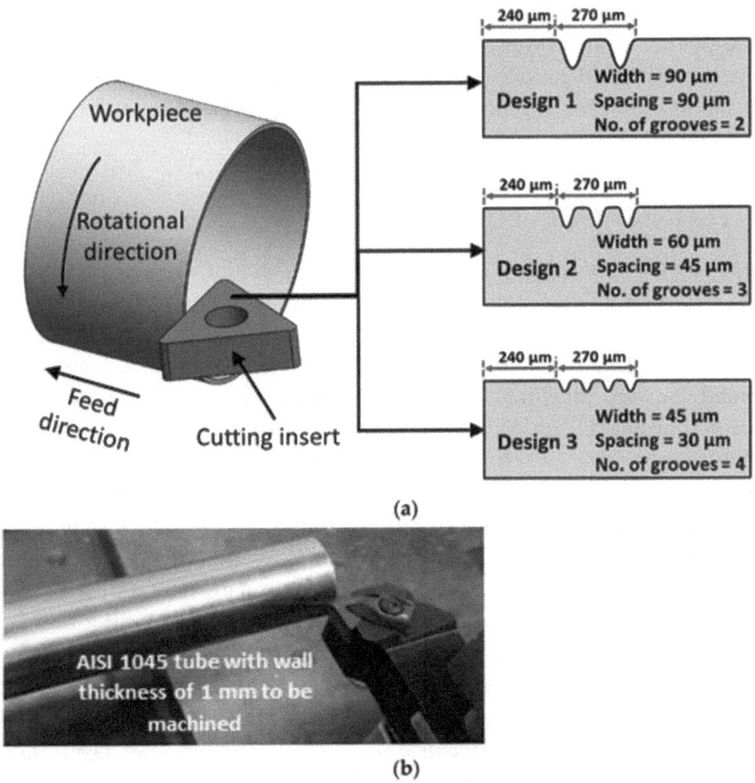

FIGURE 6.10 (a) Schematic diagram of pipe turning experiments with multiple textured structures and (b) test setup [17].

more understandable. The AISI 1045 tube used in these orthogonal cutting tests has a diameter of 30 mm and a wall thickness of 1 mm. In addition, the cutting length is 12.5 m in each test [17].

The effect of cutting speed on the tool life and surface roughness was investigated during continuous dry turning of 17–4 PH stainless steel to evaluate the cutting performance of graded cermet composite cutting inserts. In this study, materials for graded metal-ceramic composite cutting tools are developed and manufactured. The mechanical characteristics and microstructure of the tool materials are gradients. The substrate has high flexural strength, the subsurface layer has high adhesive strength, and the surface layer is very hard. Three graded cermet composite cutting inserts with a graded structure are created on three different tool faces. The effectiveness of ceramic graded inserts in dry turning has been demonstrated through increased tool life, improved wear resistance, and superior surface quality, making them a promising alternative to conventional inserts for high-speed dry machining. Designed for same cutting conditions, the ceramic-metal composite cutting inserts have shown longer tool life and better machining quality than conventional Ti(C,N) ceramic-metal inserts. Cutting speed significantly has affected the failure mode of cutting inserts and tool life. It is found that high surface hardness, good wear resistance of the surface layer, and an appropriate gradient structure provide good cutting performance of graded cermet composite cutting inserts. Finally, design 3 has four 45 μm wide microgrooves with 30 μm gap between each pair of adjacent microgrooves.

6.16 TECHNICAL DEVELOPMENTS

The use of ceramic tools is crucial to increasing manufacturing productivity and efficiency, especially when it comes to cutting metals for machining steel. The following is a summary of their importance and how they compare to carbides and other hard alloys. In addition, research on increasing the intrinsic factor of cermets is presented. Several machinability parameters are evaluated that control the performance of cermet. Ultimately, the performance of cermets as a cutting tool for hard machining is restored. To maximize productivity and efficiency, hardened ferrous metals have been processed since the 1980s using hard machining technologies. Grinding is a machining method used to machine hardened black steel before hard finishing is introduced. Due to the need for grinding equipment, longer setup times and lower metal removal rates are required. The efficiency and productivity of grinding operations are lower. This ultimately affected the cost and duration of production. Despite the long history of ceramic and metal cutting tools, tool manufacturers continue to improve them. One such innovation from Kyocera [16] is a combination of technologies used in insert casting and sintering processes to create a gradient of metal-ceramic material. Because they have a harder outer layer and greater strength in the center, we call them "hybrid cermets". Hardness and strength are combined in Kyocera ceramic inserts. The machine tool used for milling experiment is VMC-850E vertical machining center (Shenyang Shenyi Lathe Manufacturing Co., Ltd., Shenyang, China) with a spindle speed range of 50–8000 rpm. Figure 6.11 illustrates the experimental conditions [16].

FIGURE 6.11 Milling machine for experiments. ([16] [Open access].)

The solid carbide shank of Walter Prototyp MC275/MC075 ceramic cutters is soldered to the ceramic head. In addition, the tool manufacturer offers a design with replaceable ceramic heads that screw into the shank. The purpose of application is the milling of high-temperature alloys. While the MC075 has a geometry for high feeds when machining difficult materials, the MC275 offers a general geometry for a variety of applications. Both models offer cutting speeds up to 3,300 ft^2/min. Walter also produces sintered inserts for turning.

However, nickel-bonded cermet cutting tools are not suitable for machining nickel-based superalloys since superalloys often contain nickel. "You get a lot of build-up on the cutting edge, the cobalt leaches out of the cutting edge, the carbide grains become loose, the edge chips, and the tool doesn't last very long", Benkochi explained. If there is a cobalt in cutting grade and there is a cobalt in workpiece, this indicates that when the adhesive film is formed, the temperature at tool-chip interface gradually increases. When the milling length reaches 30 m, a sticky layer begins to form. The temperature at tool-chip interface gradually decreases after the adhesive layer stabilizes (Figure 6.12) [16].

A literature review showed that coated cemented carbide cutting tools are more commonly used for hard machining than coated cermet [1,2], as demonstrated in Figure 6.13a and b. The crater wear in cutting-edge area *A* and the adhesive layer deposited on tool surface *B* at 20 m form the rake face wear morphology. The crater A also has molten solid particles attached to it, and the tool matrix with a porous structure is visible in the non-adhesive part of the crater. The formation of the adhesive layer is discussed layer by layer. The crater becomes much larger at 40 m, and the rake face begins to crack. On the tool surface, the adhesive layer becomes thicker [16].

The crack trajectories of each microstructure with different fracture energy ratios are reported. Transgranular cracking is observed in the intergranular microstructures

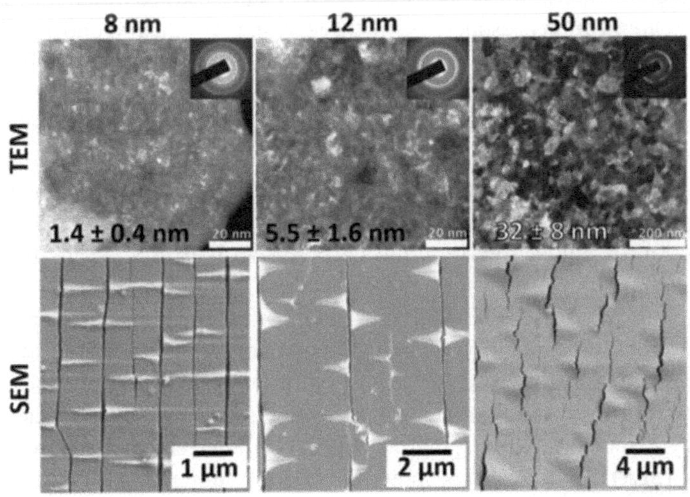

FIGURE 6.12 Wear morphology of front surface at (a) 2.5 m, (b) 20 m, (c) 30 m, and (d) 40 m at 0.07 mm/rev. ([16] [Open access].)

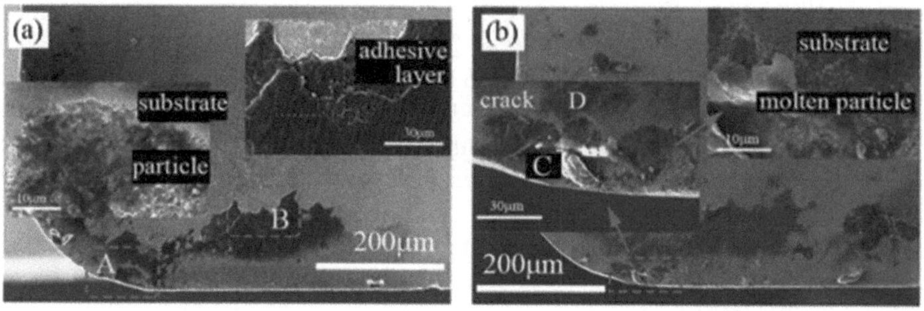

FIGURE 6.13 Wear morphology of the front surface tool at 1200 m/min is as follows: (a) $L=20$ m and (b) $L=40$ m. ([16] [Open access].)

with strong interfaces ($D=3:2:1$). When the intergranular crack comes into contact with the nanoparticles at the interface, it is likely to propagate into the matrix because fracture energy of interface is larger than matrix grain. Comparing the transgranular fracture phenomenon is more prominent, and the pinning and deflection effect of nanoparticles at the interface on intergranular crack is stronger when the interface bond strength is higher [76].

6.17 CONCLUDING REMARKS

Ultra-high-temperature ceramics (UHTCs) fill a gap in many advanced industrial manufacturing applications. It can be achieved with super-hard ceramic materials that withstand the high temperatures of high-speed cutting tools and molten metal crucibles due to oxidation resistance of TiCN/TiN-based cermet. However, the main disadvantages are poor sintering and insufficient mechanical stability of properties. Despite its high-temperature properties, TiCN/TiN are brittle, which may limit their use in some applications, especially in the aerospace industry and in high-speed cutting tools. Manufacturing processes are the most important factor in using and standardization of TiCN/TiN composites. TiN, one of several compounds used as an additional reinforcement to improve the properties of TiCN, can impart special properties to TiCN and TiN that make them very attractive as advanced technologies.

When many carbon allotropes are combined, TiCN performs better. Enhancing TiCN's qualities such as its oxidation resistance and fracture toughness is crucial. The production of binder-free titanium carbonitride is examined in further detail in this review in order to clarify the impacts of several additions such as WC, WC-Cr_2C_3, and TiC. It demands a great deal of attention in research labs and academic institutions. Additionally, the computed wear factors are found to be incredibly low on scale: 10–8. For both coatings, abrasive wear is the primary source of wear, though. Furthermore, some plastic deformation is found by the wear trace profiles. Energy dispersive X-ray spectroscopy (EDS) analyses and scanning electron microscopy (SEM) investigations show that coating integrity is maintained during lubrication conditions. Secondary carbides have important properties that can influence the overall performance of composite materials. However, methods for producing binder-free TiCN using metallic powders remain largely unexplored in the scientific literature. Key aspects such as the selection of starting powders, control of impurities in the powder, dispersion methods, equipment requirements (e.g., mixing and milling), sintering methods and parameters, and the most effective methods for characterizing the resulting composites have not been adequately addressed. Last, but not least, it is a very important to characterize the mechanical behavior (varicosity and hardness) of obtained titanium carbonitride cermet without binder.

PROBLEMS

1. Explain the nitride titanium (TiN) and its manufacture.
2. List several applications for treatment techniques.
3. Identify the challenges and restrictions of TiN and TiCN in manufacture.
4. Describe the coating of TiCN and TiN.
5. Explain the titanium carbonate nitride and TiN without a binder using sintering techniques.

REFERENCES

1. Ma, D., Deng, Q., Liu, H., & Leng, Y. (2021). Effect of ion energy on the microstructure and properties of titanium nitride thin films deposited by high power pulsed magnetron sputtering. *Coatings*, 11(5), 579.
2. Samoilova, O., Shaburova, N., Pashkeev, K., Samodurova, M., & Trofimov, E. (2022). Al0.25CoCrFeNiV high entropy alloy coating deposited by laser cladding on stainless steel. *Materials*, 15, 7058.
3. Gao, M., Li, S., Guan, W., Xie, H., Wang, X., Liu, J., & Wang, H. (2022). Excellent thermal shock resistance of NiCrAlY coatings on copper substrate via laser cladding. *J. Mater. Sci. Technol.*, 130, 93–102.
4. Blixt, K. H. & Hallberg, H. (2022). Grain boundary and particle interaction: Enveloping and pass-through mechanisms studied by 3D phase field crystal simulations. *Mater. Des.*, 220, 110845.
5. Jia, L., Rong, X., Zhao, D., Zhang, X., He, C., & Zhao, N. (2022). Microstructural characteristic and mechanical properties of the in-situ $MgAl_2O_4$ reinforced Al matrix composite based on Al-Mg-ZnO system. *J. Alloy Compd.*, 891, 161991.
6. Zhang, Y., Miyamoto, G., & Furuhara, T. (2022). Enhanced hardening by multiple microalloying in low carbon ferritic steels with interphase precipitation. *Scr. Mater.*, 212, 114558.
7. Zeng, W., Gan, X., Li, Z., & Zhou, K. (2017). Effect of TiC addition on the microstructure and mechanical properties of TiN-based cermets. *Ceram. Int.*, 43(1), 1092–1097.
8. Yadav, S., Paul, C. P., Jinoop, A. N., Rai, A. K., & Bindra, K. S. (2020). Laser directed energy deposition based additive manufacturing of copper: Process development and material characterizations. *J. Manuf. Process.*, 58, 984–997.
9. Li, W. Y., Yang, X. F., Xiao, J. P., & Hou, Q. M. (2021). Effect of WC mass fraction on the microstructure and friction properties of WC/Ni60 laser cladding layer of brake discs. *Ceram. Int.*, 47, 28754–28763.
10. Gu, D., Shi, X., Poprawe, R., Bourell, D. L., Setchi, R., & Zhu, J. (2021). Material-structure-performance integrated laser-metal additive manufacturing. *Science*, 372, eabg1487.
11. Torbov, V., Troitskii, V., & Rakhmatullina, A. (1979). Hot pressing and vacuum sintering of fine titanium nitride powder. *Sov. Powder Metall. Met. Ceram.*, 18(12), 878–882.

12. Sabzi, H. E., Hernandez-Nava, E., Li, X.-H., Fu, H., San-Martín, D., & Rivera-Díaz-Del-Castillo, P. E. (2021). Strengthening control in laser powder bed fusion of austenitic stainless steels via grain boundary engineering. *Mater. Des.*, 212, 110246.
13. Lee, J., Beach, J., Bellon, P., & Averback, R. S. (2017). High thermal coarsening resistance of irradiation-induced nanoprecipitates in Cu-Mo-Si alloys. *Acta Mater.*, 132, 432443.
14. Scattergood, R. O. & Bacon, D. J. (1975). The Orowan mechanism in anisotropic crystals. *Philos. Mag.*, 31, 179–198.
15. Zhang, Z., Zhou, H., Yao, P., Fan, K., Liu, Y., Zhao, L., … & Deng, M. (2021). Effect of Fe and Cr on the macro/micro tribological behaviours of copper-based composites. *Materials*, 14, 3417.
16. Meng, X., Lin, Y., Mi, S., & Zhang, P. (2023). The study of tool wear mechanism considering the tool–chip interface temperature during milling of aluminum alloy. *Lubricants*, 11(11), 471.
17. Salem, A., Hegab, H., & Kishawy, H. A. (2023). Experimental investigation of the derivative cutting when machining AISI 1045 with micro-textured cutting tools. *Metals*, 13(9), 1587.
18. Taheri, R., Jalali, M., Al-Yaseri, A., & Yabesh, G. (2022). Improving TC drill bit's efficiency and resistance to wear by graphene coating. *Pet. Res.*, 7, 430–466.
19. Aamir, M., Tolouei-Rad, M., Giasin, K., & Nosrati, A. (2019). Recent advances in drilling of carbon fiber–reinforced polymers for aerospace applications: A review. *Int. J. Adv. Manuf. Technol.*, 105, 2289–2308.
20. Wang, X., Shen, X., Zeng, C., & Sun, F. (2018). Combined influences of tool shape and as-deposited diamond film on cutting performance of drills for CFRP machining. *Surf. Coat. Technol.*, 347, 390–397.
21. Li, Z. M., Li, X. N., Hu, Y. L., Zheng, Y. H., Yang, M., Li, N. J., … & Zhang, X. W. (2021). Cuboidal γ' phase coherent precipitation-strengthened Cu-Ni-Al alloys with high softening temperature. *Acta Mater.*, 203, 116458.
22. Aliofkhazraei, M., Walsh, F. C., Zangari, G., Köçkar, H., Alper, M., Rizal, C., … & Allahyarzadeh, M. H. (2021). Development of electrodeposited multilayer coatings: A review of fabrication, microstructure, properties and applications. *Appl. Surf. Sci. Adv.*, 6, 100141.
23. Bachmaier, A., Pfaff, M., Stolpe, M., Aboulfadl, H., & Motz, C. (2015). Phase separation of a supersaturated nanocrystalline Cu-Co alloy and its influence on thermal stability. *Acta Mater.*, 96, 269–283.
24. Zhao, Y. H., Zhan, Z. J., Lv, X. Z., & Cao, H. Y. (2022). Microstructure and properties of ZrB_2-SiC reinforced copper matrix composite coatings prepared by laser cladding. *Materials*, 15, 67–77.
25. Montanari, R. & Varone, A. (2021). Flat-top cylinder indenter for mechanical characterization: A report of industrial applications. *Materials*, 14, 1742.
26. Nagaraja, K. C., Rajanna, S., Prakash, G. S., Koppad, P. G., & Alipour, M. (2020). Studying the effect of different carbon and glass fabric stacking sequence on mechanical properties of epoxy hybrid composite laminates. *Compos. Commun.*, 21, 100425.
27. Dong, C. & Davies, I. J. (2014). Flexural and tensile strengths of unidirectional hybrid epoxy composites reinforced by S-2 glass and T700S carbon fibres. *Mater. Des.*, 54, 955–966.
28. Tan, C. L., Azmi, A. I., & Mohamad, N. (2014). Performance evaluations of carbon/glass hybrid polymer composites. *Adv. Mater. Res.*, 980, 8–12.
29. Hu, D., Fu, Q. G., Zhou, L., Liu, B., Cheng, C. Y., Li, X. X., & Sun, J. (2021). Self-healing improvement strategy of thermally sprayed $MoSi_2$ coating at 1773 K: From calculation to experiment. *Corros. Sci.*, 189, 109–599.

30. Tan, C. L., Azmi, A. I., & Muhammad, N. (2016). Delamination and surface roughness analyses in drilling hybrid carbon/glass composite. *Mater. Manuf. Process.*, 31, 1366–1376.
31. Krishnaraj, V., Prabukarthi, A., Ramanathan, A., Elanghovan, N., Senthil Kumar, M., Zitoune, R., & Davim, J. P. (2012). Optimization of machining parameters at high speed drilling of carbon fiber reinforced plastic (CFRP) laminates. *Compos. Part B Eng.*, 43, 1791–1799.
32. Gaitonde, V. N., Karnik, S. R., Rubio, J. C., Correia, A. E., Abrão, A. M., & Davim, J. P. (2008). Analysis of parametric influence on delamination in high-speed drilling of carbon fiber reinforced plastic composites. *J. Mater. Process. Technol.*, 203, 431–438.
33. Campos Rubio, J., Abrao, A. M., Faria, P. E., Correia, A. E., & Davim, J. P. (2008). Effects of high speed in the drilling of glass fibre reinforced plastic: Evaluation of the delamination factor. *Int. J. Mach. Tools Manuf.*, 48, 715–720.
34. Selmy, A., Abd El-baky, M. A., & Hegazy, D. A. (2019). Mechanical properties of inter-ply hybrid composites reinforced with glass and polyamide fibers. *J. Thermoplast. Compos. Mater.*, 32, 267–293.
35. Attia, M. A., Abd El-baky, M. A., & Alshorbagy, A. E. (2017). Mechanical performance of intraply and inter-intraply hybrid composites based on e-glass and polypropylene unidirectional fibers. *J. Compos. Mater.*, 51, 381–394.
36. Rawat, S. & Attia, H. (2009). Wear mechanisms and tool life management of WC–Co drills during dry high speed drilling of woven carbon fibre composites. *Wear*, 267, 1022–1030.
37. Navarro-Mas, M. D., Meseguer, M. D., Lluch-Cerezo, J., & García-Manrique, J. A. (2020). Comparison of different parameters to evaluate delamination in edge trimming of basalt fiber reinforced plastics (Bfrp). *Materials*, 13, 5326.
38. Mkaddem, A., Soussia, A. B., & El Mansori, M. (2013). Wear resistance of CVD and PVD multilayer coatings when dry cutting fiber reinforced polymers (FRP). *Wear*, 302, 946–954.
39. Meinhard, D., Haeger, A., & Knoblauch, V. (2021). Drilling induced defects on carbon fiber-reinforced thermoplastic polyamide and their effect on mechanical properties. *Compos. Struct.*, 256, 173–183.
40. Heisel, U. & Pfeifroth, T. (2012). Influence of point angle on drill hole quality and machining forces when drilling CFRP. *Proc. CIRP*, 1, 471–476.
41. Poulachon, G., Outeiro, J., Ramirez, C., André, V., & Abrivard, G. (2016). Hole surface topography and tool wear in CFRP drilling. *Proc. CIRP*, 45, 35–38.
42. Franz, G., Vantomme, P., & Hassan, M. H. (2022). A review on drilling of multilayer fiber-reinforced polymer composites and aluminum stacks: Optimization of strategies for improving the drilling performance of aerospace assemblies. *Fibers*, 10, 78.
43. Gemi, L., Köklü, U., Yazman, Ş., & Morkavuk, S. (2020). The effects of stacking sequence on drilling machinability of filament wound hybrid composite pipes: Part-1 mechanical characterization and drilling tests. *Compos. Part B Eng.*, 186, 107787.
44. Abd El-baky, M. A. (2017). Evaluation of mechanical properties of jute/glass/carbon fibers reinforced hybrid composites. *Fibers Polym.*, 18, 2417–2432.
45. Findenig, G., Buchegger, C., Lengauer, W., Veitsch, C., & Demoly, A. (2017). Investigation of the main influencing parameters on the degassing behavior of titanium carbonitrides using mass spectrometry. *Int. J. Refract. Metals Hard Mater.*, 63, 38–46.
46. Suib, S. L. (2017). A review of nanoceramic materials for use in ceramic matrix composites. In *Sol-gel Based Nanoceramic Materials: Preparation, Properties and Applications*; Mishra, A. K. Ed.; Springer: Cham, Switzerland, pp. 185–230.

47. Nguyen, T. K. (2015). *Fundamental Tool Wear Study in Turning of Ti-6Al-4V Alloy (Ti64) and Nano-Enhanced Minimum Quantity Lubrication (MQL)* Milling; Michigan State University: East Lansing, MI, USA.
48. Lagarde, Q., Wagner, V., Dessein, G., & Harzallah, M. (2021). Effect of temperature on tool wear during milling of Ti64. *J. Manuf. Sci. Eng.*, 143, 071007.
49. Veiga, C., Davim, J. P., & Loureiro, A. (2012). Properties and applications of titanium alloys: A brief review. *Rev. Adv. Mater. Sci.*, 32, 14–34.
50. Pramanik, A., Islam, M. N., Basak, A., & Littlefair, G. (2013). Machining and tool wear mechanisms during machining titanium alloys. *Adv. Mater. Res.*, 651, 338–343.
51. Ulutan, D. & Ozel, T. (2011). Machining induced surface integrity in titanium and nickel alloys: A review. *Int. J. Mach. Tools Manuf.*, 51, 250–280.
52. Amin, A. K. M. N., Ismail, A. F., & Khairusshima, M. K. N. (2007). Effectiveness of uncoated WC–Co and PCD inserts in end milling of titanium alloy—Ti–6Al–4V. *J. Mater. Process. Technol.*, 192–193, 147–158.
53. Sahoo, P., Singh, V. K., & Patra, K. (2023). Influences of tribological and mechanical properties of (Ti/Al)N and diamond like carbon coated tungsten carbide tool on machining and dynamic stability in micro milling of Ti_6Al_4V. *J. Manuf. Process.*, 85, 915–934.
54. Rahman, M., Wang, Z., & Wong, Y. (2006). A review on high-speed machining of titanium alloys. *JSME Int. J. Ser. C Mech. Syst. Mach. Elem. Manuf.*, 49, 11–20.
55. Jianxin, D., Yousheng, L., & Wenlong, S. (2008). Diffusion wear in dry cutting of Ti–6Al–4V with WC/Co carbide tools. *Wear*, 265, 1776–1783.
56. Hosseini, A., Kishawy, H. A., & Hussein, H. M. (2014). Machinability of titanium and its alloys. In *Machinability of Advanced Materials*; Paulo Davim, J. Ed.; John Wiley & Sons, Inc.: Hoboken, NJ, USA, pp. 95–118.
57. Umbrello, D. (2008). Finite element simulation of conventional and high speed machining of Ti_6Al_4V alloy. *J. Mater. Process. Technol.*, 196, 79–87.
58. Kumar, K. M., Mathew, N. T., & Baburaj, M. (2022). Sustainable milling of Ti-6Al-4V super alloy using AlCrN and TiAlN coated tools. *Mater. Today Proc.*, 50, 1732–1738.
59. Hosseini, A. & Kishawy, H. A. (2014). Cutting tool materials and tool wear. In *Machining of Titanium Alloys*; Davim, J. P., Ed.; Springer: Berlin/Heidelberg, Germany, pp. 31–56.
60. Nabhani, F. (2001). Machining of aerospace titanium alloys. *Robot Comput. Integr. Manuf.*, 17, 99–106.
61. Soroka, O. B. (2010). Evaluation of residual stresses in PVD-coatings. Part 1. Review. *Strength Mater.*, 42, 287–296.
62. Abadias, G., Chason, E., Keckes, J., Sebastiani, M., Thompson, G. B., Barthel, E., ... & Martinu, L. (2018). Review Article: Stress in thin films and coatings: Current status, challenges, and prospects. *J. Vac. Sci. Technol. A*, 36, 20801.
63. Patro, S. S., Roy, S., & Sahoo, A. K. (2023). Investigation of influence of design on residual stress at free edge of multilayer Ti/AlN coatings by PVD using FEM. *Int. J. Interact. Des. Manuf. JIDeM*, 17, 2987–2998.
64. Zhang, J., Ling, L., Luo, D., Deng, C., Huang, X., Tao, G., & Cao, H. (2023). Cutting performance and surface quality of Ti-6Al-4V by longitudinal ultrasonic vibration-assisted high-speed dry milling with coated carbide tools. *Int. J. Adv. Manuf. Technol.*, 126, 5583–5596.
65. ISO 8688-2:1989. (1989). *Tool Life Testing in Milling—Part 2: End Milling.* ISO: Geneva, Switzerland.
66. Kumar, T. S., Prabu, S. B., Manivasagam, G., & Padmanabhan, K. A. (2014). Comparison of TiAlN, AlCrN, and AlCrN/TiAlN coatings for cutting-tool applications. *Int. J. Miner. Metall. Mater.*, 21, 796–805.

67. Liu, J., Wang, Y., Liu, G., Hua, J., & Deng, X. (2023). Properties and performance of TiAlSiN and AlCrN monolayer and multilayer coatings for turning Ti-6Al-4V. *Coatings*, 13, 1229.
68. Feng, Y., Zhang, L., Ke, R., Wan, Q., Wang, Z., & Lu, Z. (2014). Thermal stability and oxidation behavior of AlTiN, AlCrN and AlCrSiWN coatings. *Int. J. Refract. Metals Hard Mater.*, 43, 241–249.
69. Kim, D., Swan, S. R., He, B., Khominich, V., Bell, E., Lee, S.-W., & Kim, T.-G. (2021). A study on the machinability of advanced arc PVD AlCrN-coated tungsten carbide tools in drilling of CFRP/titanium alloy stacks. *Carbon Lett.*, 31, 497–507.
70. Akinribide, O. J., Obadele, B. A., Akinwamide, S. O., Bilal, H., Ajibola, O. O., Ayeleru, O. O., … & Olubambi, P. A. (2019). Sintering of binderless TiN and TiCN-based cermet for toughness applications: Processing techniques and mechanical properties: A review. *Ceram. Int.*, 45(17), 21077–21090.
71. Mishra, S. K., Ghosh, S., & Aravindan, S. (2018). Characterization and machining performance of laser-textured chevron shaped tools coated with AlTiN and AlCrN coatings. *Surf. Coat. Technol.*, 334, 344–356.
72. Reiter, A. E., Derflinger, V. H., Hanselmann, B., Bachmann, T., & Sartory, B. (2005). Investigation of the properties of Al1–xCrxN coatings prepared by cathodic arc evaporation. *Surf. Coat. Technol.*, 200, 2114–2122.
73. Gao, M. Y., Li, S. C., Guan, W. M., Xie, H. B., Wang, X. X., Liu, J. B., & Wang, H. T. (2022). Excellent thermal shock resistance of NiCrAlY coatings on copper substrate via laser cladding. *J. Mater. Sci. Technol.*, 130, 93–102.
74. Zhu, L., Chen, P., Cai, Z.-M., Feng, P.-Z., Kang, X.-Q., Akhtar, F., & Wang, X.-H. (2022). Fabrication of $MoSi_2$ coatings on molybdenum and its high-temperature anti-oxidation properties. *Trans. Nonferrous Met. Soc.* China, 32, 935–946.
75. ISO 19252:2008. (2008). *Plastics—Determination of Scratch Properties, Technical Committee ISO/TC 61, Plastics, Subcommittee SC 2, Mechanical Properties*. ISO: Geneva, Switzerland.
76. Zhou, T., Meng, L., Yi, M., & Xu, C. (2023). Simulation of microscopic fracture behavior in nanocomposite ceramic tool materials. *Lubricants*, 11(11), 489.
77. Yetim, T., Turalioğlu, K., Taftali, M., Tekdir, H., Kovaci, H., & Yetim, A. F. (2022). Synthesis and characterization of wear and corrosion resistant Ni-doped Al_2O_3 nanocomposite ceramic coatings by sol-gel method. *Surf. Coat. Technol.*, 444, 128659.
78. Sonkar, V., Abhishek, K., Datta, S., & Mahapatra, S. S. (2014). Multi-objective optimization in drilling of GFRP composites: A degree of similarity approach. *Procedia Mater. Sci.*, 6, 538–543.

7 Cutting Tool Materials for High Speed and Advancements in Nanocoating Technology

7.1 INTRODUCTION

The need for improved performance under challenging conditions has led to significant advances in cutting tool materials for high-speed machining (HSM). Hardness, wear resistance, and thermal stability have improved dramatically as a result of combining advanced materials such as coated carbides and ceramics with advances in nanocoating technology. By reducing friction and improving durability, nanocoatings improve these tools and extend their life and efficiency. To meet the challenges of modern machining applications, research and development in materials science and coating technologies will be essential as manufacturing processes continue to advance. This will ultimately lead to increased productivity and cost-effectiveness in the industry. A sophisticated and newly developed processing method called high-speed machining (HSM) is widely used to produce complex parts with increased productivity, better quality, environmental friendliness, and cost-effectiveness. HSM is first created for the processing of aluminum and rocket and aircraft components based on its alloys. With continuous advances in HSM [1], newly introduced machine tools can machine hard metals, difficult-to-cut materials, complex 3D geometries, and micro/nano features with great precision and accuracy [1]. In addition, the modern machine used today can cut burr-free edges and eliminate built-up edges (BUE), providing an exceptional surface finish. HSM is a metal cutting technology that uses cutting speeds 10–15 times faster than traditional machining. Researchers have used some parameters that characterize HSM and classify it into many areas including high-throughput, HSM, high-feed machining, and high-speed spindle machining. Many factors including work material, tool material, cutting conditions, and machine tools influence the cutting speed range in HSM. On the other hand, cutting speed is highly dependent on the material being processed. For example, aluminum has a significantly higher cutting speed than titanium alloy. On the other hand, cutting speed is highly dependent on the material being processed. For example, a new method of high-speed rotary ultrasonic machining (HRUM) for the microstructure of hard and brittle materials was reported. Since microstructure machining usually involves small workpieces, the ultrasonic vibration can directly act on the workpiece. To assist in HRUM, an HRUM machine (spindle speed $\geq$ 8000 rpm) was established.

DOI: 10.1201/9781003604648-7

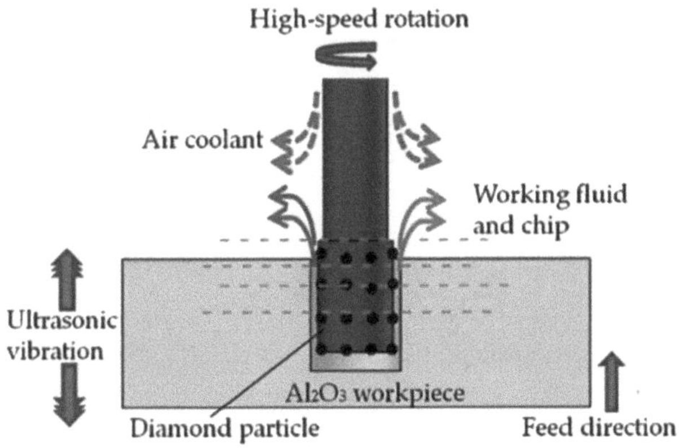

FIGURE 7.1 Illustration of HRUM. ([2] [Open access].)

In HRUM, as shown in Figure 7.1, the Al_2O_3 workpiece vibrates longitudinally at ultrasonic frequencies, while the sintered diamond tool rotates rapidly under the action of an electric spindle [2].

7.2 APPLICATIONS OF HIGH-SPEED MACHINING

HSM has emerged as a revolutionary method for some production applications, offering significant benefits in terms of efficiency, accuracy, and surface quality. HSM reduces tool wear and thermal distortion while enabling faster material removal rates through the use of advanced cutting tools and improved manufacturing standards. In industries where accuracy and productivity are critical such as aerospace, automotive, and mold making, this technology is highly beneficial. HSM is likely to become more widely used as advances in machine design, process improvements, and tool materials continue to drive gains in manufacturing capabilities and industry competitiveness. The effectiveness of HSM is determined by the following machine subsystem components: spindles, axes, motor drives, tool holders, and cutting tools. Since the spindle is arguably the most critical component [1], runout must be kept as low as possible to maintain optimal accuracy and performance. With a cutter, the shorter the interval between changing inserts, the smaller the runout. The runout size is important for accuracy, especially in HSM applications [1,3].

The deposition of carbon nanotubes (CNTs) using plasma-enhanced chemical vapor deposition (PECVD) on a high-speed steel tool was studied. The mechanical properties, surface quality, cutting parameters, temperature, cutting forces, friction coefficient, and tool life of coated tool are evaluated. The CNT-coated tool is found to be a good choice for machining harder materials according to the experimental results studied for three different cutting conditions. Moreover, it is observed that the superior mechanical and thermal properties of CNTs resulted in a significant reduction in cutting tool tip temperature and cutting forces. As shown in Figures 7.2 and 7.3, this indicates that CNT-coated tools perform better in terms of application than DLC-coated tools [2].

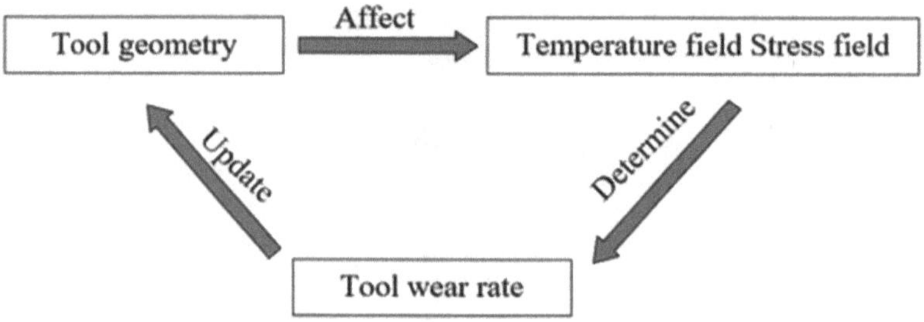

FIGURE 7.2 Temperature readings at levels 1, 2, and 3. ([2] [Open access].)

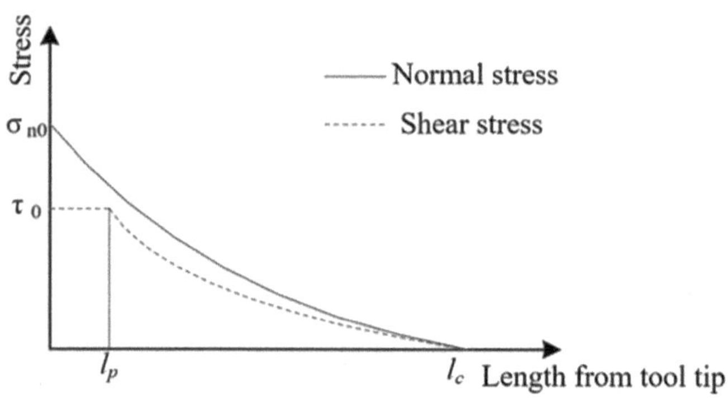

FIGURE 7.3 Flank wear results under cutting conditions of levels (a), (b), and (c). ([2] [Open access].)

At the cutting edge, the total indicator reading (TIR) should not exceed 10 μm [2]. An additional 10 μm of runout results in 50% reduction in tool life. There can be multiple sources of imbalance, even in cases where the tool, tool holder, and spindle are perfectly balanced and unpredictable. The contact between the tool holder and spindle contact is one of the instability causes. With this grip, there is usually a noticeable gap. In addition, chips or dirt may form inside the cone. There is usually no need to balance 20,000 rpm end mill to anything greater than 20 and 5 g/mm is suitable for much higher speeds.

7.3 MACHINE TOOLS FOR HSM—REQUIREMENTS

Machine tool designers and engineers have begun to build machines for HSM, although high-speed alternatives to spindles for conventional machining centers have been available for some time. As mentioned earlier, HSM requires the adjustment of many machine-related factors.

7.3.1 Hard Material Cutting

Cutting hard materials presents unique challenges that require advanced tooling solutions and tactics. To achieve precise and efficient cutting performance, the creation of cutting tools specifically designed for machining hard materials—such as high-speed steels, carbides, and advanced ceramics—has proven to be critical. Coatings and tool engineering advances improve thermal stability and wear resistance, allowing for longer tool life and improved surface polish. Continued study of material properties and cutting processes will be essential to drive progress and maintain market competitiveness as companies seek higher levels of precision and efficiency when machining hard materials. The new cubic boron nitride (CBN) cutting tools with textured flank surfaces are developed considering the stabilization effect of textured surfaces. Hard material cutting refers to machining techniques specifically designed for materials with a high degree of hardness such as ceramics, titanium alloys, and hardened steels. Specialized cutting tools and techniques are required due to the hardness and wear resistance of these materials.

Important elements include:

- Tool materials: Due to their long life and hardness, carbides, ceramics, and high-speed steels are frequently used.
- Cutting conditions: To minimize tool wear and produce proper surface finishes, appropriate cutting speeds, feeds, and depths must be used.
- Cooling and lubrication: By reducing friction and dissipating heat, effective cooling techniques extend the life and functionality of tools.
- Advanced coatings: The use of coatings such as titanium nitride (TiN) or diamond-like carbon increases wear resistance and extends tool life.
- Process optimization: Ongoing research aims to improve cutting parameters and create new tool designs to increase production and efficiency.

This area is essential for sectors including aerospace, automotive, and manufacturing that require precision machining of hard materials. A femtosecond laser is used to stabilize the adhesion layer on the flank surface and prevent its peeling. As shown by a series of cutting experiments in Figures 7.4 and 7.5, the surface roughness on the flank surface is a promising method to improve the toughness of CBN during

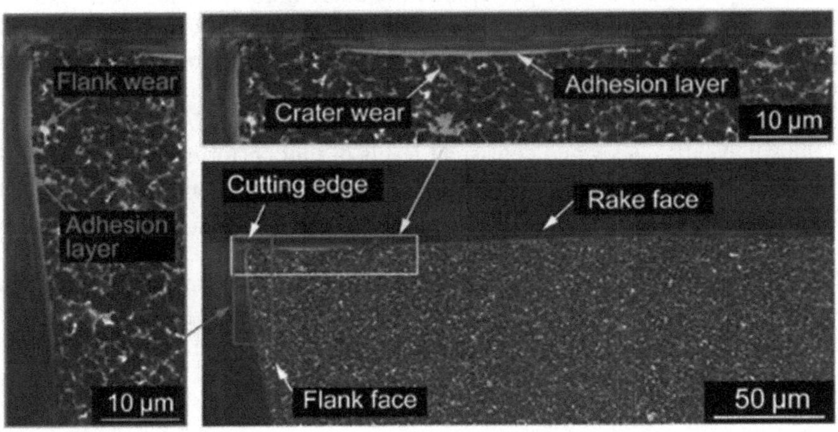

FIGURE 7.4 After cutting for 25 m, the cross section of the CBN cutting tool is observed. ([2] [Open access].)

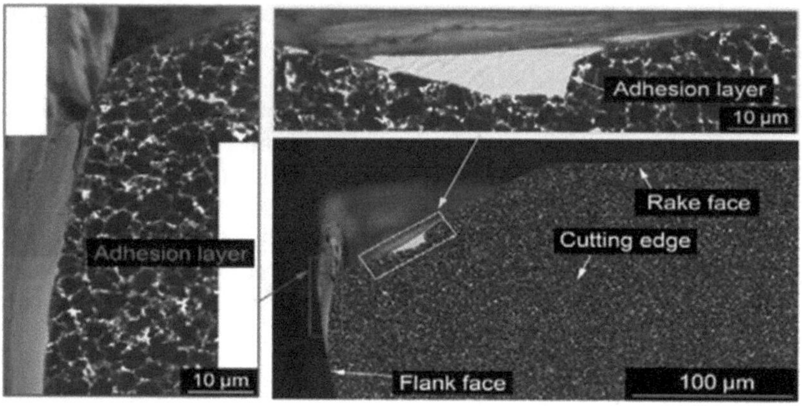

FIGURE 7.5 After cutting for 200 m, the cross section of the CBN cutting tool is observed. ([2] [Open access].)

high-speed milling of Inconel 718. The series of cutting experiments have shown that the CBN-ORE cutting tool significantly reduces the tool waste compared with the conventional CBN cutting tool without surface texture [2].

Hard cutting is the process of cutting hard materials such as hardened steel, with hardness values greater than 47 HRC. This manufacturing method includes all cutting operations such as turning, drilling, milling, broaching, and other processes with geometrically defined cutting edges (as opposed to abrasive processes). In the past, the hard materials are only processed using abrasive methods and worked into their final shape unless they needed repair. In the 1980s and 1990s, suitable cutting tools, machines, and processing methods are developed [1,2,4].

Today, most hard materials can be cut, and since the 1990s, cutting hard materials has become an industry standard, particularly in instrumental materials. The main technological advances have been high hardness and sufficient strength at elevated temperatures, mainly polycrystalline cubic boron nitride (PCBN), as well as ultra-fine-grain carbide and mixed oxide ceramics (Al_2O_3/TiCN) with improved toughness. Alumina ceramics with a very negative effective rake angle have been used to successfully machine steels with hardness greater than 60 HRC as early as the mid-1960s; however, it is not until the development of CBN/TiCN composites in the 1980s that satisfactory and consistent levels of tool performance were achieved. Physical vapor deposition (PVD) coatings have been successfully applied to PCBN tools over the last 10 years or so, allowing for increased cutting speeds and, in some cases, more than doubling tool life.

Hardened steels account for a significant portion of hard materials that are cut in the industry, and case-hardened steels account for the majority in the automotive sector. The main technological advances have been high hardness and sufficient strength at elevated temperatures, mainly in PCBN, as well as ultra-fine-grain carbide and mixed oxide ceramics (Al_2O_3/TiCN) with improved toughness. Alumina ceramics with very negative effective rake angles have been used to successfully machine steels with hardness greater than 60 HRC. The three main types of abrasive failure are shedding, fragmentation, and wear. Since the cutting force applied to a single diamond grain is less than the bond strength of the bond, the tip area becomes more visible and wears continuously. The abrasive granules shift and increase the cutting force when the cutting force exceeds the bond strength. When the bond collides with the workpiece, the cutting force depends on the bond strength until new diamond abrasive grains are formed. A super depth of a field video microscope (DVM6, Leica, Wetzlar, Germany) was used for the initial observation to compare the end faces of the diamond tool after ductile grinding mode (DGM) and hybrid rotary ultrasonic machining (HRUM). Figure 7.6a and b shows the results. The end face has a more uniform distribution of exposed abrasive grains compared to those without ultrasonic vibration. The sharp edges of the abrasive grains are blunted by the friction between the abrasive grains and the surface of the difficult-to-cut material, which increases the abrasive wear. At the same time, the diamond abrasive grains form a smaller plane. However, the ultrasonic vibration effect can reduce the friction coefficient and grinding force in HRUM. The Al_2O_3 ceramics are removed

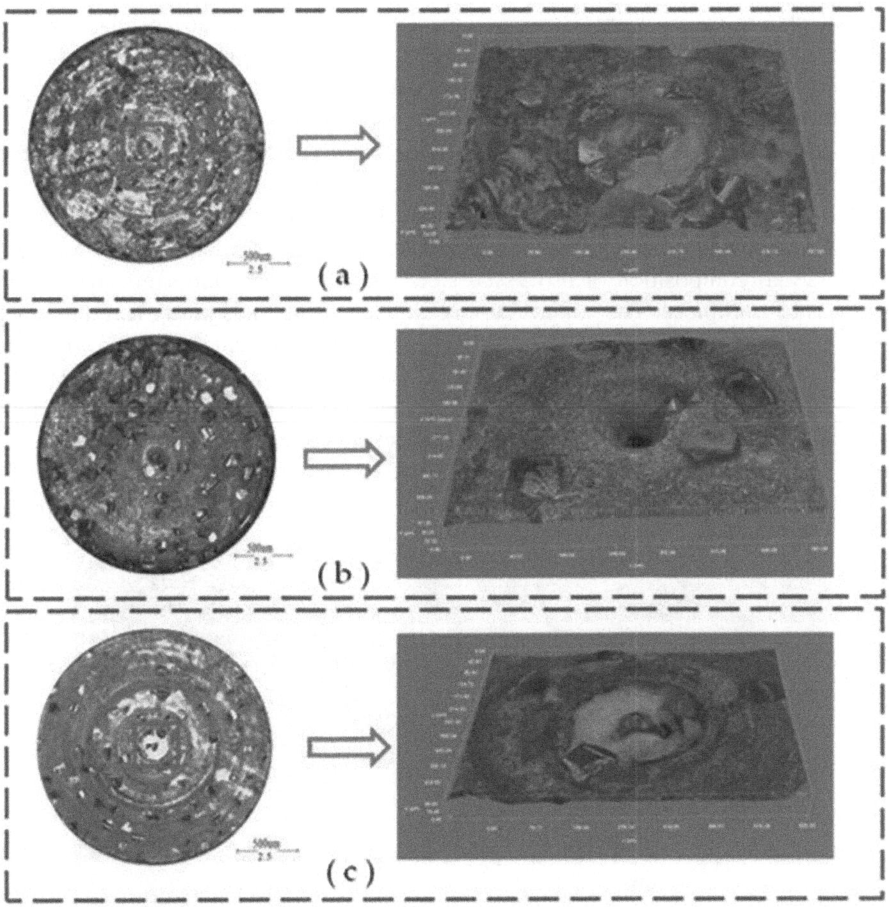

FIGURE 7.6 Micrograph images of the tool end face: (a) DGM and (b) HRUM for 160 mesh abrasive at the last hole; (c) DGM 200 mesh abrasive at the last hole. ([1] [Open access].)

as chips during tool grinding, as shown in Figure 7.6c, but they also tend to stick to the end face of the tool [1].

Low-alloy steels are not only susceptible to thermochemical wear mechanisms but also experience more aggressive abrasive wear due to higher concentrations of carbide-forming components. The size of carbide grains in conventional steels causes significantly higher wear rates than smaller carbides found in steels produced from PM. PCBN, mixed oxide ceramics (SC), and fine-grained tungsten carbides (TC) are the best tool materials; however, the latter should only be used in special circumstances. PCBN ceramics is used for low-volume operations with few breaks in the workpiece, while heavy roughing and finishing turning processes (H05–H30) are used for higher volumes. Cemented carbides are primarily used in milling applications, where more geometrically complex tools can be used with them; however, PCBN is used in very high-volume applications such as ball milling with constant velocity joints. Classic hole-making defects of hard and brittle ceramic materials, such as cracks at the entry and exit of holes, are shown in Figure 7.7. In contrast, hole exit defects are more severe. Burrs and cracked edges are less severe than rupture defects as the number of holes produced increases. The sharpness of diamond abrasive grains is almost directly related to the material removal by ultrasonic vibration. Therefore, hole exit defects are caused by the action of the macroscopic force of all diamond abrasives. Due to friction, diamond abrasive grains are worn during the machining process, which increases the tip area and thus the cutting force [1].

Unlike straight lines when cutting mild steel, unusual curves show that there is usually a narrow ideal speed range for hard turning. Tests revealed differences in the hardness and composition of processed steels, which are partially reflected in the hardness of workpiece. The material used to make cutting tools must meet these two requirements: it must be more resistant to fracture and plastic deformation. In addition, the material used to make cutting tools must be heat-resistant and chemically stable and have an ideal strength-hardness ratio [5]. Many manufactured cutting tool materials have been found to differ in two specific properties: toughness and hardness. These two qualities must present in an ideal ratio in material of cutting tool. The evolution of cutting tool materials has revealed numerous trends and changes.

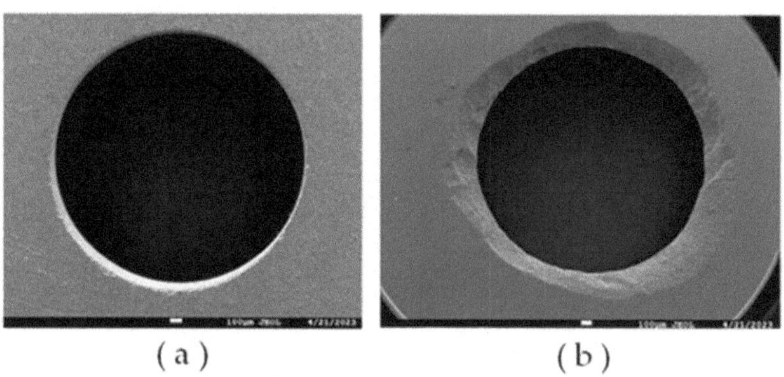

FIGURE 7.7 Quintessential hole-making defects in hard and brittle ceramic material: (a) entrance defect and (b) exit defect. ([1] [Open access].)

The initial and sustained transition occur from high-speed steel to carbide. The second change concerns the tool material: from carbide to polycrystalline diamond (PCD) [6].

Figure 7.8 illustrates the solid end mills used in HSM. There are tools with different coatings, helix angles, number of grooves, and structures available for a wide variety of applications. Before selecting cutting tools for HSM, it is important to consider the following factors: contact conditions, cutting tool material, cutting tool geometry, tool design, and associated processing methods [7].

The economic and environmental benefits shown in Figure 7.8 should be a typical interpretation that does not apply to every component or process. When working with case-hardening steels, the drill must be able to reliably cut through both soft and hard areas of the part. As shown in Figure 7.9, there is a limited processing window for hard drilling of hardened materials, with the two main characteristics being low feed rates and relatively low cutting speeds. Typically, short tungsten carbide drill bits are used. The ideal cutting material is ultra-fine-grained tungsten carbide. Boring tools with replaceable tips, mainly TiN coating, are used for large diameter (>12 mm) TiAlN coatings. Such coatings reduce friction between the tool and hole surface, as well as between the tool and chips and increased wear resistance. In finishing

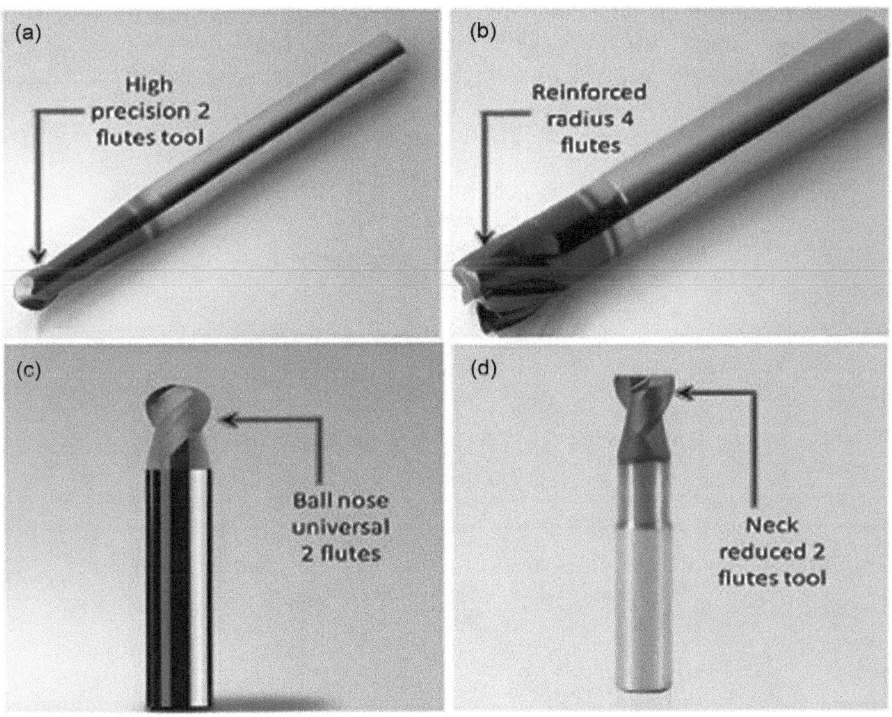

FIGURE 7.8 The following tools: (a) solid end mill with hard surface for high-precision operations, (b) hard-coated toroidal end mill to ensure high precision, (c) hard-coated short flute tool, and (d) a CVD-coated two-flute tool. (Image taken from Seco Tools website. [7] [Open access].)

conditions, solid carbide drills can provide diameter quality from IT 7 to IT 9, as well as surface roughness Rz ¼ 2–4 mm and, under certain circumstances, Rz ¼ 1 mm. Hard turning is a common finishing method that replaces grinding processes. When switching from laser polishing to continuous laser polishing, where the cutting action is performed by the cutting blade and the abrasive material is worn away, the laser radiation helps remove chips and the metal binder from the blade, allowing the diamonds to emerge evenly during the cutting process (Figure 7.9) [8].

Due to its photovoltaic and acousto-optic properties, lithium niobate ($LiNbO_3$) is becoming important in nonlinear optics and advanced photonics [9]. Optimal process settings for chip- and crack-free surfaces result in the formation of edge waveguides with a depth of 40 μm and widths of 8 and 10 μm on the surface of a planar waveguide, as shown in Figure 7.10, using ion implantation and fine-cutting with a diamond blade. The edge waveguide is visible in Figure 7.10d as a bright horizontal region at the top of the cross section of the structure; after annealing, the propagation loss of the 8 μm-wide edge waveguide was 2.2 dB/cm [8].

Important information about the properties of single-layer graphene (SLG) generated on copper (Cu) and transferred to gold (Au) substrates can be learned from examining the properties of sample before biological interaction. Scanning electron microscopy (SEM) has provided visual validation of the coating process, while Raman spectroscopy successfully characterized the quality and integrity of SLG in both configurations. Differences in surface wettability between the untreated and graphene-coated samples, which are highlighted by contact angle analysis, suggest

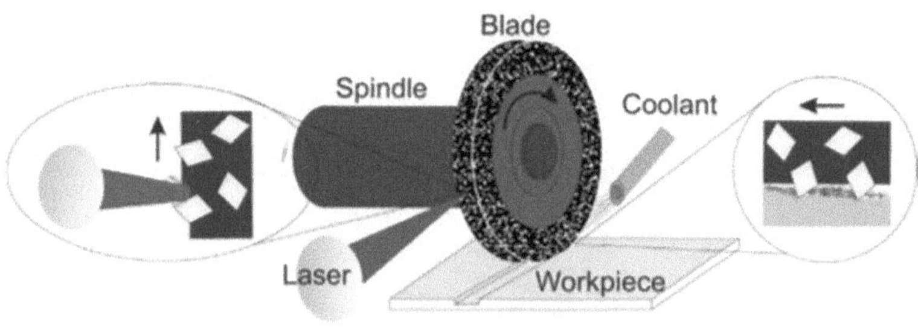

FIGURE 7.9 Principle of continuous laser dressing. ([8] [Open access].)

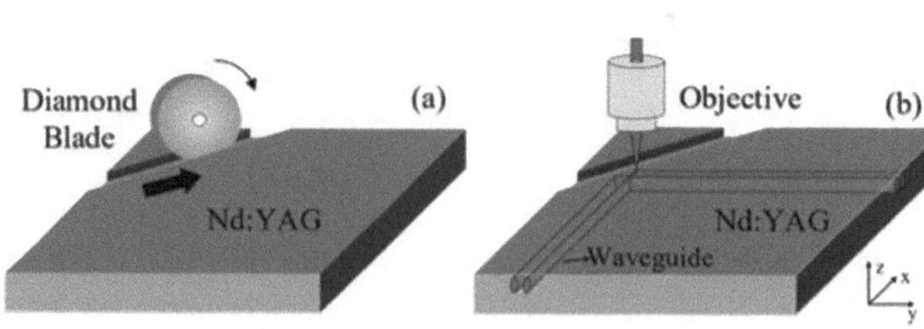

FIGURE 7.10 Schematics of (a) 3 MeV oxygen ion implantation, (b) diamond blade dicing, (c) cross sections of the ridge waveguides, and (d) transmission optical microscopy image showing a close-up of the ridge waveguide. ([8] [Open access].)

the potential for improved biological interaction properties. Overall, these results highlight the importance of surface modification and substrate selection when tailoring graphene for biosensing and biocompatibility applications.

Raman spectroscopy is used to evaluate the properties of samples before biological interaction, especially SLG produced on copper (Cu) and transferred to gold (Au). This method has provided an important new information about the electrical properties and structural quality of graphene.

Different peaks in Raman spectra such as the G and 2D bands show the quality of graphene. By distinguishing between SLG on copper and SLG transferred on gold, the intensity ratios of these peaks provided information about the degree of disorder and defects present in the material. Understanding how the physical properties of graphene affect its interactions in biological applications depends on this characterization. Taken together, these measurements provide the basis for evaluating the effectiveness of graphene-based materials.

To confirm the graphitic nature of SLG prepared on Cu and transferred onto Au, Raman spectroscopy is used (Figure 7.11c). Graphene prepared on Cu exhibits the expected 2D/G intensity ratio of 2.47 for monolayer graphene with a strong first-order G bond stretching band centered at around 1589 cm^{-1} and 2D phonon band centered at around 2683 cm^{-1}. SEM is used to characterize the morphology of samples (Figure 7.11b and c). According to literature [10], the uncoated gold samples showed a continuous coating consisting of irregular leakage grains with flat tops. Monolayer graphene deposited on gold shows a smoother surface with wrinkles across it in SEM. This micrograph does not reveal whether the graphene covers the grains in the pits between the gold or coating broke off during the transfer process [10].

More isotropic features emerge when smaller areas of graphene-coated gold samples are subjected to scanning tunneling microscope (STM) analysis (Figure 7.12). In particular, graphene ripples are evident in the spaces between large wrinkles (Figure 7.4b). Figure 7.4d depicts these nanometer-scale ripples, which are comparable to those observed on free-standing graphene and have been documented in STM studies of graphene on SiO_2 [10]. In 2D materials, partially decoupled bending and stretching modes lead to ripples, which in turn provide stability [10]. These ripples exhibit a characteristic honeycomb pattern of graphene over the scan area of $20 \times 20\,nm^2$ (Figure 7.12e). The root means square roughness on the flat grain surface (Figure 7.12b) is 6 Å (Figure 7.12c) when ripples and wrinkles are taken into account, while it is approximately 1.5 Å (Figure 4d) when ripples are considered alone (Figure 7.12f). The high-resolution method is used to detect even the smallest traces of polymethyl methacrylate (PMMA) contamination. In a number of industries such as materials research, pharmaceuticals, and semiconductor manufacturing, where even minute levels of impurities can have a significant impact on productivity and quality, high-resolution analysis methods are critical [10].

Low-alloy hardened steels have been widely used in manufacturing for decades due to their outstanding characteristics. Tool wear has an impact on surface's physical state (Figure 7.13). These materials are increasingly being used, requiring high-throughput processing methods, resulting in ongoing challenges. Numerous studies have been carried out to improve the machining characteristics of low-alloy steels [1,2,4,11]. Hard machining is the machining process of low-alloy hardened

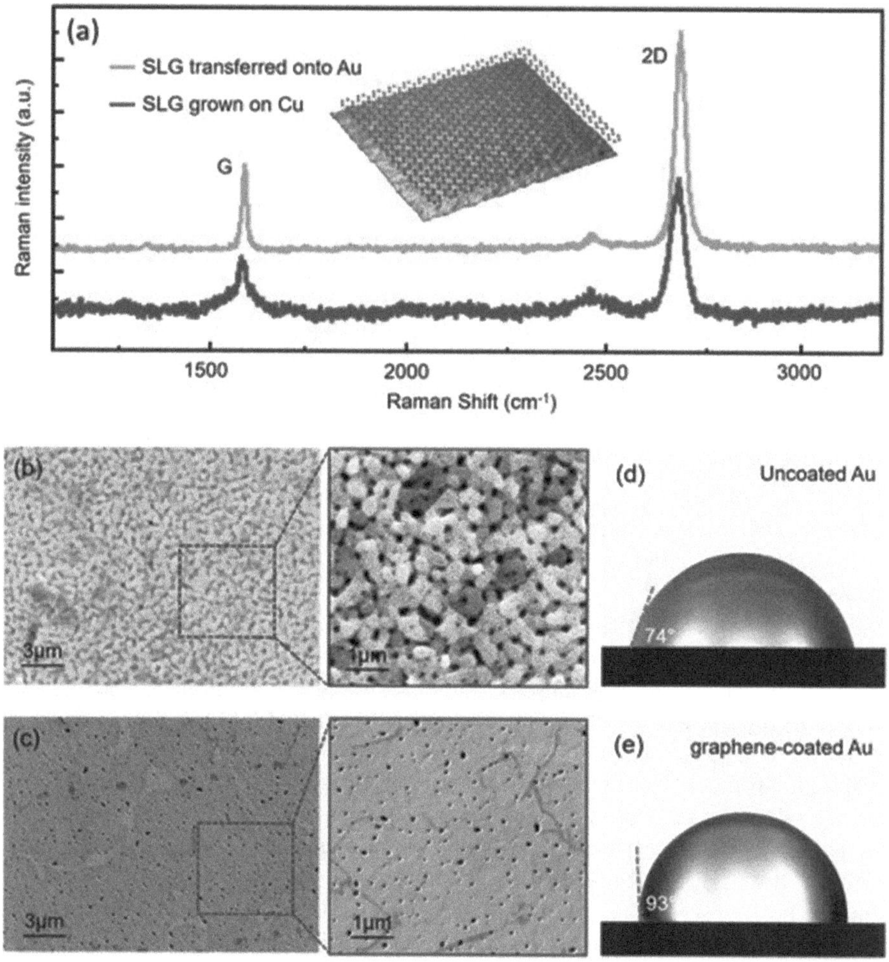

FIGURE 7.11 The characteristics of samples before biological interaction. (a) SLG grown on Cu and SLG transferred onto Au samples are measured by Raman spectra. The Au sample is grown on mica before graphene transfer. The SEM image of SLG-coated (c) and untreated (b) Au. Contact angle analysis between untreated (d) and graphene-coated (e) samples. ([10] [Open access].)

steels (>45 HRC), especially when using single-point turned bearings [3]. Bearing steel, defined as high-chromium, low-alloy carbon steel AISI 52100, is characterized by high strength, abrasion resistance, and dimensional stability [3,12]. Instead of grinding, bearing steel can be hard machined, providing a cost-effective alternative with high material removal rates. However, due to its extreme strength, AISI 52100 bearing steel is a difficult material to machine. High temperatures between the workpiece, tool, and chips also reduce tool life and result in poor surface finish. Coolant is a useful tool because it controls the temperature generated during machining, reducing surface roughness, cutting force, and tool wear. However, flood

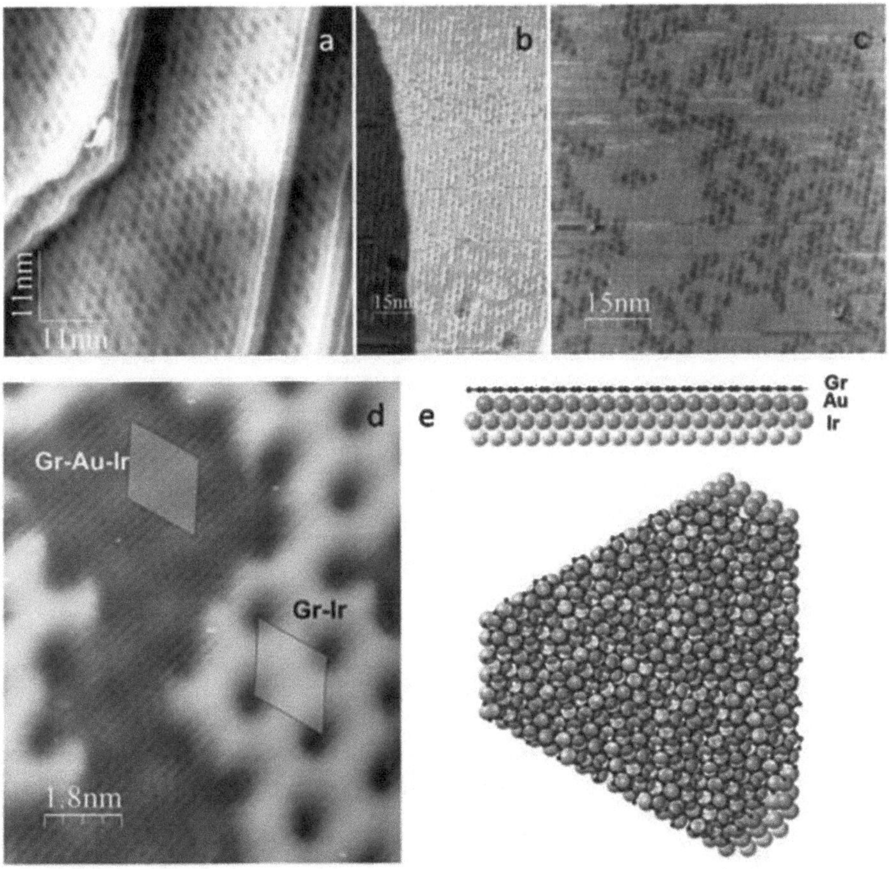

FIGURE 7.12 A more detailed investigation of roughness of graphene deposited on gold samples using STM images at different scales. (c) Roughness analysis of both wrinkled and rippled regions (Figure 7.4b); (d) $180 \times 180\,nm^2$, $I = 0.03\,nA$, VBIAS = 0.9 V; and (e) $20 \times 20\,nm^2$, $I = 0.03\,nA$, VBIAS = 0.9 V. (f) Roughness analysis of rippled regions only (Figure 7.4d). (a) $750 \times 750\,nm^2$, $I = 0.1\,nA$, VBIAS = 0.6 V; (b) $375 \times 375\,nm^2$, $I = 0.1\,nA$, VBIAS = 0.6 V. ([10] [Open access].)

cooling methods can be hazardous to human health and the environment. They may require additional steps such as post-processing cleaning and increase production costs [10]. In addition, green-cutting technology is a new idea that adapts well to strict environmental laws [13]. The industry can reduce carbon emissions and help create a cleaner, healthier environment by using green-cutting technology. Adopting green-cutting technology can improve the quality of life of industrial workers by reducing their exposure to hazardous pollutants and improving workplace safety. All things considered, green-cutting technology represents a viable option for businesses looking to maintain sustainable practices and meet environmental standards. Another machining method that has gained popularity is dry machining, which has numerous environmental and economic benefits [9]. By reducing machine downtime

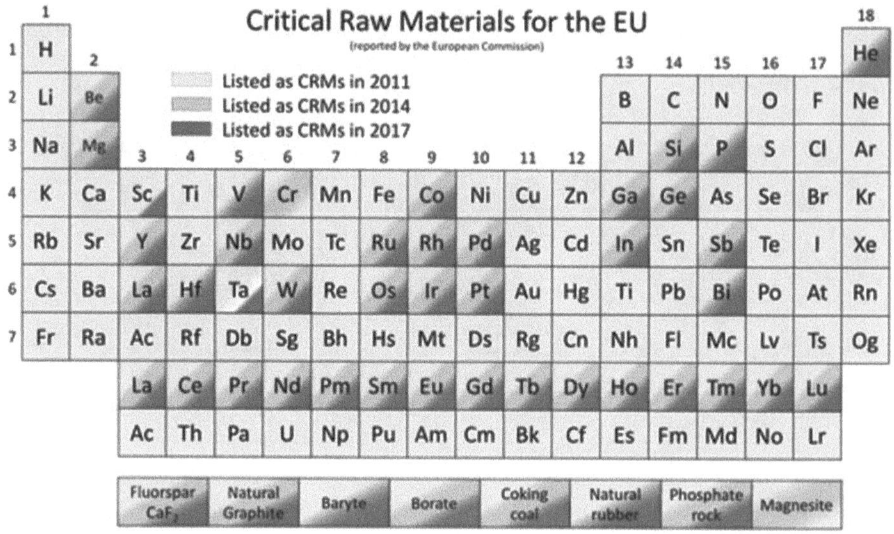

FIGURE 7.13 Periodic table of elements with a list of critical raw materials for the period of 2011–2017. ([14] [Open access].)

associated with fluid changes and cleaning, dry machining can improve productivity and efficiency. Tungsten and cobalt, two essential components of carbide tools discussed above, are [3] recognized as being among the 14 essential raw materials (see Figure 7.13) required by EU companies [14]. Due to their economic importance and risk of supply disruption, they have remained on this list more recently, both in 2014 and in 2017 [3,12].

The main obstacles facing metal cutting today are the creation and implementation of creative solutions designed to firstly reduce cycle times and costs to improve profitability and productivity, as well as increasing productivity and most importantly, and reduce emissions and energy and resources used. The use of hard machining modes (HMC and HPC) is limited by machine performance or tool wear and workpiece machinability such as when cutting lightweight materials or high-temperature alloys. Generally speaking, using high speed can result in reduced cutting forces, power and energy, which is beneficial for machine and tool design. The primary focus of CIRP's activities lies in the domain of high-speed machining (HSM) and related advanced manufacturing technologies. Figure 7.9 demonstrates the range of various characteristics of tool material such as temperature hardness and toughness, which are related to cutting speed and toughness, respectively. The wear marks formed on the surfaces of specimens after the friction test are shown in Figure 7.14 using 3D LSCM and SEM images. Very rough surfaces are formed on the wear marks that are lubricated with pure deionized (DI) water and were dry. It is obvious that the repeated sliding motion damages the graphene layer and insufficient development of lubricating film aggravates the wear. Smoother wear marks and light scratches are formed on the wear mark surfaces when the lubricants containing graphene and DI water are used instead of dry lubrication and pure DI water conditions [13].

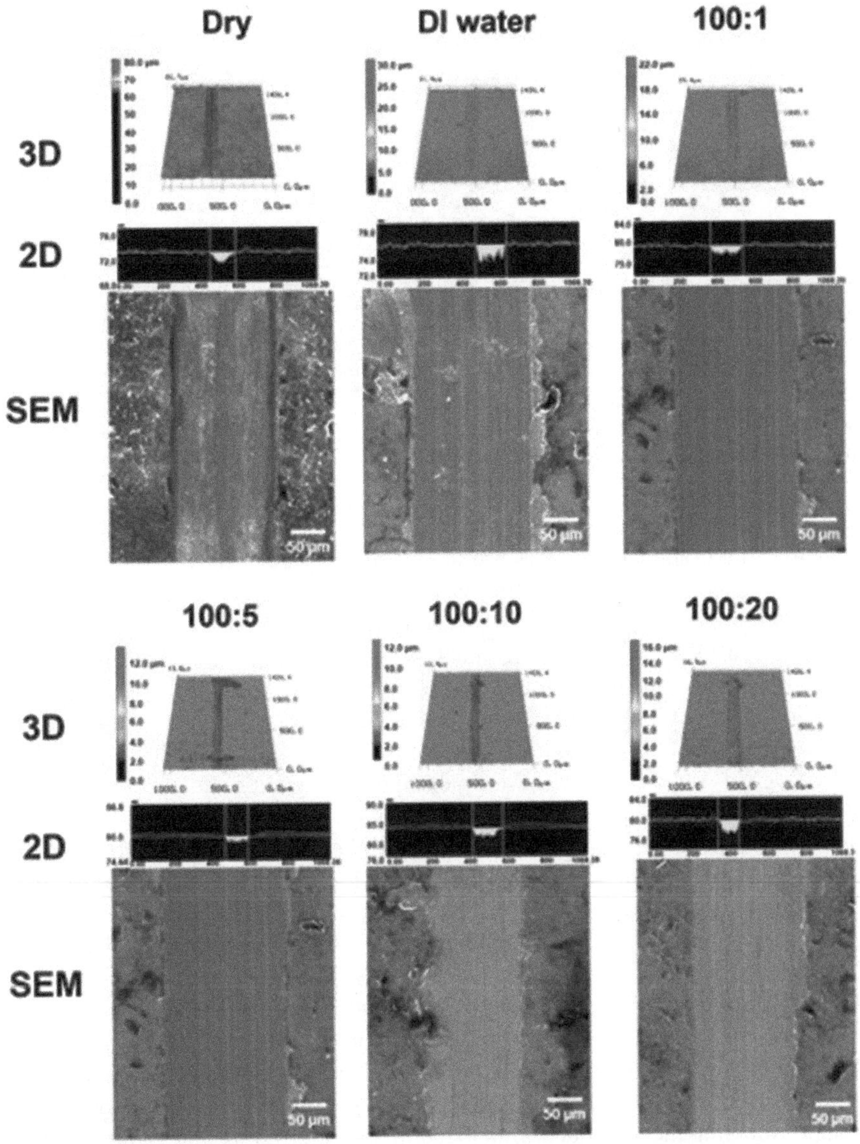

FIGURE 7.14 After 10,000 cycles, wear patterns under 200 mN lubrication conditions are observed using 3D laser scanning confocal microscope and SEM images. ([13] [Open access].)

After sliding motion for 100,000 cycles, wear marks are formed on the graphene-coated surface, as shown in Figure 7.15. When using a DI water-based graphene lubricant, the formed wear mark is the smallest and scratches are formed in the sliding direction. In contrast, the surface inside wear mark is very rough under the dry and pure DI water lubrication conditions. The graphene particles in the graphene solution, which are sprayed onto the surface to generate the graphene-coated sample,

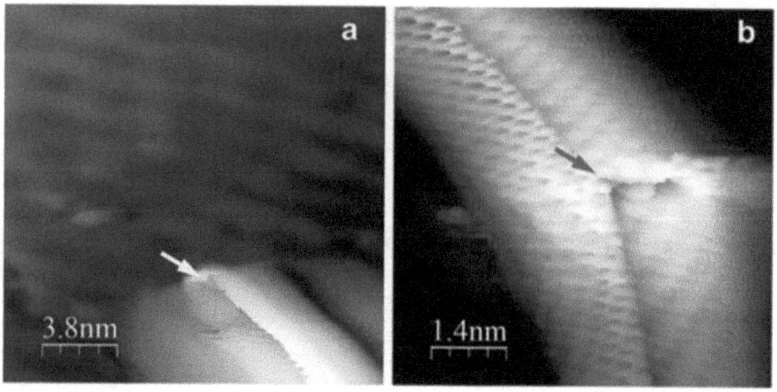

FIGURE 7.15 Photographs of wear scars under 200 mN load obtained over 100,000 cycles using laser scanning confocal microscope and SEM. ([13] [Open access].)

were uniformly distributed. On the other hand, in the case of dry and pure DI water lubrication, the graphene particles are aggregated on the surface of the wear mark and form a discontinuous lubricating layer when they are separated from the graphene coating during repeated sliding with the counter-tip [13].

Typically, after earlier precision machining procedures such as grinding, honing and superfinishing (sometimes called short-stroke honing), abrasive machining operations using bonded abrasive tools are used. However, superfinishing uses fine-grit abrasive tools. The overall goal of honing and superfinishing is to improve the tribological properties of the workpiece, surface quality, dimensional accuracy, and shape. The latter is usually achieved by increasing the surface bearing ratio, which is the ratio of the bearing contact area to the total area of the workpiece and surface roughness parameter R_{vk}, or reduced cavity depth, which affects the lubricity of surface. In addition, subsurface damage to the workpiece caused by previous manufacturing operations such as grinding can be eliminated through cleaning and enhancement.

Bonded honing and superfinishing tools typically use the same identification system as grinding tools, which can be manufactured using super abrasives or conventional abrasive grains. Before initial use, these tools must be prepared, that is, sharpened and profiled. Standard tool preparation procedures for bonded honing and superfinishing include machining using diamond grinding and grinding wheels, silicon carbide bonded grinding wheels or corundum grinding wheels and hardened steel or abrasive coating alignment sleeves. Like grinding, honing involves fixing abrasive grains (or grits as they are most commonly known) in a bonded tool. The main purpose of honing is to give the cylinder bore a smooth surface. The honing stones opposite the hole are subjected to radial outward pressure. Two things differentiate honing from grinding.

First, the abrasive tool moves slowly relative to the workpiece during honing. The surface speed is usually between 0.2 and 2 m/s. The vibrating and rotating actions of the tool work together to provide average material removal across the entire surface of workpiece and create the characteristic "hatch" pattern that is preferred for

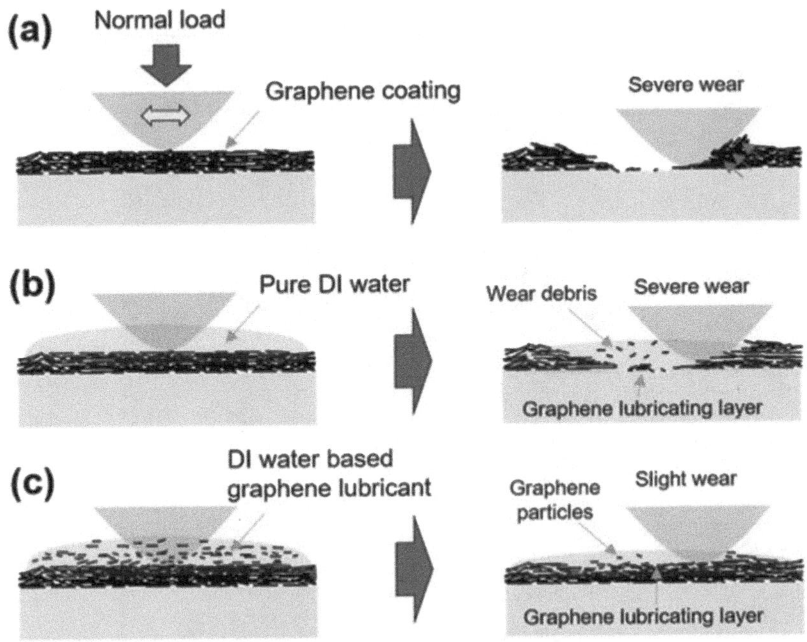

FIGURE 7.16 Schematic representation of the graphene coating corrosion mechanism in the following scenarios: (a) dry, (b) with pure DI water, and (c) with DI water-based graphene lubricant. ([13] [Open access].)

retaining oil in engine cylinder bores. Another difference between grinding and honing is that the tool is freely positioned with the workpiece surface. This indicates that the hole eccentricity relative to outer diameter cannot be adjusted. On the other hand, in the case of clean and dry DI water lubrication, graphene particles accumulate on the wear track surface and form a discontinuous lubricating layer when they are separated from the graphene coating by repeatedly sliding with counter-tip. Figure 7.16 illustrates the wear mechanism of graphene coating in clean, dry and mixed DI water lubrication scenarios. Honing is a crucial step in the production of highly functional surfaces, especially for hydraulic components and automotive industry. The shape of cylinder block holes, dimensional accuracy, and surface quality (Figure 7.16) have a significant impact on engine characteristics such as fuel consumption [13].

Abrasive machining techniques called honing (long-stroke honing) and superfinishing (short-stroke honing) are used to improve the dimensional and shape accuracy, surface quality, and tribological performance of the workpiece. By increasing the surface bearing coefficient and reducing the cavity depth Rvk, which affects the lubrication properties of the surface, the tribological properties can be improved. In addition, honing and superfinishing can be used to eliminate subsurface damage to the workpiece such as grinding burns caused by previous manufacturing operations. Thus, honing can be used to maximize the surface integrity of the workpiece, including microhardness, residual stresses, and lattice structure. Honing is

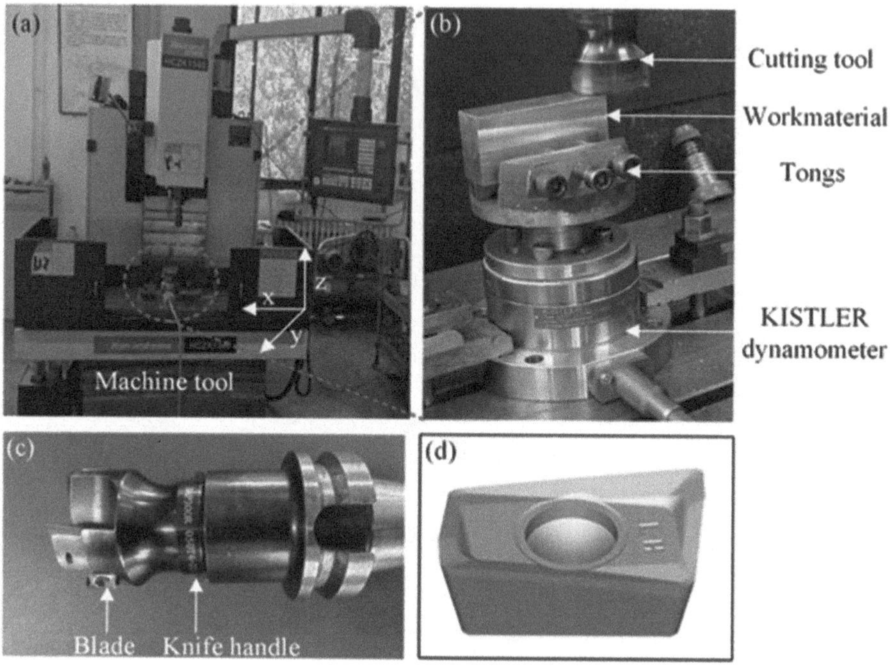

FIGURE 7.17 Schematic diagram of the test setup. Tool handle (c), blade (d), KISTLER dynamometer (b), and machining center (a). ([15] [Open access].)

divided into sub-processes such as external and internal honing, surface honing, gear honing (internal and external) and form honing, as well as the distinction between long-stroke and short-stroke honing (superfinishing). Some techniques have been developed to improve honing efficiency such as high-speed honing and spiral slide honing. The computer numerical control (CNC) machine tool is developed by Zhengdang Precision Machinery Co., LTD., Shenzhen City, Guangdong Province, China. As shown in Figure 7.17a, the tool handle is made of Winon BAP300R 050T4–22 WINIO CNC cutting head from Taiwan Province, China; as shown in Figure 7.17c, the tool is made of APKT11T308-LH milling cutter from Diamond Factory, Zhuzhou City, Hunan Province, China; and as shown in Figure 7.17d, the force gauge is installed on the CNC machine tool and connected to the workpiece via a fixture, as shown in Figure 7.17b [15].

The growing need for faster and more efficient supply networks has led to the development of robotics and automation as critical supply chain management (SCM) technologies. By reducing long-haul costs, increasing work and utilization efficiency, reducing errors, reducing repetitive inventory checks, updating orchestration, taking care of time and gathering information in problem and dangerous regions, robotics and automation are improving SCM. Through the use of computer software, other technologies and autonomous physical devices, automation helps perform jobs that are often done by humans. Although artificial intelligence and machine learning are widely recognized as decision support tools, SCM has rarely ever used them. This

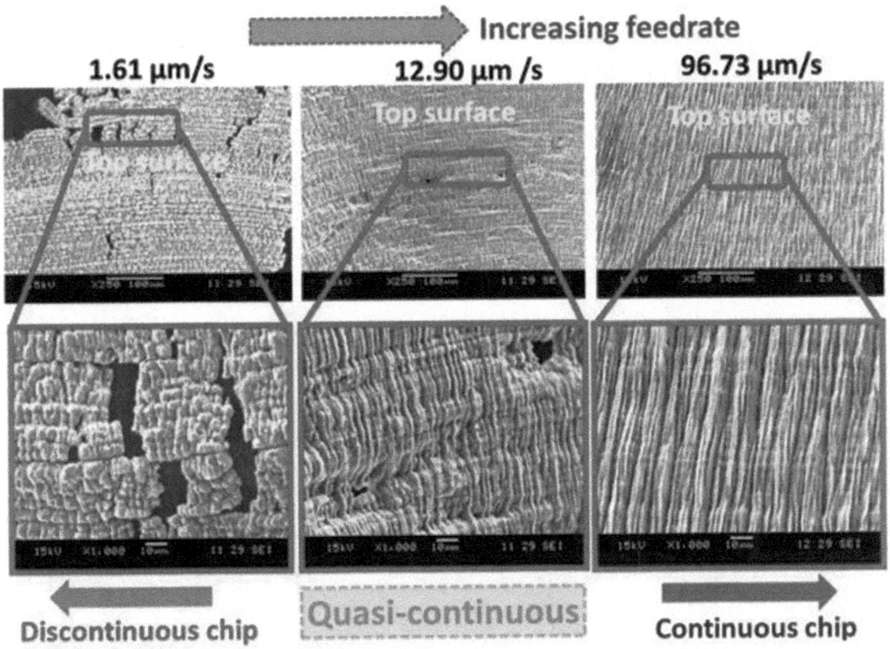

FIGURE 7.18 Chip macroform. (a) Chip variation with the cutting speed. (b) Chip variation with feed. (c) Chip variation with the axial depth of the cut. (d) How chip changes as radial notch depth increases. ([15] [Open access].)

study covers several major subfields of artificial intelligence and machine learning. Using the Japanese KEYENCE ultra-deep field microscope to view the chips, the stability of milling process is confirmed in each milling experiment by observing the chip shape and tool wear. Figure 7.18 illustrates how changing the settings changes the macroscopic shape of the device [15].

They are ideally suited to solve practical problems associated with SCM to maximize the potential benefits of artificial intelligence and machine learning for SCM. In doing so, this study examines the history of successful implementations of artificial intelligence and machine learning in SCM and identifies the most productive areas of SCM for technological applications. ZL205A aluminum alloy will produce chips continuously when the cutting requirement of test structure is met. The macroscopic change of chips is not obvious even with increasing of cutting speed (Figure 7.19a). The cutting speed and the amount of mass removed increase as feed increases (Figure 7.19b) and cutting depth (Figure 7.19c and d). Therefore, the chip length increases, and length deformation ratio decreases when the protruding chip length in the cutting zone increases (Figure 7.19b). The distribution of cutting force and chip formation behavior in cutting zone depend on the increasing of cutting force and heat generated with increasing of axial cutting depth. The tool wear is shown in Figure 7.19, but Figure 7.19a compares the new blade with blade used in milling test (Figure 4b). All four blades are stable [15].

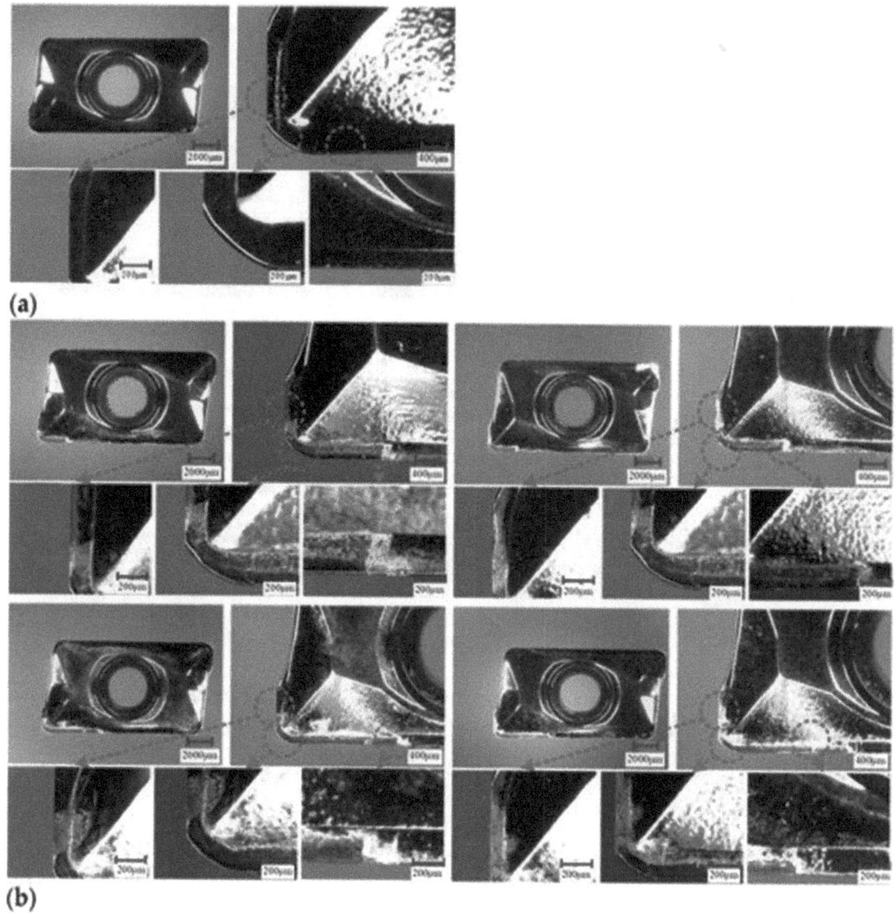

FIGURE 7.19 Tool wear: (a) new blade and (b) four blades on a cutter blade. ([15] [Open access].)

7.3.1.1 Hybrid Cutting

To improve the accuracy of temperature prediction in machining processes, a new hybrid cutting model has been developed to predict the workpiece temperature distribution during dry grinding of compacted graphite cast iron (CGI), which combines several methods. The hybrid model consists of a finite element analysis (FEA) based on the thermal model and analytical force model based on the mechanistic approach. The advection heat distribution model is used to calculate the amount of heat generated during the milling process and transferred to the workpiece. The temperature field's development during laser cutting is shown in Figure 7.20 [16].

The cutting forces are used to calculate the temperature distribution of the workpiece, which is obtained using thermal loads and used as boundary conditions in FEA. The constructed power and thermal models are tested experimentally and good agreement between the calculated and measured results that is obtained. Energy and

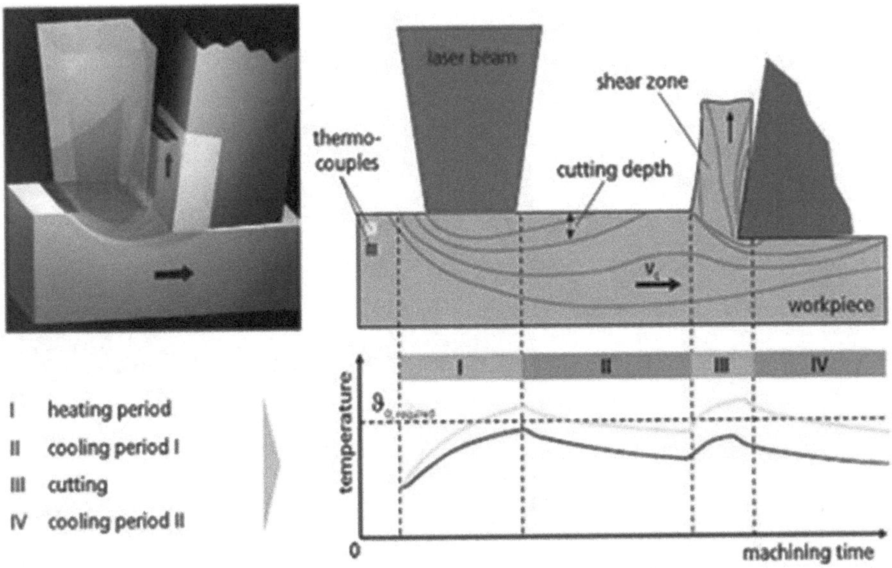

FIGURE 7.20 Development of the temperature field during laser cutting. ([16] [Open access].)

active force calculations show that by doubling the feed during CGI milling, nearly doubling the cutting forces is achieved, although only about 10% of the energy is used [3]. However, thermal cutting tools such as arc plasma torches inevitably produce radioactive aerosols and gaseous vapors when cutting radioactively contaminated metal structures, increasing the risk of radiation exposure to workers [10]. The most common type of drill bit for deep geological drilling of hard rocks is the impregnated diamond drill bit, which meets the requirements of highly abrasive formations due to rock-breaking and self-sharpening processes [10,9,13,16–19].

Its rock-breaking method is comparable to the action of grinding wheel on a piece of material. Figure 7.21 shows the rotational model of rock breaking by the impregnated diamond drill bit. It involves crushing and breaking the rock at the bottom of borehole with a large number of small diamond particles on the surface of drill bit edge [16]. In addition, molten materials can interfere with the cutting process [12]. In addition, strong arc currents, 200–900 when cutting thick metals [12,13], are detrimental to the service life of plasma torch, increasing the cost and duration of dismantling. Thus, using only an arc plasma torch for dismantling radioactively contaminated metal structures is ineffective [1–4,11]. The stress cloud pattern indicates a significant stress concentration on the front and back surfaces of diamond blade as it moves from point 2 to point 8, which dissipates the stress waves and enlarges rock-crushing hole.

A complete annular crushing pit is formed at point 9, where the stress value decreases significantly and pit size gradually increases with rotation (Figure 7.22). The next section discusses the energy used by diamond to crush a unit volume of rock per revolution and physics of rock crushing to shed light on this phenomenon,

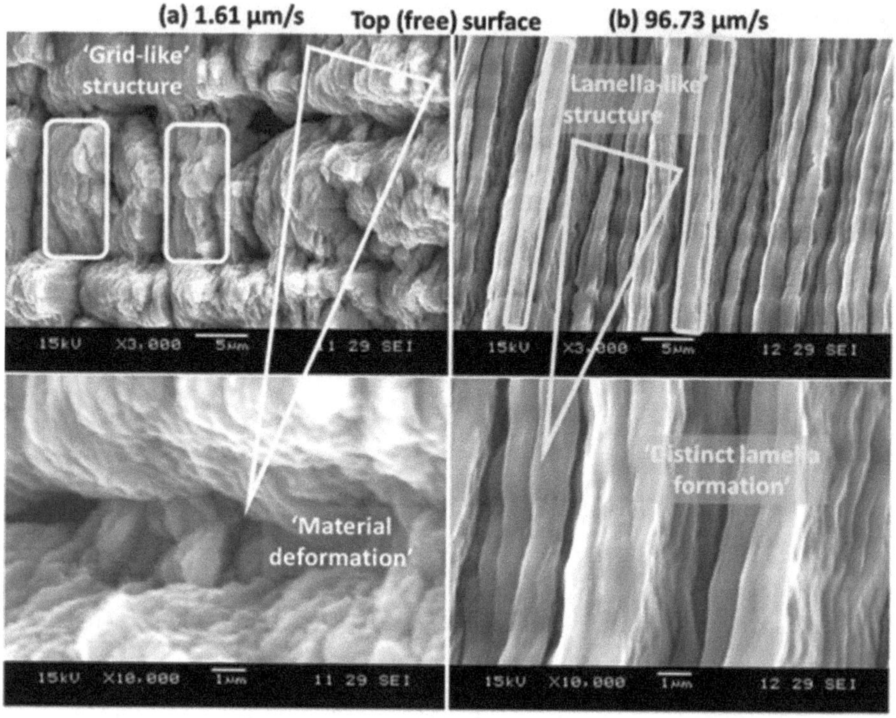

FIGURE 7.21 Schematic diagram of the rotary drilling technology using an ID bit. ([16] [Open access].)

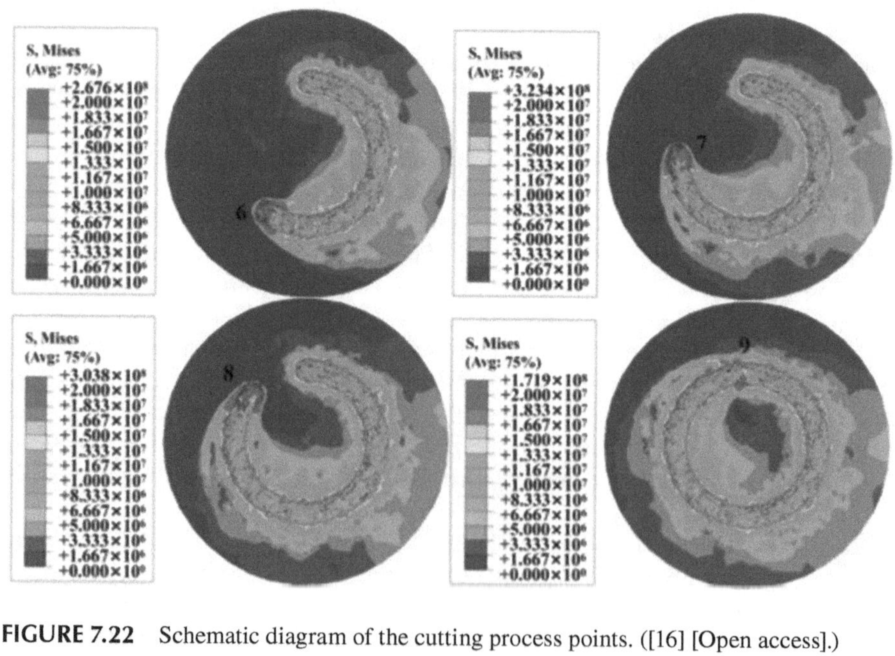

FIGURE 7.22 Schematic diagram of the cutting process points. ([16] [Open access].)

taking into account the significant decrease in mean squared error (MSE) in stage II (Figure 7.22a) [16].

Milling cutters, such as end mills, emit very little aerosols compared to arc torches and other mechanical tools except for easily removed chips making them suitable for safely cutting metal structures contaminated with radioactivity [4]. Additionally, since the final cutting pass splits the inner surface of workpiece [4], these mechanical cutting tools can be useful for segmenting metal structures such as pipes and tanks whose internal layers are radioactively contaminated. Despite these clear advantages, cutters have not found widespread use in nuclear power plant dismantling except for a few specialized applications such as pre-preparation due to high time and tooling costs, especially when using tools without cutting fluid [9].

7.3.1.2 Chemical Vapor Deposition

Chemical vapor deposition (CVD) reactors must be built with the ability to control several aspects of the film, including composition, surface morphology, thickness and crystal structure. The direction of gas flow to the substrate holder (susceptor) can be used to distinguish between horizontal and vertical systems. Reactors designed for air and low-pressure deposition have been specially designed. Using radio frequency, which is conductive, the substrate holder is heated by radiation. Rotating substrate holders are A typical CVD system as shown in Figure 7.23, where reactive gases, known as primary gases, which are introduced into a chamber at an appropriate temperature. The gases flow through the reactor and come into contact with the heated substrate. After their reaction with the substrate, a solid layer of ceramic coating with a specific property such as Al_2O_3, Cr_2O_3, or SiC is deposited on it. It should be noted that argon or another neutral gas is used as a diluent in the CVD process [20].

As mentioned earlier, the two limiting parameters in this process are the deposition temperature and pressure designed to improve film uniformity and flow dynamics. Large-scale CVD production is possible, so reactors such as barrel reactors, pancake reactors, and spinning disk reactors have been built that allow simultaneous deposition onto multiple substrates. The selected deposit is created on the surface of the substrate as a result of reaction or destruction of one or more volatile precursors when the substrate is exposed to them by CVD.

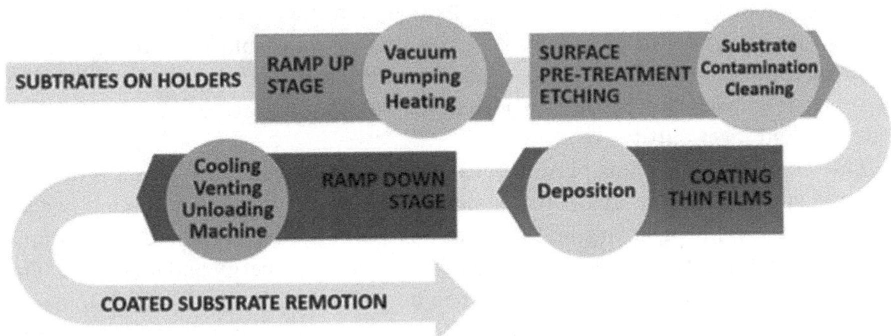

FIGURE 7.23 Physical vapor deposition flow chart. ([20] [Open access].)

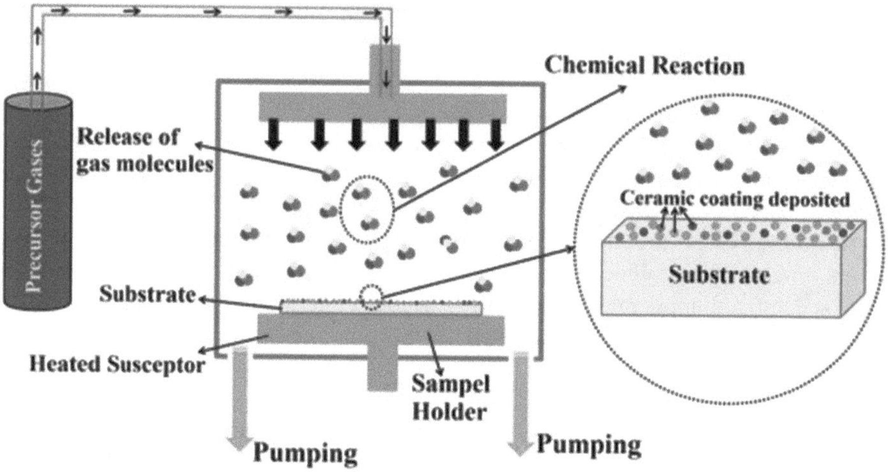

FIGURE 7.24 Illustration of CVD used to coat ceramic nanocomposite materials. ([17] [Open access].)

In addition, volatile by-products are formed and ejected from the reaction chamber by gas stream [7]. System conditions play an important role in determining the process success. During this procedure, the temperature in the area varies from 800°C to 2000°C. Thanks to this method, excellent adhesion and coating thickness of up to 20 μm are achieved [3,11]. Characterization techniques such as SEM and X-Ray diffraction (XRD) concluded that the pore surface characteristics changed as a result of CVD nanocoating (Figure 7.24) [17].

7.3.1.3 Plasma Arc

The electrically conductive material is heated, melted, and cut into unique patterns and designs using a plasma torch. This manufacturing technology is known as plasma cutting, sometimes known as fusion plasma cutting or plasma art cutting. Plasma cutting can cut a wide range of metal materials such as copper, aluminum, alloy steel, and structural steel, which can cut materials that are 0.5–180 mm thick. Rubber, plastic, and other non-conductive materials are not suitable for plasma cutting. There are different types of plasma cutting such as water injection plasma arc cutting, secondary plasma arc cutting, and standard (or traditional) plasma arc cutting. Alternatives to plasma cutting include oxy-fuel, waterjet, and laser cutting.

The technique, known as plasma cutting, uses a jet of hot plasma that is accelerated and cuts electrically conductive materials. When an electrically conductive gas, such as oxygen, nitrogen, argon, or a combination of gases, is ionized and forms a plasma arc, it binds to the material being cut, transferring energy and melting it. This is exactly how plasma cutting works. When a gas is heated to such a high temperature that electrons are stripped from the atoms, creating a mixture of free electrons and ions (atoms that have gained or lost electrons), a unique state of matter known as plasma is formed. Ionization is the process by which atoms lose their electrons.

Three states of matter are most mentioned: solid, liquid, and gasoline. However, plasma, which is found both inside and outside the galaxy (for example, in the centers of stars and form of lightning), is sometimes called the fourth state of matter. From a technical view point, plasma is the most common state of matter, both by volume and by mass. When more energy is given to a gas, the molecules can move faster and collide with greater force, resulting in plasma.

Molecules can be separated into their constituent atoms by collisions, and these individual atoms can be separated from their outer shell electrons to form ions. The gas reaches a critical point where the number of atoms gaining free electrons and the number of atoms losing electrons balance out as more atoms lose their electron [13]. Figure 7.25 shows a schematic illustration of the plasma arc process. Sylvain Coulomb and Cao Qin demonstrated how to create organic layer-coated metal nanoparticles. Comparing substrates with low critical process spray parameter (CPSP) and high CPSP coatings, the latter was found to exhibit the least wear resistance [18].

The indirect plasma arc system is a type of arc plasma. Without being connected to the workpiece, the copper nozzle functions as cathode to positive terminal of DC power supply. The electrodes themselves create the electric arc. Through the nozzle, plasma gas creates an arc and discharges at high temperature and velocity. The plasma jet can reach a maximum temperature of 11,726°C and a maximum velocity of 600 m/s. This system uses different electrode geometries to create the plasma, with a hollow cylinder or at least two coaxial cylinders serving as anodes and a cylindrical rod serving as the cathode. A water-cooling system attached to the electrodes maintains cooling. The output powers of indirect arc torches range from 1 to 6 MW. The reactor system is shown in Figure 7.3b, while a non-transferred arc system is shown schematically in Figure 7.25a [18].

To further reduce zirconium at low power and short processing time, a low-power plasma arc transfer cutting torch is used. Low power plasma arc transfer method is a welding or surface coating technology that uses a low power plasma arc to transfer molten material (such as metal powder or wire) to the surface of a workpiece to modify the surface of the material or improve its properties, such as wear resistance or hardness. Since the recovery of the base metal has economic value (i.e., cost), plasma arcs have been used for the recovery of titanium scrap and metal refining since the 1970s. Electric arc furnaces (EAFs) are widely used in the steel industry, but they pollute the environment. Commercial applications of plasma process include recovery of iron and zinc from electric arc furnace dust, aluminum from aluminum production slag, and copper alloys from copper foundry slag [21,22].

The same basic ideas that govern plasma welding apply to plasma cutting. A stream of hot ionized plasma is ejected from a converging nozzle at high speed to melt the base metal. The molten metal is then ejected to create the cut. This chapter will briefly cover several pieces of plasma cutting equipment including the control panel, torch, plasma gas, and shielding gas selection. The basic operating principles of a programmable automation controller (PAC) will be briefly covered later.

The basic terms associated with plasma cutting are shown in Figure 7.26. When the electrode arc passes through an orifice, it converges. When the arc and orifice gas come into contact, it quickly raises temperature to high level. Consequently, it expands to larger volume. It accelerates as it passes through the converging nozzle.

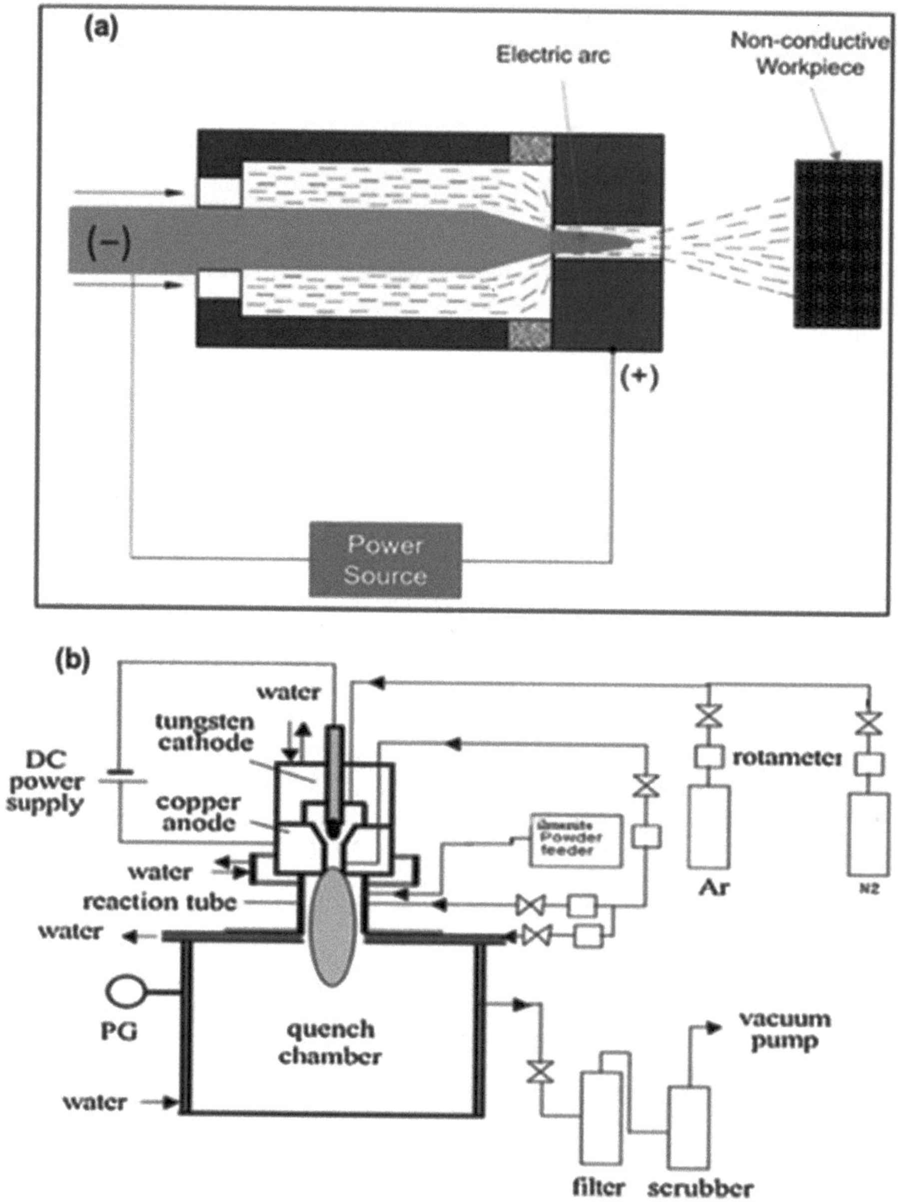

FIGURE 7.25 Schematics of the non-portable system (a) and non-portable plasma reactor (b). ([18] [Open access].)

We call this a plasma gas. The diameter and shape of the converging nozzle, the gas inlet pressure, the current at the electrode inlet, and other factors affect the temperature and velocity of plasma gas. Thermal plasma finds application in the production of nickel-based powders, niobium-based superconducting alloys, titanium alloys, and various other special metal applications [23,24].

It consists of water-cooled crucibles and copper melting hearth. The red clay is melted in a 35 kW DC expanded thermal plasma reactor since it contains 30%–50% iron. For a 350 g batch, the maximum iron recovery is 71% with an optimum melting time of 15 minutes [25]. A SEM image of iron trapped in the reduced red clay slag during melting is shown in Figure 7.26. Expanded graphite flakes are scattered throughout the iron matrix in the as-cast microstructure (Figure 7.26a). The SEM image of etched cementite-pearlite interdendritic structure is shown in Figure 7.26b [18].

7.3.1.4 Flame Synthesis

An industrially scalable method for creating nanomaterials is called flame spray pyrolysis (FSP), which can produce a wide range of metal or metal oxide nanoparticles with specific physicochemical properties [17]. The complex process architecture, including precursor atomization, combustion, and nanoparticle production is the basis of FSP technology [18]. This enables rapid, one-step synthesis and eliminates the need for post-production modifications that are necessary with other approaches.

The importance of FSP has increased in the creation of nanodevices, in which nanoparticles can be inserted in situ into complex functional arrays [19]. Consequently, FSP is not only transforming the manufacturing of nanomaterials, but with recent advances, it is opening the door to development of next-generation nanodevices [26]. So, we offer a revised summary of the current state of the art in nanodevice development, new materials synthesis, and innovative reactor and process projects in FSP technology.

In the field of nanotechnology, the synthesis of nanostructures is an important area of research that includes a wide range of techniques in addition to FSP. These various methods, including electrospinning, sol-gel processing, and CVD, provide specific qualities in terms of shape, chemical composition, and particle size control. The choice of an appropriate synthesis technique depends on several factors directly related to the intended use of the nanostructures. When assessing the suitability of the method for a variety of applications, from solar devices to drug delivery systems, it is important to consider factors such as temperature range, synthesis time, environmental impact, and material adaptability. Several synthesis methods used to create nanomaterials (including FSP) and describe the specific standards applicable to their use. This careful consideration ensures that the synthesis procedure meets the functional specifications of the end application, thereby optimizing the efficiency and practicality of generated nanostructures. An alternative to FSP is atomic layer deposition (ALD) [23], especially for thin film technology. ALD provides atomic-level precision and high levels of consistency, resulting in extremely stable and flawless films. Disadvantages include the need for expensive controlled environment equipment, slower deposition rates, and possibly significant precursor prices [18].

7.3.1.5 Laser Ablation

One method for creating several types of nanoparticles is laser ablation. These include core-shell nanoparticles, semiconductor quantum dots, CNTs, and nanowires. Using this method, particles subjected to laser evaporation in a background gas are nucleated and multiplied to form nanoparticles. It is beneficial to produce

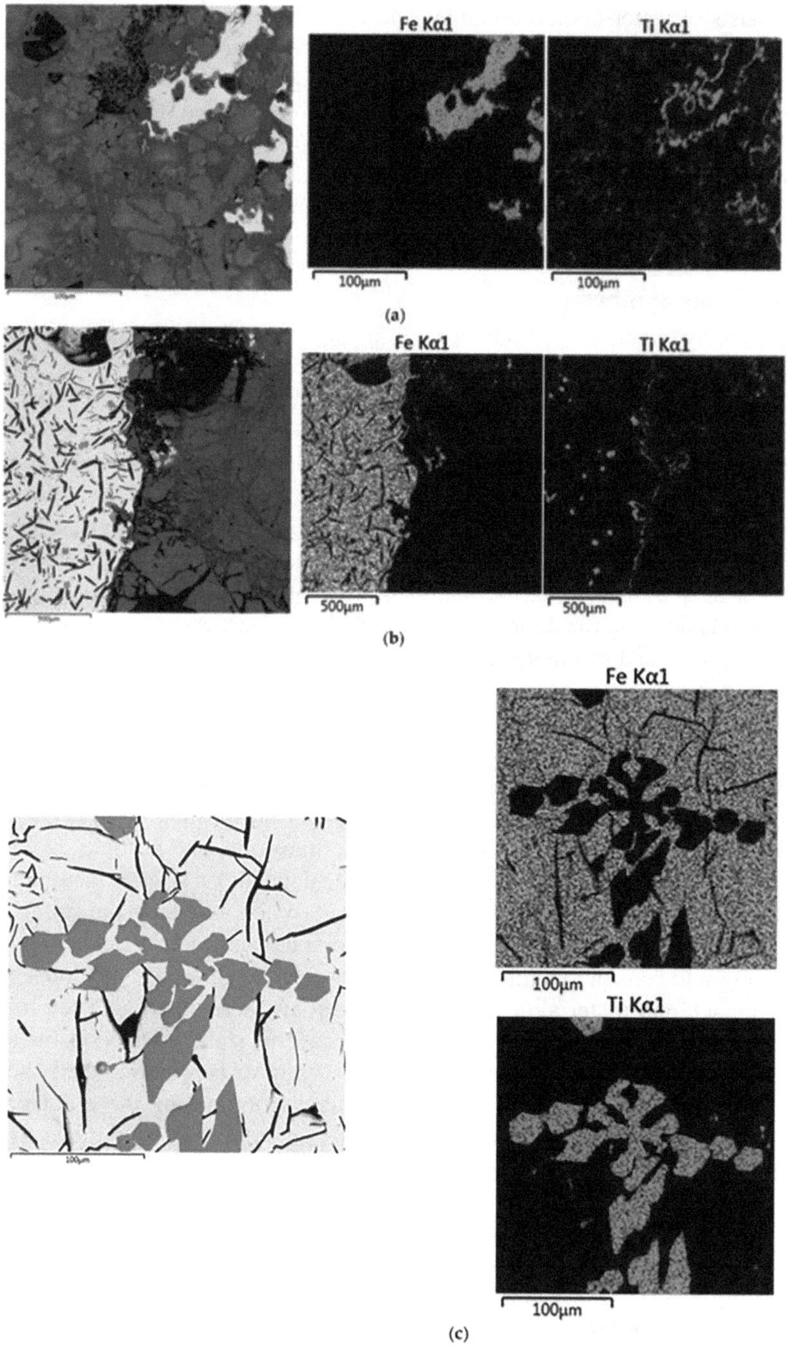

FIGURE 7.26 SEM images of slag and cast iron samples: (a) titanium carbide in cast iron, (b) reduction smelting at 1750°C, and (c) reduction smelting at 1650°C. Fe and Ti distributions are shown in the maps. ([18] [Open access].)

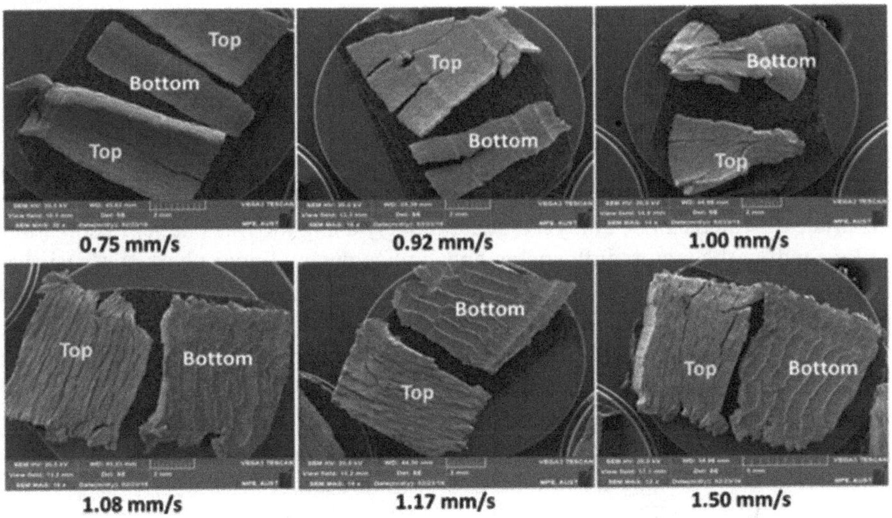

FIGURE 7.27 SEM surface structure. Three-dimensional morphology: (a2) 42CrMo, (b2) HT, and (c2) UB samples; and (a1) 42CrMo, (b1) HT, and (c1) UB samples. For UB samples, the EDS spectrum is located to the right of (a1,b1,c1), respectively. ([27] [Open access].)

high-purity nanoparticles in the quantum size range (<10 nm) by extremely rapid steam cooling. This provides an overview of process of creating nanoparticles using laser ablation. Next, we review recent developments in nanoparticle size control and challenges of synthesizing functional nanoparticles using modern laser ablation technologies. Figure 7.27 illustrates the 3D morphology of three prepared samples and SEM images of the surface [27].

Three specimens [control, heat-treated (HHT), and ultrasonic bombardment (UB)] are subjected to visual inspection and surface roughness measurement. The control and HHT specimens, which exhibit distinct peaks and valleys indicating noticeable cut marks, are shown in Figure 7.27a and b. The UB specimen, on the other hand, has a flat surface without discernible cut lines (Figure 7.3c), resulting in noticeably lower roughness values of R_a1.19, R_a1.12, and R_a0.17, respectively. Interestingly, compared to the HHT specimen, the roughness is reduced by 79.2% after UB treatment. The tip of the polishing tool, which facilitates the movement of near-surface components from the peaks to valleys, is responsible for the smoothing effect of the UB specimen. In addition, sporadic black spots found on the surface of each sample are identified as pores and carbides by energy dispersive X-ray spectroscopy (EDS) analysis, showing a higher concentration of carbon near these locations. It is important to note that the elemental composition of heat treatment (HT) and UB processed samples did not change, as UB uses plastic deformation to improve the surface without changing the composition of material. Here, the segregation of carbides is mainly responsible for the decreasing in carbon content in the UB-treated specimen [22]. At the same time, due to the material flow of UB, there are much fewer black spots in the UB-treated specimen than in the HT-treated specimen because some pores on the nearby surfaces that formed as a result of laser cladding are filled [5].

7.3.2 Method of Liquid Phase

Another method for creating phosphorus compounds based on silicates is liquid-phase synthesis. This procedure can produce approximately 90% more transformation than the high-pressure process, which produces 20%. Solvents are widely used in liquid-phase synthesis as reaction media. The response speed and system can vary significantly depending on the range of solvents. The effect of solvation is the term for this process, which can be analyzed quantitatively using the free energy of solvation. Understanding the factors influencing appearance during liquid-phase synthesis has received much less attention than gas-phase synthesis methods. Using the method of liquid-phase precursors, a green-emitting phosphor $(Ba_{1-x},Sr_{1-x})SiO_4{:}2xEu^{2+}$, where $x=0.03$, 0.05, 0.1, and 0.15, is created [28]. When phosphors are produced using a liquid-phase precursor, their emission intensity is reported to be significantly higher than other synthesis methods. The most widely used film is SiO_2, which is easily hydrolyzed by metal, boric acid, or aluminum. In addition to basic solutions, extremely acidic solutions are formed when creating a thin film.

7.3.2.1 Electrochemistry

Electrochemistry is the study of chemical processes involving the separation of charges, usually in liquid media such as solutions. Charge transfer, which can occur heterogeneously on the surface of electrodes or homogeneously in solution between different chemical compounds, is associated with charge separation. Thus, it is immediately obvious that its application is very wide and will be discussed later.

At least two charge transfer half-reactions, oxidation (loss of electrons or increase in the oxidation state) and reduction (gain of electrons or decrease in the oxidation state), occur simultaneously and in opposite directions to ensure electrical neutrality. Oxidation and reduction half-reactions are spatially separated at electrode surfaces and occur at different electrodes immersed in solution within the same cell or in different compartments of cell. The electrodes are connected by conductive paths that allow the transfer of charge and completion of an electrical circuit both externally (via electrical wires) and internally (via ion transport). Batteries and fuel cells can be used to capture the electrical energy released when the total number of Gibbs energy changes at both electrodes is negative. If the difference is positive, external electrical energy can be used to counteract the positive Gibbs energy and drive electrode reactions that convert chemicals (electrolysis). Therefore, as in industrial electrolytic reactions, the products of two-electrode reactions can be separated if the cell design allows it [19].

7.3.2.2 Precipitation

The most popular method for extracting dissolved (ionic) metals from solutions such as hazardous metal-containing process wastewater is chemical precipitation. A chemical reaction between the soluble metal compounds and precipitating reagent converts the ionic metals into an insoluble form (particle). By settling and/or filtration, the particles resulting from this process are removed from solution. Because metal hydroxides are amphoteric, they become more soluble at both high and low pH levels. Thus, the metal may dissolve or begin to reenter the solution depending on ideal pH for that metal to precipitate. Precipitation of these various metals as hydroxides can be a complex procedure since most process wastewaters contain mixed metals.

Another way to think of interdomain deposition is a crystal stacking defect. The interface between two consecutive concentric domains, where the growth cycle repeats to initiate a new domain after deposition is called the "interdomain" region. Atomic force microscopy (AFM) phase images and height profiles along white line corresponding to interdomain defects in extinction bands of P12T thin films at two thickness levels, 1.5 and 3.5 μm, are shown in Figure 7.28. At the beginning of the stacking cycle, terrace-like single-crystal lamellae are packed as thick scale-like structure before depositing into nearly empty valley near substrate glass surface. The sharp/narrow optical extinction is caused by the empty valley, which is devoid of any observable crystals. The end of the extreme edge about to precipitate is one side of the interface of the interband, which is densely packed with twisted single-crystal lamellae, while the empty valley about to start a new cycle on the other side of the interface is smooth and free of crystals. In the step-shaped lamellae at the beginning of the banding cycle, height differences of 10–20 nm (Δz) are observed. However, the vertical ranges of precipitation height across two successive bands at the end of

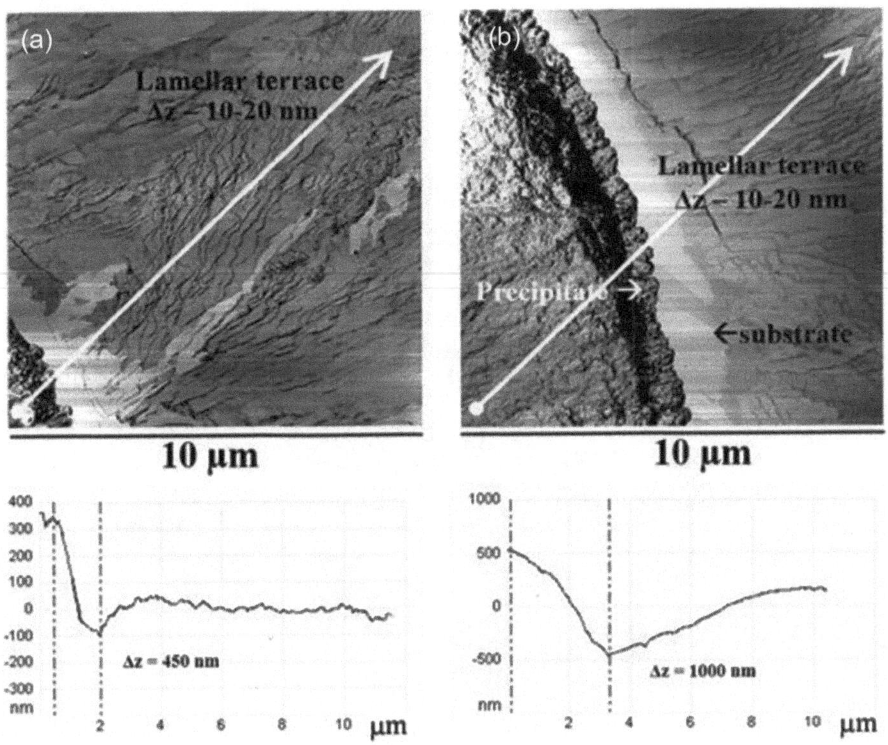

FIGURE 7.28 Thin film of P12T crystallized at 110°C yielded AFM phase images and interdomain defect height plots in extinction bands. Film thickness is (a) 1.5 μm and (b) 3.5 μm. ([29] [Open access].)

banding cycle are determined to be 450 nm (for the 1.5 μm sample) and 1000 nm (for the 3.5 μm sample). The locations of empty valleys for two samples are shown by dotted lines in the profile images [29].

In this method, the typically required unit steps are neutralization, sedimentation, coagulation/flocculation, separation of particles from liquids and dewatering. The efficiency of the chemical precipitation process depends on several variables such as the type and number of ionic metals in the solution, precipitant used, reaction conditions (especially pH of solution), and presence of additional substances that may interfere with precipitation reaction. The most widely used chemical precipitation method is hydroxide precipitation, known as pH precipitation, which uses precipitants such as sodium hydroxide (caustic) or calcium hydroxide (lime) to create metal hydroxides. The ideal pH for the precipitation of hydroxides of each dissolved metal ranges from 7.5 for chromium to 11.0 for cadmium cyanide. Lead(II) iodide precipitate is formed when potassium iodide combines with lead(II) nitrate. Different liquid-phase methods for producing nanocoatings are shown in Figure 7.29. Nucleation is a crucial step in the deposition process. The effectiveness of agar-based aerogels as supercapacitor materials was demonstrated by preparing nitrogen-doped porous carbon aerogels. They were synthesized by a sol-gel process using a cross-linking agent that improved the structural integrity and electrochemical performance of the resulting materials [30]. The agar aerogels have showed different morphologies of their structures. In this case, the urea/potassium hydroxide-treated agar aerogel has a porous carbon

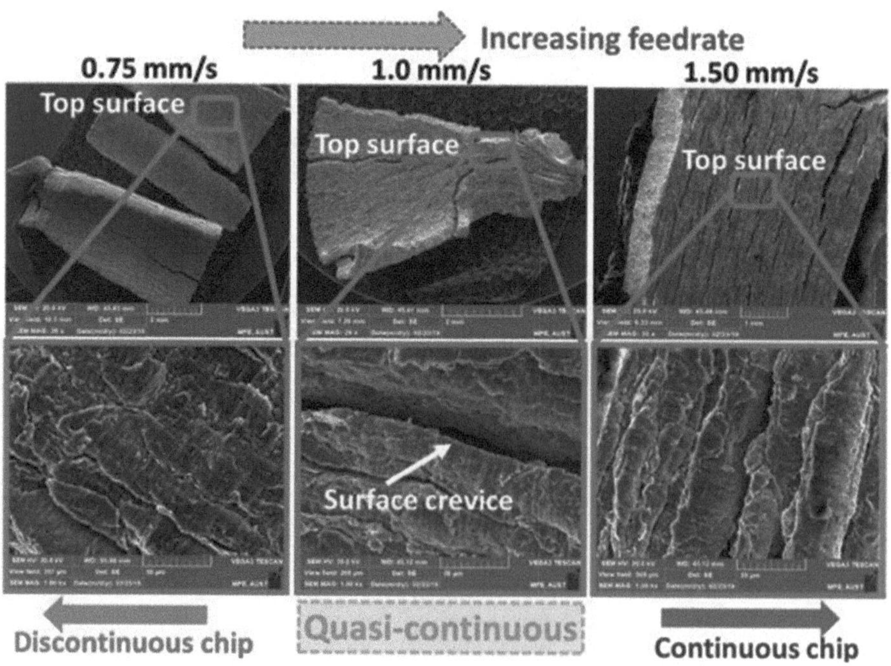

FIGURE 7.29 Varieties of liquid-phase techniques used to create nanocoatings. ([20] [Open access].)

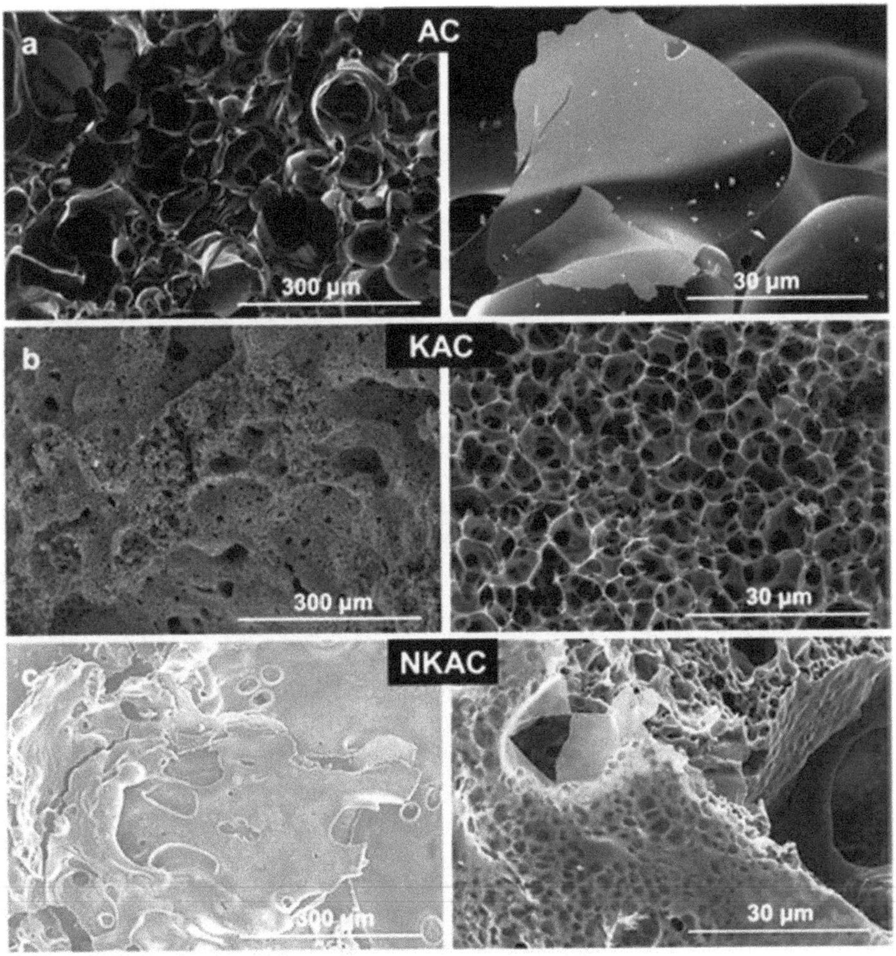

FIGURE 7.30 SEM of agar aerogel (a), urea/KOH-treated agar aerogel (b), and KOH-treated agar aerogel (c). SEM of nano-knurling-assisted cutting (NKAC). ([20] [Open access].)

network with a Brunauer–Emmett–Teller (BET) surface area of 2000 m^2/g, while the agar and potassium hydroxide-treated aerogel consisted of three-dimensional microporous cross-linked carbon networks (Figure 7.30).

7.3.2.3 Sol-Gel

One of the common wet chemical methods for applying nanocomposite films is the sol-gel procedure. Material sources that are first prepared in this process included metal oxide and alkoxide precursors, metallic nanoparticles, tetraethoxysilane for the silica matrix, stabilizers, catalysts, and other porosity additives. After this, the sol undergoes processes of polycondensation and hydrolysis, which gradually lead to the development of network resembling a gel. The fundamental morphology or structure of the solid phase can take several forms such as continuous

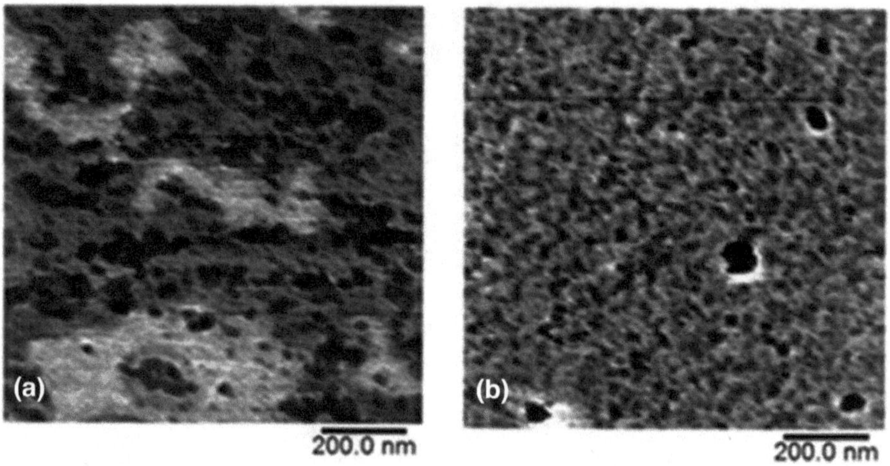

FIGURE 7.31 AFM pictures of hybrid composite nanocoating in 24-hour salt spray tests for functionalized (b) and non-functionalized (a) silica coatings. ([20] [Open access].)

polymer networks or discrete colloidal particles. Based on the obtained results, it can be concluded that the matrix of anticorrosive coating is formed by adding silicon nanoparticles, since it increases the homogeneity and reduces cracking (Figure 7.31). However, after the functionalization of silicon nanoparticles using glycidyloxypropyltrimethoxysilane (GPTMS), the glycidyl group creates a stronger interface matrix of nanoparticles, which prevents aggregation, reduces conductivity, and enhances anticorrosive properties, which is associated with improved barrier properties of the hybrid composite [20].

Figure 7.32 demonstrates the synthesis of tungsten trioxide [20] and its formulation. Cyclic voltammetry is used to study the dopamine sensing properties of fabricated WO_3, and the material demonstrated excellent dopamine sensing with outstanding stability. The chemical processes of sol-gel formation are depicted in Figure 7.24 and include the stages of hydrolysis, condensation, gelation, aging, drying, and compaction.

Then, the addition of an acid (e.g., hydrochloric acid) or a base substance (e.g., potassium hydroxide) activates the precursor. Subsequently, the activated precursors combine to form a network. With time and temperature, the network expands and ages until it reaches the size of container. From this point, the viscosity of liquid increases exponentially until gelation occurs [31]. The structure of gels obtained in two different modes is shown in Figure 7.33. Several variables, including pH, solvent, temperature, time, catalyst, and agitation, affect sol-gel chemistry [20].

7.3.2.4 Hydrothermal

Hydroxyapatite (HAp) is a naturally occurring form of calcium apatite that is an inorganic component of bones. About 80% is minerals, 22% is protein, and 8% is water. A hexagonal crystal structure is a form of calcium phosphate with a chemical formula $Ca_{10}(PO_4)_6(OH)_2$. Hydroxyapatite, a needle-like material 60 nm long and 5–20 nm wide, imparts rigidity and strength to bones and accounts for about 65% of bone mass [1,2].

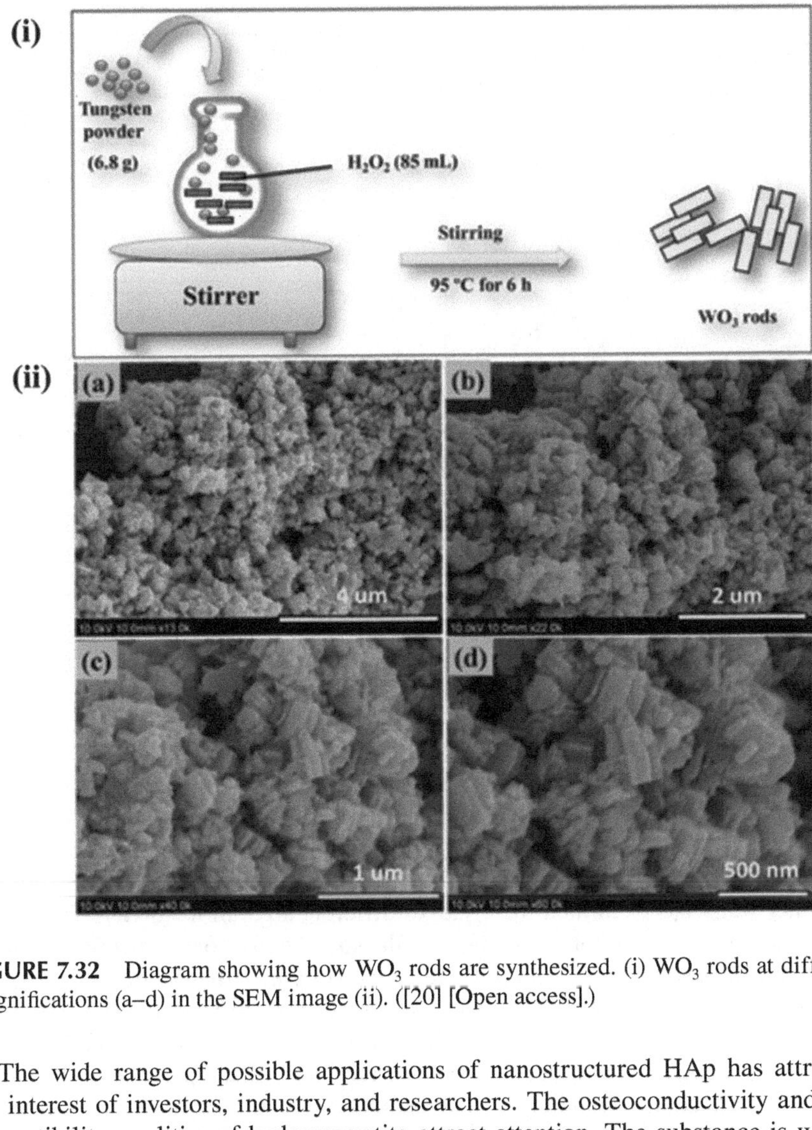

FIGURE 7.32 Diagram showing how WO_3 rods are synthesized. (i) WO_3 rods at different magnifications (a–d) in the SEM image (ii). ([20] [Open access].)

The wide range of possible applications of nanostructured HAp has attracted the interest of investors, industry, and researchers. The osteoconductivity and biocompatibility qualities of hydroxyapatite attract attention. The substance is widely and effectively used in drug delivery systems and regenerative medicine [4,11]. Hydroxyapatite particles with a nanostructure can be used in caries therapy as building blocks for damaged enamel [3]. Numerous studies show that nanosized HAp can improve its mechanical properties due to its high surface energy, which allows for faster surface renewal of implants [12]. Hydroxyapatite materials are used as sunscreen, anti-wrinkle agents, and bone replacement. They help speed up wound healing and used in toothpaste formulations. Polish scientists have created patches and ointments that use the special properties of HAp [10]. There is a great interest in hydroxyapatite to restore bone loss in animals. It has been demonstrated that HAp nanoparticles (nano-HAp) rather than micrometer-sized hydroxyapatite (micro-HAp)

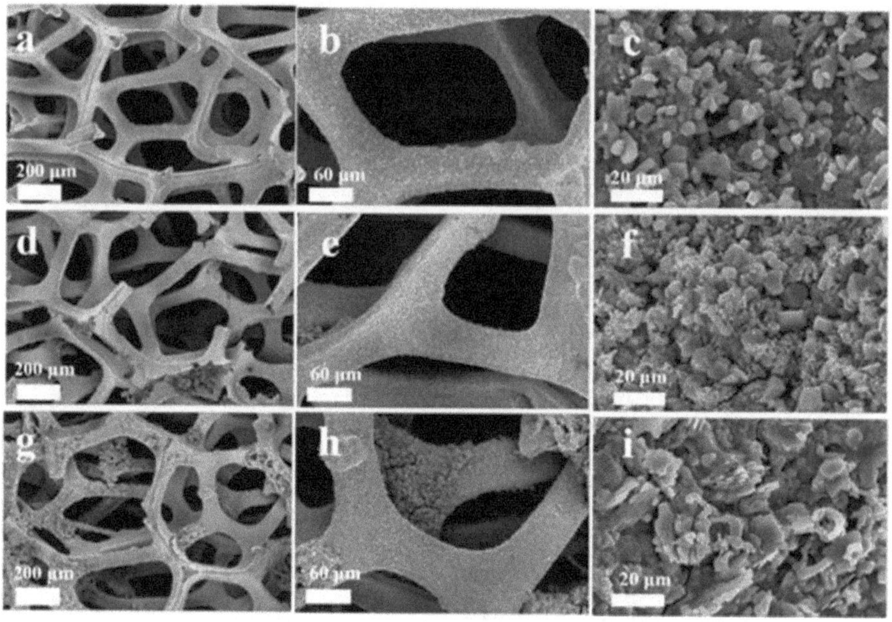

FIGURE 7.33 Simplified diagram showing the many gel processing structures. ([20] [Open access].)

are more suitable to function as bone apatite substitute in biomedical applications [13]. The effects of differently shaped nano-HAp particles on extremely aggressive melanoma cells are investigated. The results obtained have shown that the appearance of the nanoscale effect, in contrast to morphology of hydroxyapatite (HAP) particles, more effectively suppresses the development of such malignant cells [9]. A further study has examined the effect of different sizes of nano-HAp from 20 to 80 nm on the growth of bone-associated cells, specifically bone marrow mesenchymal stem cells and osteosarcoma cells [16]. A cell culture experiment confirmed that the hydrophilicity of nanophase mineral is increased by using nano-HAp compared to traditional HAp. In addition, nano-HAP is found to enhance stem cell viability and proliferation, especially for the smallest particles, 20 nm. Similarly, the smallest 20 nm particles are found to be the most effective inhibitor of osteosarcoma cell development, while each nano-HAp slowed down the growth rate. The anticancer activity of nano-HAp is found in another study [17].

7.3.2.5 Microemulsion

Nanotechnology is a branch of research that originated in the physical, chemical, biological, and engineering sciences. These domains are known for their innovative methods of manipulating individual atoms and molecules. In the context of nanotechnology, a nanoparticle is a small object with transport characteristics and possible applications that behaves as a whole. Simple or compound nanoparticles are useful building blocks for creating innovative nanodevices with a wide range of physical, biological, biomedical, and pharmacological applications. For many years, scientists have been fascinated by nanoparticles and nanostructured materials due to their applications in Raman scattering, catalysis, photonics, electronics, labeling, imaging, and sensing. They involve intriguing modifications in their optical, magnetic, catalytic, electrical, and drug delivery properties, coupled with improved physical properties such as mechanical hardness, thermal stability, or chemical passivity and one of the most challenging areas of modern scientific and technological research [27].

When comparing the optical characteristics of single crystals of the same chemical composition with nanoparticles, significant differences are found. The properties of nanomaterials differ from their bulk phases due to the influence of significant number of surface atoms and effect of quantum confinement of electronic states on nanosized crystallites. Compared to bulk materials and individual atoms or molecules, the properties of matter at this scale are very different. These nanoparticles, both organic and inorganic, have been prepared using a range of physical and chemical methods such as thermal and irradiation methods. One of the many preparation methods that can control many particle properties including size, geometry, morphology, uniformity, and surface area is the microemulsion method, which covers various aspects of microemulsions. Colloidal minerals, colloidal AgCl, nanosized Fe_2O_3, TiO_2, Al_2O_3, and high-temperature $YBa_2Cu_3O_7$ oxides are synthesized by microemulsion method. In addition to summarizing the latest studies on synthesis of organic and inorganic nanoparticles using this technique, this study aims to provide a clear overview of using microemulsions to produce nanoparticles and control their particle size and shape [25].

7.4 ADVANCEMENTS IN NANOCOATING TECHNOLOGY

In every field, including mechanical, electrical, and electronic devices and computer applications, which are mainly used to improve performance, prevent corrosion, prevent weathering, and prevent water damage, nanotechnology is the fastest-growing field. Thus, it is used everywhere nowadays. In addition, as they become more accessible, we are trying to use nanotechnology in turning and other machining operations. In this case, the nanoparticles will be coated using traditional methods and the hard, difficult-to-cut material will be turned. Before going deep into this project, we will first learn about nanotechnology, the materials that we will use, the place of our study, and types of tools that will be used, as we can see one by one. In nanotechnology, one billionth of meter (10^9m) is called a nanometer; it is approximately 100,000 times smaller than diameter of human hair, 1000 times smaller than red blood cell, and half the size of DNA molecule. Nanotechnology is defined as the creation and application of structures, devices, and systems that, due to their small size, have unique properties and functions, and the ability to control or manipulate matter on an atomic scale and research and development of technologies at the atomic, molecular, or macromolecular levels using a length scale of approximately 1–100 nm in any dimension. Engineering at the molecular level (groups of atoms) is known as nanotechnology. It is a broad term used to describe a wide range of methods, tools, and techniques for manipulating matter at the nanoscale, typically between 1 and 100 nanometers. On such an extremely small scale, the classical rules of physics and chemistry are not easily applied for two reasons. First, compared to their larger counterparts, very small particles can have completely different electrical characteristics. Second, the material's characteristics change because the surface atoms are the most reactive, resulting in a much larger surface area-to-volume ratio [7].

7.5 APPLICATION OF NANOCOATINGS

Typically, single- or multi-phase solid structures called nanocoatings, about 100 nm thick or less, are deposited onto an immediate surface to impart specific properties or purposes to the substrate material. Different materials have different characteristics at nanoscale. The particle size of material has a significant impact on overall properties such as hydrophobicity, electrical conductivity, magnetic permeability, melting point, and reactivity. Intermolecular forces such as hydrogen bonds, van der Waals interactions, and electrostatic forces control material properties at nanoscale. The height profile and 3D morphology of graphene oxide (GO) sheets are shown in Figure 7.34i and j. The GO nanosheets have a lateral size of 400 nm and a thickness of 1 nm. To quantitatively evaluate the adhesion of coating and coating on the substrate, the surface roughness, which is an important parameter, is evaluated [14,24]. The surface morphology and roughness values of bare magnesium sample and EPD-GO-coated magnesium samples are shown in Figure 7.34a–h. For bare magnesium, EPD-GO-A, EPD-GO-E, and EPD-GO-W, the corresponding average roughness (R_a) is 34.10, 26.41, 11.54, and 9.65 nm [24].

The length scales over which these intermolecular forces act vary from few angstroms to several nanometers. When the dimensions of a material reach the nanoscale, or the same order of magnitude as intermolecular force distances, the properties of the material change significantly. Tapping mode AFM is used to study the morphology of thin layers of graphene oxide, known as graphene oxide sheets, which have several applications including coatings. Information on the environmental impact of product's life cycle is provided through the electrophoretic deposition (EPD) techniques. Thus, many new applications can be explored by improving the physical and chemical properties of nanocoatings made from nanoparticles or multiple nanoscale layers. The main industrial applications of nanocoatings are wear-resistant coatings, self-cleaning (non-stick) surfaces, corrosion prevention, and thermal protection. These applications are particularly important to the aviation sector as they allow aircraft engine, cabin and frame components to be modified to improve performance, reduce operating costs, and reduce fuel efficiency [21]. It is reported that the solar absorber used in the collector panel determines how well solar radiation is converted into thermal energy. The significant surface roughness of the Mg substrate did not allow these domains to merge perfectly to form a smooth GO coating. As a result, cracks are formed [24]. Figure 7.35a and b shows the high- and low-magnification SEM images of GO sheets. The energy dispersive X-ray spectroscopy (EDX) analysis of uncoated and EPD-GO-E coated parts of the Mg substrate is shown in Figure 7.35c. The deposited coating is composed of GO due to the presence of carbon and large proportion of oxygen on the coated side [24].

A solar absorber with low emissivity and high absorption coefficient is ideal. It has been found that electrochemical deposition of selective nickel-based coatings can

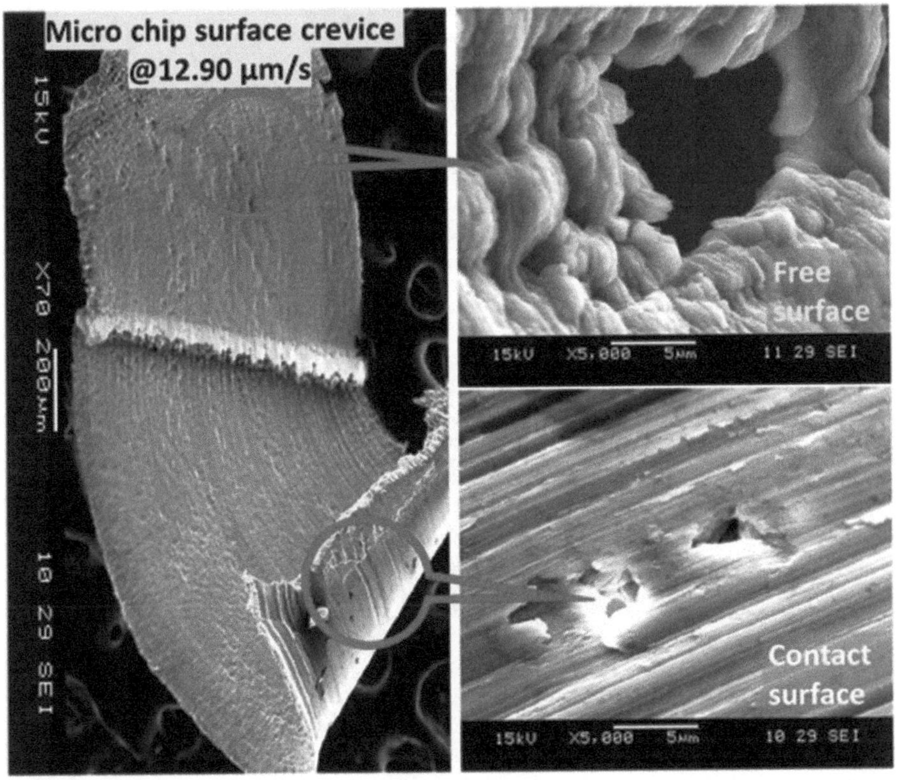

FIGURE 7.34 AFM images of (a) and (b) 2D and 3D roughness profiles of pure magnesium, (c) 2D and 3D roughness profiles and 3D profiles of the EPD-GO-A-coated magnesium sample, (e) 2D and 3D roughness profiles and 3D profiles of the EPD-GO-E-coated magnesium sample, (h) 2D and 3D roughness profiles of the EPD-GO-W-coated magnesium sample, and 2D and 3D profiles of GO sheets with a height profile. ([24] [Open access].)

provide both cost-effective and good optical performance required for solar energy absorbers [22]. However, scientists are moving toward a PVD method to be used to coat medium- and high-temperature solar photovoltaic panels due to the poor thermal stability of the CVD method and environmental concerns. Sputtering is one of the PVD processes that has been successfully used to coat receiver tubes and flat plate solar collectors [24]. These defects may be due to the nucleation of GO at different locations on magnesium substrates, which creates multiple GO domains. These domains then expand and merge to form GO coating. The base material is CuCrZr alloy (Cr: 0.2%–1.2%, Zr: 0.03%–0.3%, other: max. 0.2%, Cu: residual). A wire-cutting machine is used to cut the copper alloy substrate into samples with a size of 100 mm × 100 mm × 10 mm. Before drying for use, the surface of the base material is first polished with 240 mesh sandpaper, then polished with 600 mesh sandpaper, and

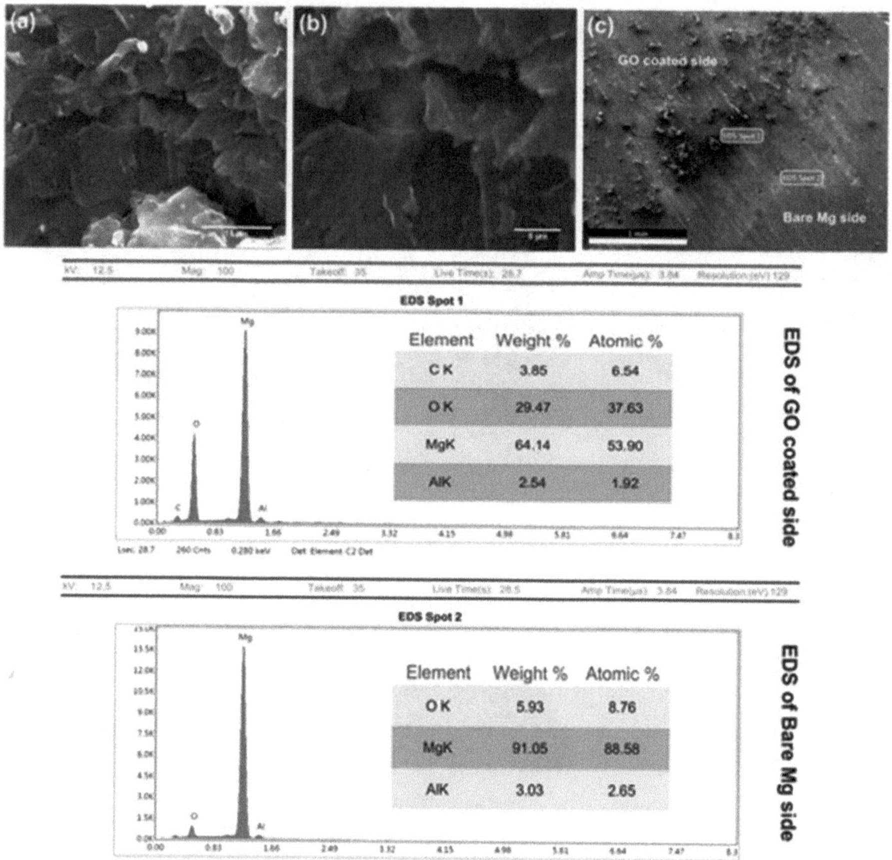

FIGURE 7.35 SEM images of bare magnesium and the EPD-GO-E coated area: (a, b) GO sheets and (c) EDX spectrum. ([24] [Open access].)

finally ultrasonic-cleaned in 75% alcohol solution for five minutes. Spherical copper powder and $MoSi_2$ powder (Shanghai Nai'ou Nano Technology Co., LTD., Shanghai, China) are used as the raw materials for the shell, as shown in Figure 7.35. The raw materials have purity 3N and a particle size of 100–300 mesh [30].

PROBLEMS

1. Explain the applications of high-speed machines.
2. List several applications for cutting hard materials.
3. Learn about hybrid cutting in manufacturing.
4. Describe the liquid-phase method.
5. Explain developments in nanocoating technology.

REFERENCES

1. Zhang, S., Gong, M., Lian, H., Wu, J., Zhu, W., & Ou, Z. (2023). Design of a high-speed rotary ultrasonic machining machine tool for machining microstructure of brittle materials. *Micromachines*, 14(8), 1544.
2. Dabees, S., Mirzaei, S., & Sobola, D. (2022). Characterization and evaluation of engineered coating techniques for different cutting tools—Review. *Materials*, 15, 5633.
3. Hon, K. & Baharudin, B. H. T. (2006). The impact of high speed machining on computing and automation. *Int. J. Auto. Comput.*, 3, 63–68. https://doi.org/10.1007/s11633-006-0063-3
4. King, R. (1985). *Handbook of High-Speed Machining Technology* (1st ed.). Chapman and Hall. https://doi.org/10.1007/978-1-4684-6421-4
5. Blixt, K. H. & Hallberg, H. (2022). Grain boundary and particle interaction: Enveloping and pass-through mechanisms studied by 3D phase field crystal simulations. *Mater. Design*, 220, 110845.
6. Wan, Y., Wang, T., Wang, Z., Jin, Y., & Liu, Z. (2017). Construction and characterization of micro/nano-topography on titanium alloy formed by micro-milling and anodic oxidation. *Int. J. Adv. Manuf. Tech.*, 98, 1–7. https://doi.org/10.1007/s00170-017-0323-0
7. Start, Secotools.com. Available from: https://www.secotools.com/.
8. Yuan, Z., Riaz, A., & Chohan, B. S. (2023). Precision machining by dicing blades: A systematic review. *Machines*, 11(2), 259.
9. Sabzi, H. E., Hernandez-Nava, E., Li, X.-H., Fu, H., San-Martín, D., & Rivera-Díaz-Del-Castillo, P. E. (2021). Strengthening control in laser powder bed fusion of austenitic stainless steels via grain boundary engineering. *Mater. Design*, 212, 110246.
10. Parra, C., Aristizabal, J., Arce, B., Montero-Silva, F., Lascano, S., Henriquez, R., … & Barrera de Brito, A. (2021). Graphene coating as an effective barrier to prevent bacteria-mediated dissolution of gold. *Metals*, 11(1), 147.
11. Ekinovic, S. & Ekinovic, E. (2003). Some advantages of high speed machining in aspect of machined surface quality. In *3rd Research/Expert Conference with International Participations "Quality"*; Zenica, B&H, pp. 453–460.
12. Smith, S. & Tlusty, J. (1997). Current trends in high-speed machining. *J. Manuf. Sci. Eng.*, 119, 664. https://doi.org/10.1115/1.2836806
13. Lee, S. J., Sohn, Y. C., & Kim, C. L. (2022). Tribological effects of water-based graphene lubricants on graphene coatings. *Materials*, 16(1), 197.
14. Rizzo, A., Goel, S., Luisa Grilli, M., Iglesias, R., Jaworska, L., Lapkovskis, V., … & Valerini, D. (2020). The critical raw materials in cutting tools for machining applications: A review. *Materials*, 13(6), 1377.
15. Cui, J., Shen, X., Xin, Z., Lu, H., Shi, Y., Huang, X., & Sun, B. (2023). Thermodynamic analysis based on the ZL205A alloy milling force model study. *Lubricants*, 11(9), 390.
16. Li, Y., Zhang, K., & Wang, Y. (2024). Study of the critical speed of ultra-high-speed diamond drilling technology in marble. *Appl. Sci.*, 14(8), 3403.
17. Sabzi, M., Mousavi Anijdan, S. H., Shamsodin, M., Farzam, M., Hojjati-Najafabadi, A., Feng, P., … & Lee, U. (2023). A review on sustainable manufacturing of ceramic-based thin films by chemical vapor deposition (CVD): Reactions kinetics and the deposition mechanisms. *Coatings*, 13(1), 188.
18. Samal, S. & Blanco, I. (2022). An overview of thermal plasma arc systems for treatment of various wastes in recovery of metals. *Materials*, 15(2), 683.
19. Cao, S. L., Liang, J., & Zhou, J. S. (2022). Evolution in microstructure features and properties of Mo-containing Fe-Cr-Ni-B-Si composite coatings by laser cladding. *Mater. Charact.*, 188, 111926.

20. Myasoedova, T. N., Kalusulingam, R., & Mikhailova, T. S. (2022). Sol-gel materials for electrochemical applications: Recent advances. *Coatings*, 12(11), 1625.
21. Blixt, K. H. & Hallberg, H. (2022). Grain boundary and particle interaction: Enveloping and pass-through mechanisms studied by 3D phase field crystal simulations. *Mater. Design*, 220, 110845.
22. Zhang, Z., Zhou, H., Yao, P., Fan, K., Liu, Y., Zhao, L., … & Deng, M. (2021). Effect of Fe and Cr on the macro/micro tribological behaviours of copper-based composites. *Materials*, 14, 3417.
23. Zhuang, C., Xiong, Z., & Ding, H. (2010). High speed machining tool path generation for pockets using level sets. *Int. J. Prod. Res.*, 48, 5749–5766. https://doi.org/10.1080/00207540903232771
24. Maqsood, M. F., Raza, M. A., Rehman, Z. U., Tayyeb, A., Makhdoom, M. A., Ghafoor, F., … & Khan, M. F. (2022). Role of solvent used in development of graphene oxide coating on az31b magnesium alloy: Corrosion behavior and biocompatibility analysis. *Nanomaterials*, 12(21), 3745.
25. Su, H., Liu, P., Fu, Y., & Xu, J. (2012). Tool life and surface integrity in high-speed milling of titanium alloy TA15 with PCD/PCBN tools. *Chin. J. Aeronaut.*, 25, 784–790. https://doi.org/10.1016/S1000-9361(11)60445-7
26. Tlusty, J. (1993). High-speed machining. *Ann. CIRP*, 42, 733–738.
27. Xu, N., Jiang, X., Shen, X., & Peng, H. (2023). Improving the surface integrity and tribological behavior of a high-temperature friction surface via the synergy of laser cladding and ultrasonic burnishing. *Lubricants*, 11(9), 379.
28. Jia, L., Rong, X., Zhao, D., Zhang, X., He, C., & Zhao, N. (2022). Microstructural characteristic and mechanical properties of the in-situ $MgAl_2O_4$ reinforced Al matrix composite based on Al-Mg-ZnO system. *J. Alloy Comp.*, 891, 161991.
29. Woo, E. M., Lugito, G., Yang, C. E., & Chang, S. M. (2017). Atomic-force microscopy analyses on dislocation in extinction bands of poly (Dodecamethylene terephthalate) spherulites solely packed of single-crystal-like lamellae. *Crystals*, 7(9), 274.
30. Li, Y., Zhao, X., Zhai, P., Fan, P., Xu, J., Xu, Y., … & Xiao, L. (2023). A novel superhard, wear-resistant, and highly conductive Cu-$MoSi_2$ coating fabricated by high-speed laser cladding technique. *Materials*, 17(1), 20.
31. Zhang, Y., Miyamoto, G., & Furuhara, T. (2022). Enhanced hardening by multiple microalloying in low carbon ferritic steels with interphase precipitation. *Scr. Mater.*, 212, 114558.

8 Sculptured Surface Machining

8.1 INTRODUCTION

Surface sculpture processing is a widely used process in many industries, including the automotive, aerospace, die, and mold sectors. Nowadays, it is more important than ever to reduce production costs and time without compromising the quality of parts due to increasing competition in the market. Most ball milling models are either suitable for machining workpieces with simple and limited geometries such as 2½-axis milling cases, but not for relatively complex sculptured surfaces, or they are not capable of simulating multi-pass machining operations and cannot automatically detect areas of cutter-workpiece interaction, despite the recent wave of new scientific research on milling tools, which have focused on various aspects [1–6]. Unfortunately, there are insufficient scientific resources to select "appropriate" feed rate settings for surface finishing for sculpture based on a reliable force model. For this reason, the feed rates have been set at cautiously constant values. Computer-aided design (CAD) and computer-aided manufacturing (CAM) software are commonly used in modern manufacturing that are essential tools, especially in computer numerical control (CNC) manufacturing. This is an explanation of how they work together and how CNC code generators are used. They rely solely on geometric and volumetric analysis; they do not yet take into account the mechanics of manufacturing processes. For roughing, semi-finishing, finishing machining, CNC code generators and surface sculpting CAD/CAM programs recommend fixed feed rate settings for tool material combinations based on appropriate databases. These databases include general conservative values derived from testing and trial-and-error experience. They can be quite conservative, although the information may be useful to prevent adverse consequences such as chipping, excessive tool deflection, and tool breakage. Because manufacturing lacks scientific tools based on the mechanics of sculpting processes, engineers are unable to estimate cutting forces in most industrial applications. Figure 8.1 shows the workpiece's main dimensions. As a result, they are forced to choose a conservative feed rate during the process planning phase. Additionally, when machining a sculptured surface, where tool-to-work contact is constantly changing throughout tool path, maintaining a constant feed rate will result in reduced productivity [1].

These modifications open up the possibility of using variable and higher feed rates at certain portions of tool path. For the mentioned reasons, the feed rate should be varied so as not to affect the amount of cutting force that results. Thus, understanding the selection of different feed rates along tool path depends on cutting force levels and having a reliable model of force system. The main characteristics of four classes defined for industrial applications are as follows:

DOI: 10.1201/9781003604648-8

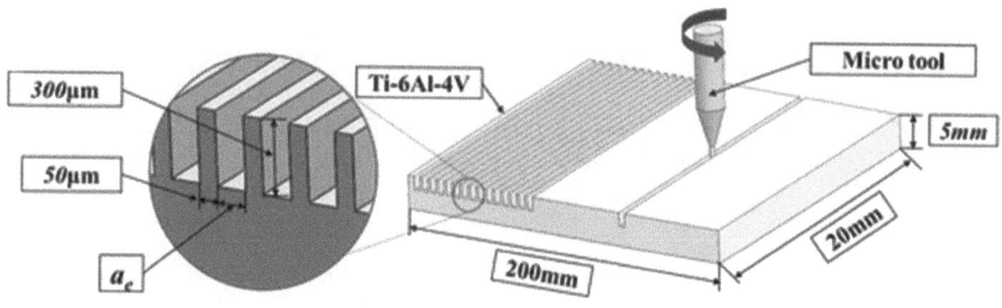

FIGURE 8.1 Workpiece's main dimensions. ([1] [Open access].)

- Forging dies: Construction materials consist of machined steels specifically designed for hot operations and having a hardness level of 30–60 HRC. Both final roughness and tolerances are quite large, deep, and narrow zones.
- Stamping dies: Ductile iron castings are constituent materials. However, the use of advanced high-strength steels (AHSS) for automotive bodies has increased over the past 5 years. As a result, a higher percentage of hardened iron, 60 HRC plate blocks, and stronger forming tool surfaces must be machined. Additionally, the final processing step can take many hours. Because of this, many users choose weaker but more durable tool materials (such as carbide) rather than more resilient but brittle tool materials (such as PCBN); in the latter case, an uncontrolled allowance can lead to unexpected destruction of the tool, which will have catastrophic consequences for die.
- Molds for casting plastics or aluminum alloys: The mold is made of hardened steel with a hardness of 50–55 HRC. The injection procedure requires good finishing and precision of form. The complexity of plastic components leads to mold designs featuring extremely deep and narrow areas, some of which push the limits of modern electrical discharge machining (EDM) technology.
- Ruled surfaces: These surfaces can be found in some energy or aircraft engine-related parts such as workpiece material Ti6Al4V, Inconel 718 (or related superalloys), or aluminum alloys. These applications require high precision in the interface areas. A certain type of side milling [1], a finishing procedure, attempts to apply a given geometry using tool paths where the tool generatrix (cone or cylinder) touches the workpiece guides. Otherwise, the same methods are used to process this surface as for free-form surfaces.

The unified power flow controller (UPFC) can be divided into two types depending on tool installation direction: flat edge milling, which occurs when the tool installation direction, is parallel to the spindle axis [13,14], and ultra-fine mesh milling, which occurs when the tool installation direction is set along the radial direction of the spindle [15,16]. Figure 8.2 shows some microstructures and scanning electron microscope (SEM) images of structures manufactured using surface milling [1].

Figure 8.3a–c [1] illustrates the successful preparation of a free-form nanostructured F-Theta surface and nano-micro-spherical array. An innovative processing

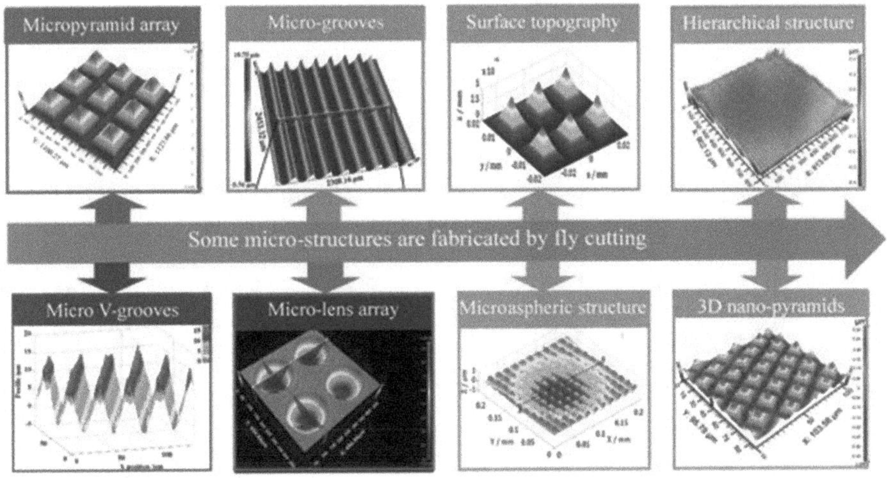

FIGURE 8.2 Fly cutting is used to create specific microstructures. ([1] [Open access].)

technology is proposed to produce microstructures with hierarchical characteristics as a result of continuous improvement of performance of the advanced fabrication control system (EFCS). By changing the cutting directions within the EFCS, this technology can generate various nano polygons as secondary structures. As shown in Figure 8.3d–f, the experimental results show that nanopyramids can be successfully fabricated on high-precision microsphere surfaces and free-form F-Theta mesh. In a breakthrough, it has presented a new micro/nano machining technology that combines diamond tool rotational spatial vibration (RSV) and workpiece servo motion to achieve the generation of multilayer nano-/microstructures. This method represents a significant improvement in the accuracy and adaptability of micromachining processes [2]. As depicted in Figure 8.3g–i [2], a series of micro-/nanostructures with complex geometry and adjustable variable feature sizes can be created by combining non-resonant triaxial vibration and servo motion.

Selecting the correct feed rate along the tool path is essential for improving cutting productivity and quality in manufacturing and industrial processes. Cutting force levels, which vary based on several variables including material quality, tool shape, and operating conditions, can affect the feed rate. Different categories or feed rate selection strategies can be created based on the characteristics of the force system to manage this complexity. Four categories are commonly used for industrial purposes with key characteristics. These four groups of feed rate selection methods provide different ways to control cutting forces and improve operating efficiency. While variable feed rates based on cutting force provide flexibility in changing situations, fixed feed rates are simple and effective for stable conditions. While material property-based optimization provides efficient operation across a wide range of materials, feed rate optimization based on tool wear focuses on maintaining tool performance. The specifics of the workflow and the complexity of the system used determine the best approach. In addition, the application of fixed tool system (FTS) and swiveling tool system

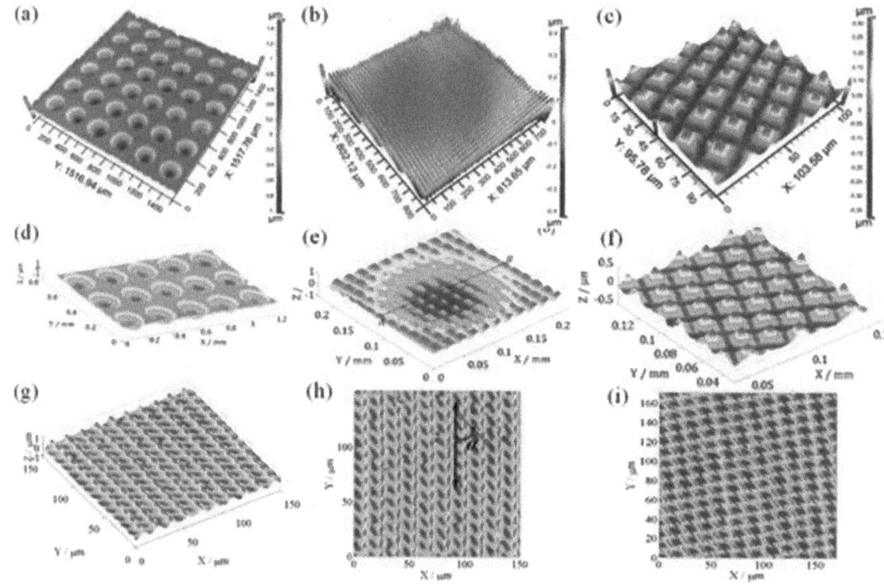

FIGURE 8.3 Surface features of free-form F-Theta nanopyramid array and 3D hierarchical structure: (a) large-area structures, (b) extracted 3D nanopyramids, and (c) surface features of an aspherical nanopyramid array. The features of following elements are described: (d) a nanopyramid array filled with nanopyramids, (e) a spherical nanopyramid array filled with individual nanopyramids, and (f) enlarged view of secondary nanopyramids. Features of nanostructures created by single-frequency rotational vibration: (g) created 3D nanopyramid array and (h, i) predicted 2D nanopyramid array. ([2] [Open access].)

(STS) to fly-head end face cutting, which could not produce high-quality large-scale microlens array (MLA) machines. In addition to achieving uniform MLA mass over a large area, higher spindle speeds can lead to a higher cutting efficiency, as shown in Figure 8.4 [2].

8.2 THE MANUFACTURING PROCESS

The finishing process technique has a significant impact on the design of complex shapes, forging dies and stamping dies. However, after the design stage, each project faces a fairly similar challenge: processing a complex surface, most often sculptural (free-form). Currently, the main differences are lead times required by the competitive market, roughness levels, and dimensional tolerances[20].

8.2.1 Technologies Involved

One well-liked non-traditional machining technique that is frequently applied to hard materials is EDM. Because EDM can manufacture any material, regardless of its hardness, it is a widely used technology. Modern engineering materials that are

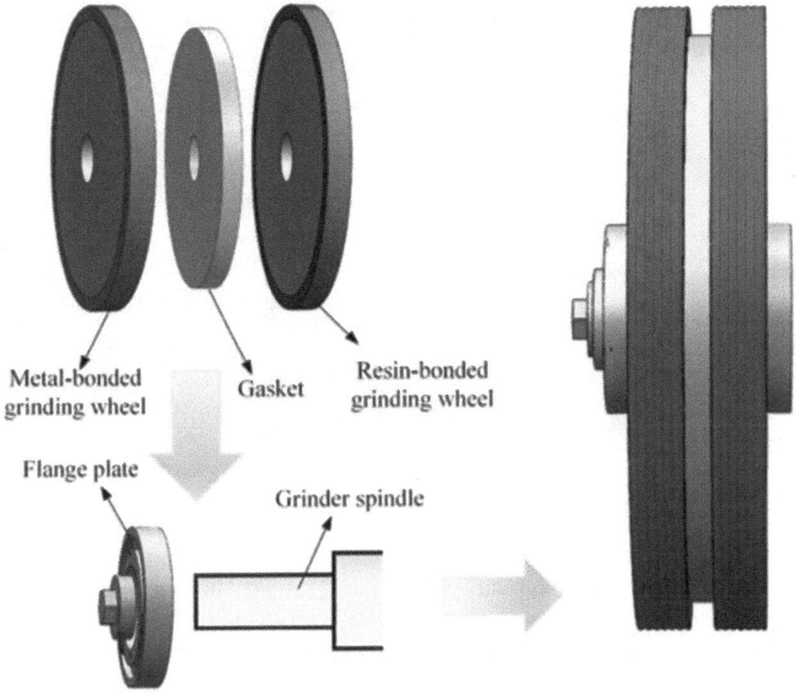

FIGURE 8.4 Characteristics of a processed microlens array: (a) large-area three-dimensional structure, (b) two-dimensional cross-sectional profile, (c) removed square region, and (d) equivalent profiles along the transverse directions of the microlens array. ([2] [Open access].)

used in harsh environments are frequently formed or produced via the EDM technique. However, this approach has disadvantages of its own. A reduced rate of material removal and considerable wear on the tools can frequently make this method's machining less efficient. To produce a more consistent and well-balanced machining setup, EDM can be coupled with other standard or nonconventional manufacturing techniques. The inherent disadvantages of EDM technique can be addressed by hybrid or combination approaches to machining. When traditional manufacturing methods are combined with the other ones, machining performance can be greatly enhanced. Three procedures that were used are as follows:

- Shape the conventional method used before development of high-speed milling
- High-speed method used in the 1990s
- Method implemented in 2000

The test setup for our typical machine is shown in Figure 8.5. The test setup included a rigid body mounted on the machine. A linear variable differential transformer (LVDT) displacement sensor with a holder and force throttle with a load cell are additional components of the test setup. The purpose of this experiment is to evaluate

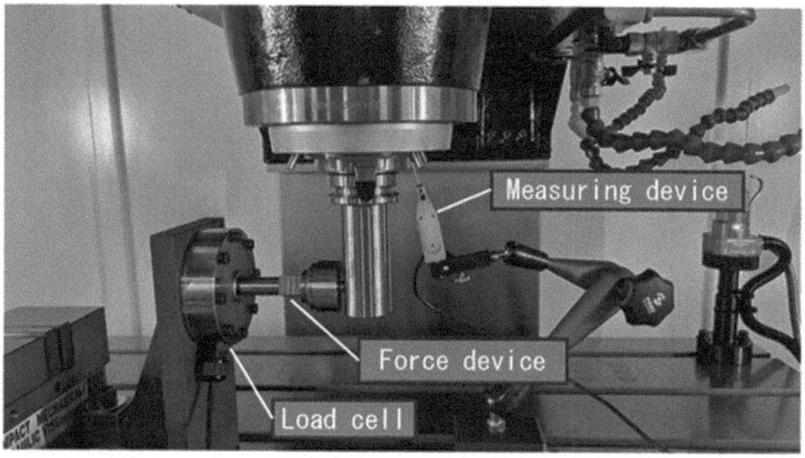

FIGURE 8.5 *X*-Axis rigidity test. ([36] [Open access].)

the overall stiffness of the machine on the *X*-axis [21]. For this purpose, the spindle is fixed by the test tool holder and force application unit is mounted on the workbench and connected to a load cell. The point of force application is 100 mm below the nose of the spindle [36].

High-speed milling machines can be used to process the graphite electrode (f1). The object is brought to final dimensions using a graphite electrode, and surface polishing can be improved using the orbital motion of the electrode. The accuracy of shape directly depends on the accuracy of the electrode. Roughing (d2) and semi-finishing (e2) between 1997 and 1999 are often performed on conventional equipment. This order is chosen for two reasons:

- Roughing, which requires less precision, is performed on machines that cost several times less per hour than high-speed machines.
- Roughing is not possible because the standard high-speed spindles available at time could not produce sufficient torque below 1500 rpm; also, tool wear is minimal due to low hardness of the workpiece material [22].

8.2.2 Five-Axis Milling

Sculpted surfaces are common in many different products, including automobiles, appliances, water, and air conditioning components. Two additional orientation axes in five-axis milling enable the processing of extremely complex parts that cannot be produced on three-axis equipment. As shown in Figure 8.6, for example, in automotive industry, five sides of an engine block can be machined in one setup and fixture. These surfaces may be required solely for aesthetic reasons or to reduce fluid resistance. Most of these surfaces are produced by machining. The designed surface is created on the cutting tool in one place using modern surface processing methods. Several closely spaced passes of the cutter are required to mill the surface to required

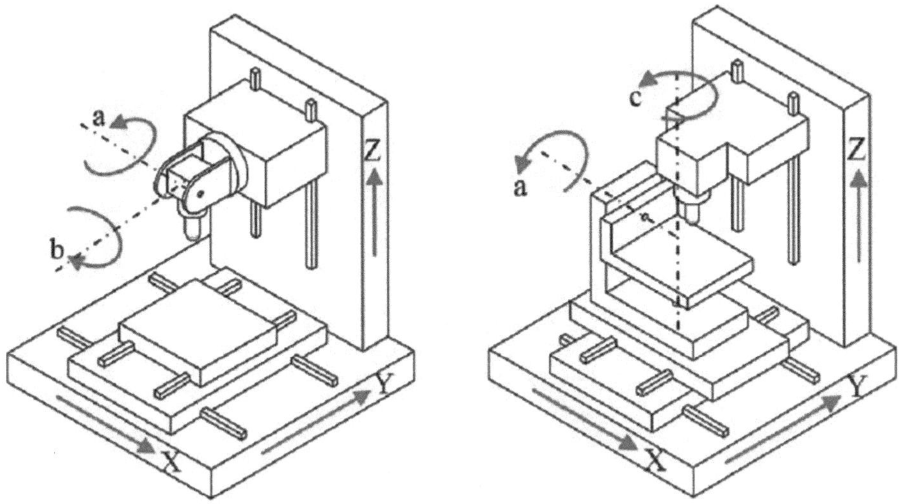

FIGURE 8.6 Axis diagram of a five-axis CNC machine with an orientable table on right and machine with an orientable spindle on left. ([3] [Open access].)

tolerance. This thesis presents a new method for five-axis machining of sculpted surfaces with flat or toroidal bottom-end mill [3].

The proposed multi-point machining tool positioning method promotes surface formation at multiple locations of the tool. Increasing the contact points between the surface and tool will speed up processing. Equations are created to describe mode 1 multi-point contact between the tool and workpiece surface. Planar, parabolic, spherical, elliptic, and hyperbolic quadratic surface shapes are five types of quadratic surface shapes with which they were studied regarding contact type. It has been found that there are two basic forms of multi-point contact; two-point contact for parabolic, elliptic, and hyperbolic surfaces and circle contact for planar and spherical surfaces. Based on these results, a precision multi-point tool positioning method for square surfaces is developed. Two algorithms are also created to adjust the multi-point tool position for non-quadric surfaces. The features of the multi-point tool positioning method are studied through cutting tests and simulations. Low and variable multi-point ridges made the machined surface appear significantly smoother than surfaces produced by competing methods. The TOPNC VMC-C50 five-axis CNC machine consists of two rotary axes (*A* and *C*) and three straight axes (*X*, *Y*, and *Z*). It is the type of five-axis machine discussed in this chapter. As shown in Figure 8.7, the cradle five-axis CNC machine has a more rigid structure than other five-axis CNC machines and can be used to produce high-quality products [8]. Consequently, the subsequent mathematical definition of every homogeneous transformation matrix is provided [14] for every machine axis.

Scallops are made with a large tool, and small inserts will be smaller than those made with small and large inserts. The direction of minimum curvature is the ideal feeding direction, and the direction of maximum curvature is the worst. Multi-point

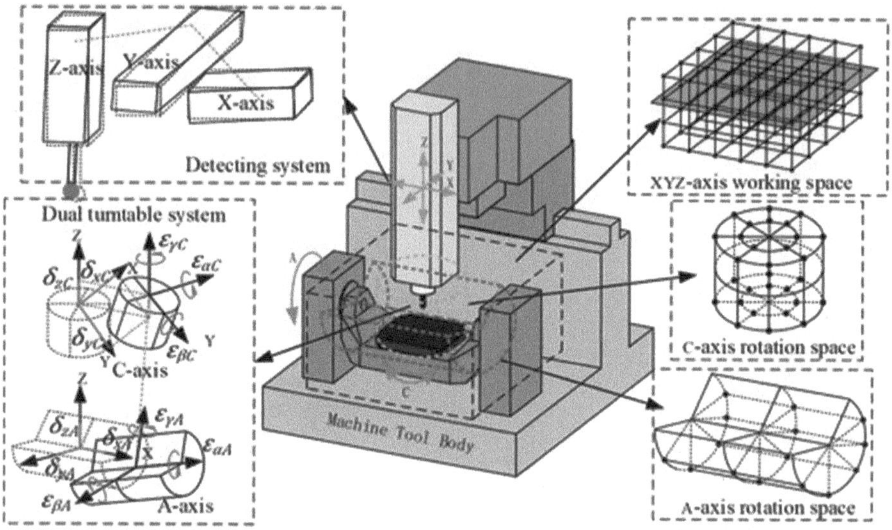

FIGURE 8.7 Schematic diagram of the spatial structure and error kinematic chain of a five-axis CNC machine tool. ([8] [Open access].)

machining provides the best results of all tool positioning methods. According to the findings of this dissertation, scallop height obtained using multi-point machining is approximately 0.0025, 0.04, and 0.5 times greater than height obtained using the other most effective tool positioning strategies; principal axis methods, and inclined tool and ball. Technologies for optimizing the tool-path and cutter orientation are necessary to ensure efficient and high-quality machining in five-axis CNC milling [17]. These systems have the following important features:

- Tool-path optimization: The path of the cutting tool during part processing is called the tool path. Finding the most efficient tool path based on the shape of part, capabilities of the cutting tool and required surface finish is called tool-path optimization. Depending on the needs of the machine tool, various tool-path methods can be used, including helical, zigzag, linear, and contour [14].
- Optimization of the cutter orientation: The angle at which the cutter is positioned relative to the workpiece is called the cutter orientation.
- The process of optimizing the cutter orientation involves determining the ideal angle at which the tool should be positioned to obtain the required surface finish and prevent collisions with the workpiece or machine tool [19].
- Collision avoidance: The complex kinematic design of a five-axis CNC router increases the likelihood of collisions between the tool and part or machine body. Real-time adjustments to the tool-path or cutter orientation are made by collision avoidance systems to identify and prevent collisions.

The need for high-precision, high-speed machine tools has increased due to growing demand from various industries, including consumer electronics, computers, communications, defense, and aviation. To achieve high speed and precision in tool manufacturing, tool manufacturing design is especially important. Nowadays, CAD is mainly used in the creation of machine tools because it eliminates the need to manufacture, process, assemble, or specify structural components, which helps to reduce development costs and speed up development time. Automobile, shipbuilding, aviation, and other equipment manufacturing industries rely heavily on CNC machine tools, which are the most important in modern manufacturing. Precision machining has always been a complex and dynamic research area [1–3]. Compared with previous multi-component structural models, the execution time is reduced by 90%, and a multi-component dynamic model for five-axis machine tools was demonstrated [4].

A topology optimization method [5] is used to optimize the shell structure by changing its design to reduce weight, improve design, and save costs. Simulation and verification technologies are used to see the machining process in action and confirm that the tool-path and cutter orientation are ideal for part. The scanning tunneling microscopy (STM) image of a sample after annealing in process 1 (at 300°C) is shown in Figure 8.8a, which shows the presence of steps with irregular edges and narrow widths on the surface of a sample. According to a line profile in Figure 8.1b (drawn along blue arrow in Figure 8.8a), the average height of the steps is 0.22 nm [26].

The bold double arrow in a side view atomic structure model (Figure 8.8c) represents the theoretical minimum separation (labeled d) of planes along Cu, which is 0.21 nm [26]. These results indicate that there is only one atomic layer in each step at this stage. After annealing in process 2 at 350°C, the step width increases, and the step edge remains irregular, as shown in Figure 8.8d. Interestingly, the step terraces have several triangular islands, some of which overlap. The average step height increases to 1.28 nm ($\approx$ 6d) in Figure 8.8e, as depicted by the height profile (plotted along the bold arrow in Figure 8.8d). The number of islands decreases sharply, and step width increases when annealing temperature increases to 450°C (process III). However, more importantly, the step edge boundary pattern takes a wavy appearance, as shown in Figure 8.8f. The average step height increases to 2.21 nm (10d), as shown in Figure 8.8g by the line profile (plotted along the blue arrow in Figure 8.8f). The shape changes from process 1 to process 3 to show that the step atomic boundaries become smoother, wider, and longer with the increase of annealing temperature [26].

Figure 8.9a illustrates the appearance of zigzag edges on steps, a significant decrease in the number of triangular islands, and appearance of few bolt dislocations (as indicated by black arrows) on the step terraces as annealing temperature increases to 600°C (process IV). However, as can be seen from the height profile in Figure 8.9b (plotted along the bold arrow in Figure 8.9a), the average step height already decreases at this point, with a value of 1.24 nm ($\approx$ 6d). Bolt dislocations appear when the triangular islands of step terrace disappear and straight edges become visible when annealing temperature reaches 650°C (process V). They remain visible, as indicated by black arrows in Figure 8.9c. The step height is shown in Figure 8.9d, and

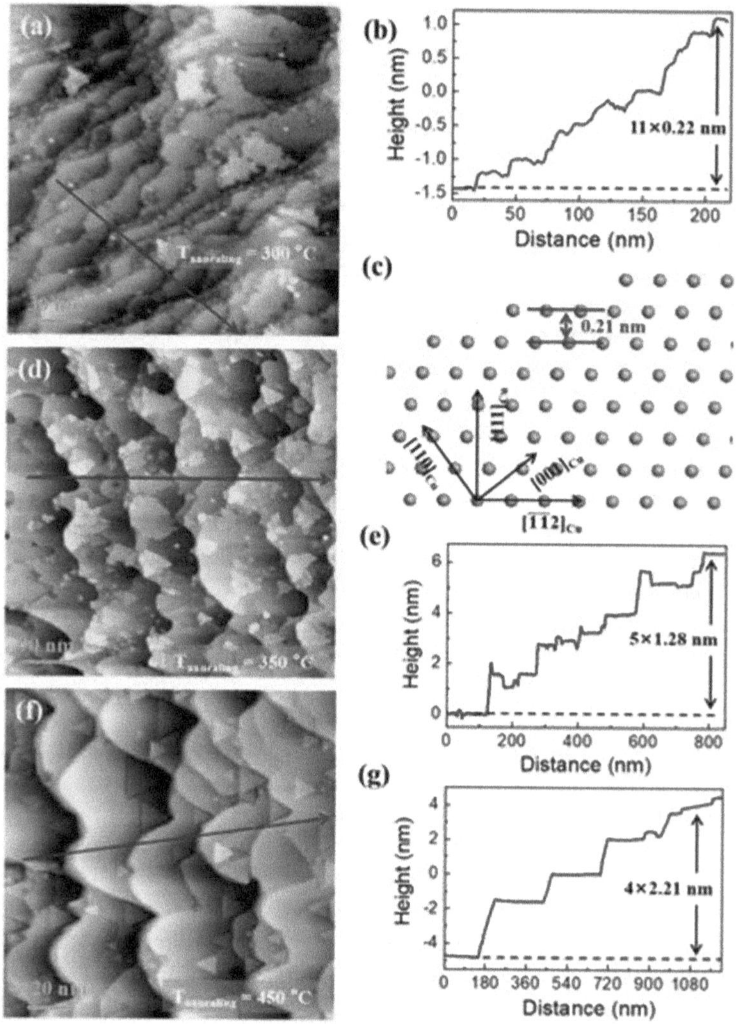

FIGURE 8.8 Shape changes from 300°C to 450°C. (a) STM image of the sample after annealing of process 1; reconstructed height profile is indicated by a bold arrow. The following images show STM images of the sample after annealing of processes 2 and 3, respectively: (b) height profile along blue arrow in (a), (c) side view of the atomic model along the [11̲0]Cu azimuth, and (e, g) height profiles along blue arrows in (d, f). In this case, the scanning bias current and voltage are 1 V and 0.1 nA, respectively. ([26] [Open access].)

the comparative height profile plotted along the blue arrow in Figure 8.9c is 1.03 nm (≈ 5d), indicating that the steps are still numerically smaller at average height. As shown in Figure 8.9e, when the annealing temperature reaches 720°C (process VI), some steps still have distinct straight edges, while others have jagged edges, and step terraces become very smooth. As can be seen from the height plot (plotted along the

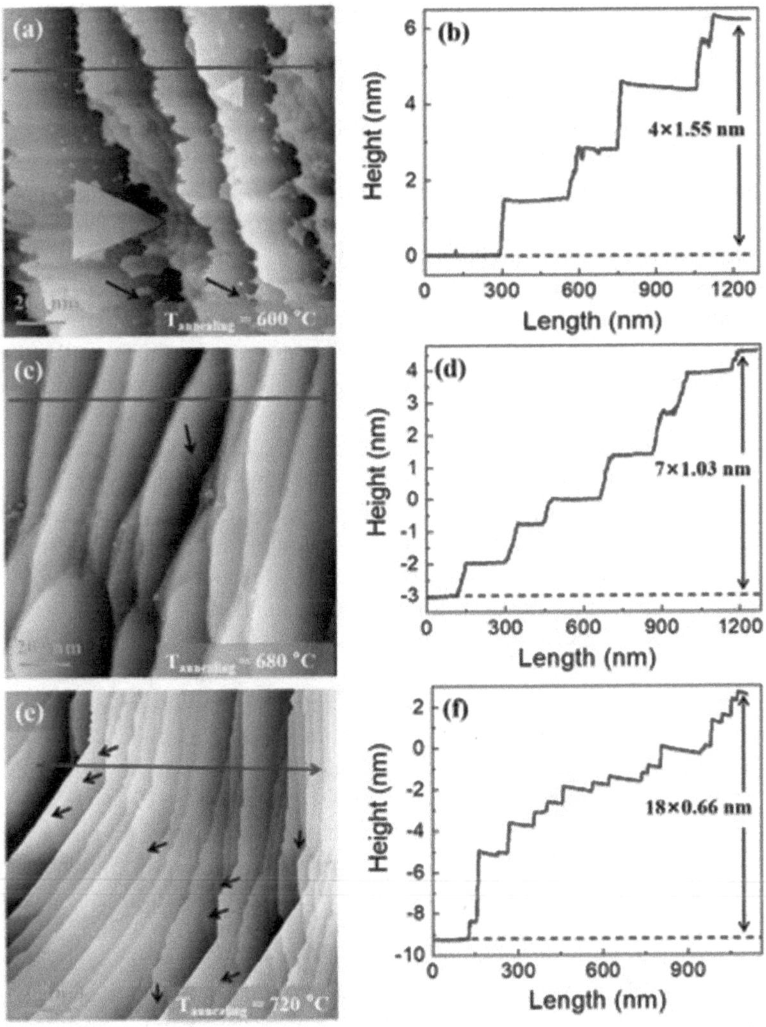

FIGURE 8.9 Evolution of the pattern from process xi to process vi. Screw dislocations are indicated by black arrows in STM images (a,c,e) of the sample after annealing at 600°C, 650°C, and 720°C, respectively; the height profiles (b,d,f) follow blue arrows in (a,c,e), respectively. In this case, the scanning bias current and voltage are 1 V and 0.1 nA, respectively. ([26] [Open access].)

bold arrow in Figure 8.9e) in Figure 8.9f, the average step height decreases to 0.66 nm (≈ 3d). It is worth noting that step terraces exhibit many torsional dislocations in this procedure, as shown by black arrows in Figure 8.9e. Previous observations of this phenomenon have been made on other cleaned metal surfaces, including Ag(111) and Au(111) surfaces [26,29,30].

8.3 COMPLEX SURFACE MACHINING (CAM) CENTER

One of the most important manufacturing methods used in manufacturing technology today is five-axis milling. When producing complex surface shapes by milling, choosing the right CNC machine is as important as the clamping technique, tool selection, and strategy. The main advantage of using five-axis milling methods is the ability to adjust the axis of milling tool relative to machined surface to ensure optimal cutting conditions [1]. Copying tools are the most commonly used tools for five-axis milling and finishing. High surface quality can be achieved with their use, but only if the influence of several important cutting process parameters is taken into consideration. These parameters include cutting conditions, the angle of inclination of tool axis to normal to surface, the accuracy of tool clamping, and finally the overall stability of process. Constant cutting conditions are not achievable when milling complex curved surfaces because the contact between the tool and workpiece varies. The effective diameter of the cutter and therefore effective cutting speed can be adjusted by adjusting the position of the tool axis. As a result, it becomes possible to mill more efficiently, while simultaneously improving surface quality, changing the force acting on the tool, extending its service life and reducing vibration in the machine–workpiece–fixture system [2,3]. Selecting the appropriate tilt angle can be especially challenging for complex components with varying surface curvatures. One way to achieve decent surface roughness and low bump height is to use a low tilt angle. A visual comparison of actual roughing, semi-finishing, and post-finishing production processes is presented in Figure 8.10. However, to prevent high roughness values in some areas of the machined surface, a significantly larger minimum inclination angle will be required. One of the indicators of process quality is surface roughness, so finding ideal conditions is critical to obtaining good results at a reasonable cost [33].

The size of an effective diameter is determined not by the tool diameter and axial depth of cut. The angle of inclination of the tool axis is another influencing factor. It is generally accepted that the variation of the tool axis vector during the milling process is one of the most significant differences between five-axis and three-axis CNC milling [4]. After creating a tool path, postprocessing introduces the problems and features described in the previous section. Defining a postprocessor is a complex task that requires a deep understanding of both the machine and postprocessor generator program. Trajectory checking is the next step after postprocessing. Although this step is highly recommended for three-axis milling, it becomes critical for five-axis machines. The surfaces that have limitations are covered and designed.

FIGURE 8.10 Visual comparison of actual manufacturing processes for roughing, semi-finishing, and post-finishing. ([33] [Open access].)

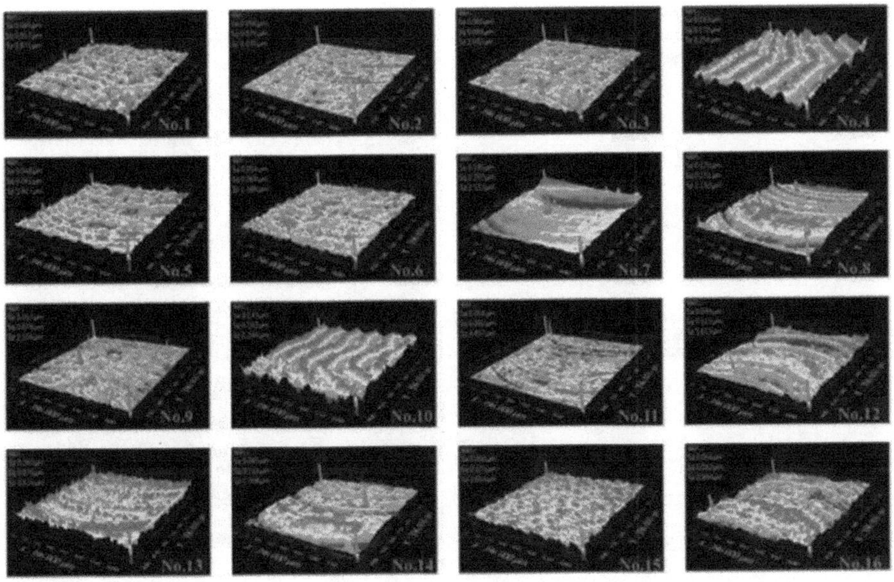

FIGURE 8.11 Surface topography of the treated surface. ([32] [Open access].)

Figure 8.11 shows the surface topography obtained using a white light interferometer. The measurement range used was 396 μm by 396 μm. The quality indicators of the micro-grinding process are Sa, Ds, and De. Similar regression models and their associated fit indices are reported in the literature [32].

It is noted that the trailing edge region of impeller blades is extremely susceptible to erosion, most likely due to a micro-cutting failure mechanism. By weighing various samples after testing, a SEM is used to determine the rate of mass loss to analyze the material removal mechanisms on the rotor passage surface. The particle contact patterns with the blade surface can be studied theoretically by examining the differences in particle impact conditions across the passages, as shown in SEM micrographs in Figure 8.12 [34].

The user can detect and correct the following problems by performing a virtual simulation before actual operation using software systems such as VericutTM, PredatorTM, and others:

- Disturbances and collisions between the tool and workpiece
- Tool holder and workpiece, or even a spindle head and machine bed
- One of the most serious problems in corner machining is problems caused by the tool cutting into the workpiece.

8.4 WORKPIECE PRECISION

Using advanced technologies, ultra-precision machining produces components with minimal surface roughness, surface/sub-surface defects, and high precision to meet the needs of consumer electronics, semiconductor technology, astronomy, and

FIGURE 8.12 Wear of pressure and suction blades under an electron microscope. ([34] [Open access].)

other fields. A 3D white light interferometer (ZYGONexView, ZYGO Connecticut, Middlefield, CT, USA) is used to measure the surface topography of workpieces to obtain the real rough surface topography data. To ensure the real reliability of test results, the instrument uses closed-loop feedback, piezoelectric ceramics, a non-contact white light interferometer, and high linear capacitance sensor, which can realize the full range of high-precision testing, 0–150 µm ± 0.1 nm. The surface roughness topography obtained by drilling, turning, smooth grinding, and polishing is shown in Figure 8.13. The corresponding values are S_a 3.031 µm, S_a 2.418 µm, S_a 0.076 µm, and S_a 0.037 µm [35].

Use in several example applications is explained to clarify specific needs for ultra-precision machining. This chapter reviews the state of the art of three commonly used ultra-precision machining processes: ultra-precision grinding, ultra-precision cutting, and finishing/shaping. Although this model does a good job of explaining quality characteristics, it ignores the influence of the tool holder and machine itself. The quality of the treated layer and wear of the cutting tool are significantly affected by the direction of carbon fiber. Material anisotropy and cutter angle have the greatest influence on the mechanical interaction between carbon fiber-reinforced polymer (CFRP) and cutting tools. To verify the simulation results, the experiments are conducted using micromachining of micro-holes in printed circuit boards using EDM with different

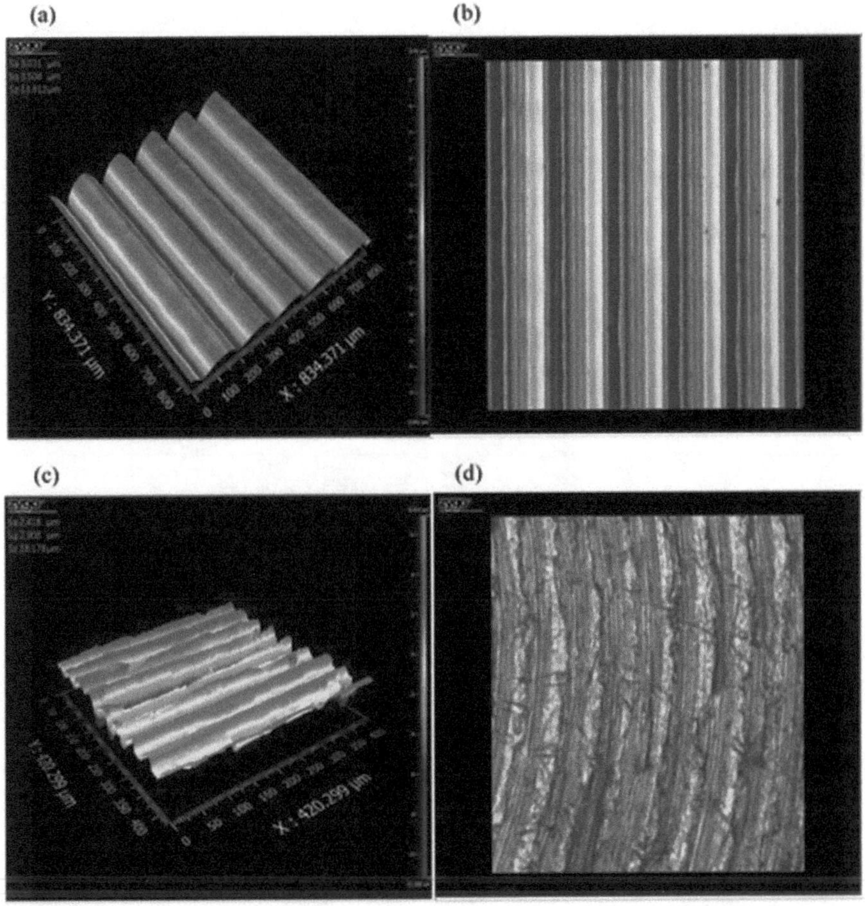

FIGURE 8.13 Surface topography for different processing methods: (a) boring, (b) actual image of boring, (c) turning, and (d) actual image of turning. ([35] [Open access].)

machining voltages, pulse widths, and spindle speeds. A SEM is used to study the morphology of microbits, and Figure 8.14 shows the experimental results [16].

The CFRP fibers carry the load, and resin facilitates the transfer of forces through fibers. As a result, the stresses to which the fibers are subjected can be used primarily to analyze the mechanisms of material failure [10,11], as depicted in Figure 8.15. According to simulation results, no gouges are created during the machining process [34].

8.4.1 Cutting Forces

Since composite materials are used in many industries, their processing is highly desirable. Due to the unique characteristics that distinguish them from conventional metallic materials, these composite materials are difficult to machine [1,2]. Cutting

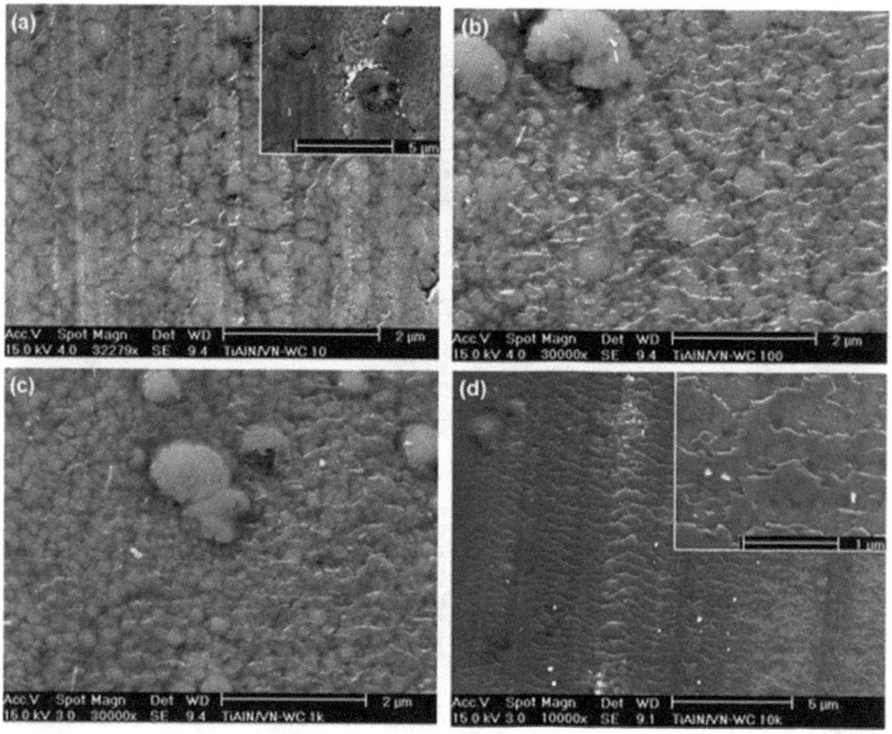

FIGURE 8.14 Micro-drill wear in EDM machining: (a) effect of machining stress, (b) effect of pulse duration, and (c) effect of spindle speed and wear. ([16] [Open access].)

FIGURE 8.15 The simulation results indicate that there are no gouges during the machining process. ([34] [Open access].)

conditions such as axial and radial depth of cut, engagement angle, feed, and feed rate profile are important factors affecting the cutting process. Cutting force, tool wear, and speed directly depend on these factors and how they change along the tool path [3–5]. Changes in the microstructure and interfacial characteristics of CFRPs have a significant impact on the types of destruction and damage patterns of these materials.

These characteristics include the presence of voids, fiber orientation and volume percentage of fiber and matrix, and cracks. In addition, the brittle polymer matrix around the fibers is subject to compression, leading to fragmentation and cracks. The tool angle largely determines the direction of distortion caused by warping and location of cracks [12–14]. An offset fly cutting-servo system (OFCS), which integrates the ideas of fly cutting and STS, is proposed to expand the application field of fly cutting. It can process straight slots of microstructure arrays on commercial machines with *X*, *Z*, and *C* axes such as pyramid arrays and triangular pyramid arrays. As shown in Figure 8.16, slots, pyramid arrays, and triangular pyramid arrays are processed based on this concept [2].

The processing process of CFRP is more complex than homogeneous materials because CFRP composites consist of two phases: continuous phase (epoxy resin) and reinforced phase (carbon fiber) [6,7]. It is not recommended to process carbon fiber-reinforced plastic at the glass transition temperature of the material matrix (T_g). The tool moves at a specified speed during milling in a direction parallel to the grain direction (0°), with the interaction between the tool face and material generating the majority of cutting force. This cutting force has two components: one perpendicular to the grain direction and the other parallel. The fibers experience compressive stress from a perpendicular force that pushes and bends them (buckling stress). Fiber rupture is caused by cracks that occur in the fibers at points of tensile stress. The case

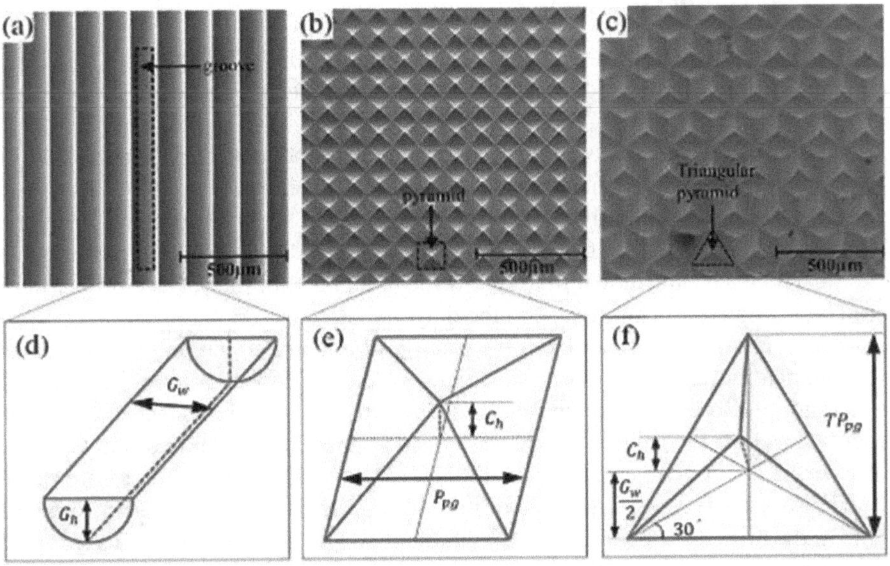

FIGURE 8.16 Microgrooves (a), micro pyramidal array (b), triangular micro pyramidal array (c), and their geometric descriptions (d–e) are electron micrographs of structured arrays. ([2] [Open access].)

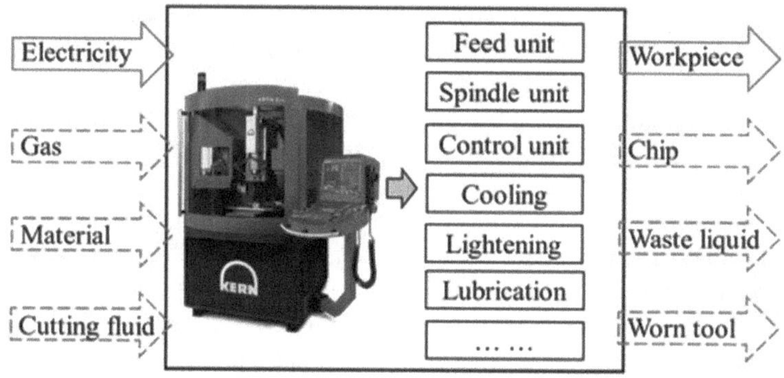

FIGURE 8.17 Maximum power consumption of a Kern Evo micro-milling machine. ([32] [Open access].)

of slope milling is presented using coordinate transformation. As a result, a force is created that acts on each element of the cutting edge. The ultimate force is calculated by adding up the numbers along the edge in contact with the material and taking into account the cutting of teeth. A technique for removing carbon fiber with a 0° fiber direction is shown in Figure 8.17, which shows the maximum power consumption of the Kern Evo micro-milling machine [32].

8.5 TOOL-PATH SELECTION USING CUTTING FORCES PREDICTION

A new approach to minimizing dimensional errors caused by tool defects when choosing milling paths for complex surfaces is presented. This leads to increased accuracy of milled surfaces. This process can be used for three- or five-axis milling. The smallest cutting force component associated with tool deflection is calculated in the case of three axes. Experiments are conducted to confirm that UPFC is superior to others in achieving deep ductile cutting regions by creating two types of free-form surfaces on silicon: microgrooves and F-Theta lenses. The surface characteristics of the F-Theta lens are shown in Figure 8.18c–e along with the surface profile measured by 3D optical surface gauge in the diagonal direction. Without brittle fracture, the shape inaccuracy is reduced to the micron level and surface roughness reaches the nanometer level. Furthermore, this approach can be applied to free-form surface processing of more brittle materials in deep cutting. This component is originally described as being located in the plane defined by the tool axis and normal vector to the workpiece surface perpendicular to the tool axis. The cutting forces are determined at a predefined grid of control points for every 15° of the tangent direction plane to the milled surface for both longitudinal and up milling. Given this information, two options are possible. The CAM operator can create an numerical control (NC) program that improves accuracy by first selecting the general tool-path direction that minimizes the average tool deflection force on the surface. Changing the milling direction at each control point is a second option. Lines of minimum force

FIGURE 8.18 The CUAT process was conducted on a traditional lathe (OKK Ramo 1000, Japan). ([38] [Open access].)

are determined on the surface of workpiece by connecting these points. As shown in Figure 8.18, the Centro Universitario de Automoción y Tecnología (CUAT) tests were conducted on a conventional lathe (OKK Ramo 1000, Yokohama, Japan). The transducer was installed in the neutral position of the horn on the tool post of the lathe. Turning was performed on workpieces with a diameter of 21–29 mm, which were made of C45 carbon steel and 201 stainless steel. With a cutting depth of 0.15 mm and a feed rate of 0.1 mm/rev, the cutting speeds were set at 20, 40, 50, 60, 100, and 120 m/min [38].

They can serve as basic guidelines for programming the tool path on the surface. The procedure is different in five-axis milling because the orientation of the tool axis relative to the workpiece surface can be changed using two rotating axes. To keep the tool deflection force below the threshold value, both tool axis orientation and machining direction can be selected for each workpiece region. A thorough analysis of results is provided as well as several case studies of different methodologies. Using this method, dimensional errors in triaxial milling are reduced from 30 μm to less than 4 μm. In most cases, the five-axis milling error can be controlled below 15 μm. This study investigates the effects of different vibration-assisted machining methods on the surface texture of workpiece. It develops a new model that takes into account the excitation of cutting forces and highlights the significant influence of spindle vibrations on surface properties. Using a dynamic 5-degree-of-freedom spindle model with hydrostatic bearings, mathematical solution is obtained. The results show the resulting surface topography, as shown in Figure 8.19a and b. The study discusses the surface formation difficulties associated with high-precision rotary machining (UPRM) under spindle vibration excitation. Spindle vibrations are described by a complex nonlinear equation, which highlights their distinct influence on the final surface design. This comprehensive study expands our knowledge of vibration-assisted machining procedures and their effects on surface quality [2]. Figure 8.19c and d shows the surface topography of spindle under horizontal cutting conditions. Furthermore, UPRM can be used in a variety of optical contexts [2].

The ability of the submicron structure surface to produce a rainbow depending on the viewing angle makes it useful for a variety of applications including stealth

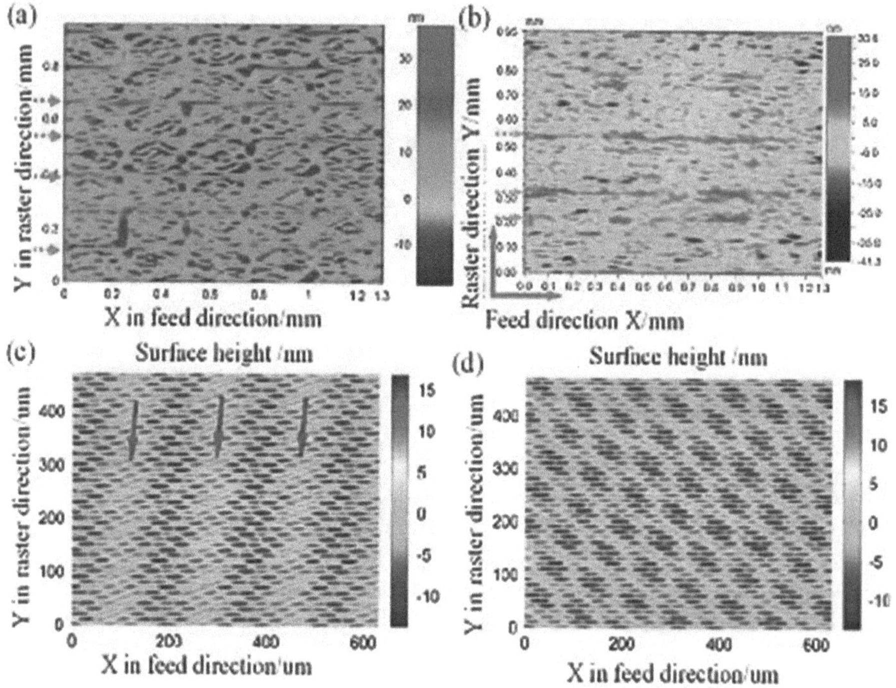

FIGURE 8.19 Surface configuration under the following cutting conditions: (a) observed topography, (b) simulated surface topography with uniform phase shift, and (c) simulated topography with a phase shift of 0.5. ([2] [Open access].)

technology, microscopic projection, and multicolor printing. In this study, two-step microstructures with first-order microscopic geometric motifs and second-order submicron grooves corresponding to diffraction gratings are fabricated. These structures are created to generate luminescence patterns according to specific shape laws. The effect of processing settings on the structure quality is demonstrated by second-order submicron grooves that are created at different feed rates, as shown in Figure 8.20a–d. The work focuses on the development of a cross-cutting method for machining micron-scale pyramids from nickel-phosphorus (Ni-P) arrays and grating milling methods for forming microstructures in brittle materials. Figure 8.20e shows the efficient fabrication of these pyramids, demonstrating the potential of the method to create microstructures with excellent quality. These efforts lead to improved micromachining methods for using in materials science and optics [2].

The surface shape of the machined convex shape and measured roughness at different inclination degrees under the optimal parameters are shown in Figure 8.21. An arithmetic roughness of 3.4 nm can be achieved for all inclination degrees [27].

Moreover, the inclination degree has little effect because the arithmetic roughness S_a is so small. This indicates that when using the ideal combination of cutting settings, surface finishes unevenness when machining geometries with large inclination degree dispersion can be mitigated using this ultrasonic tool system. Mills with a

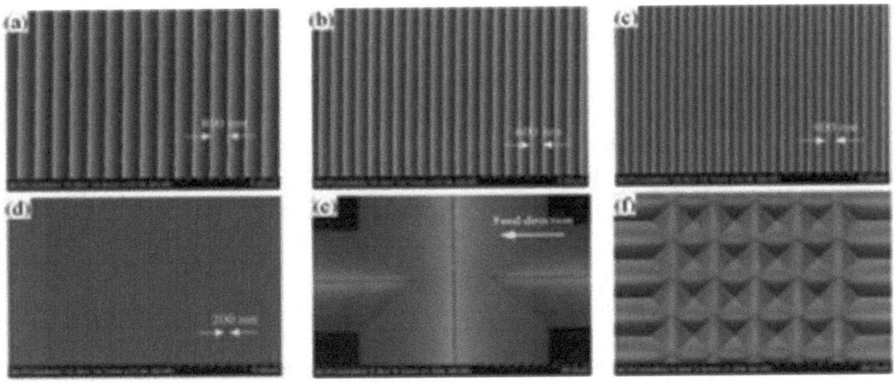

FIGURE 8.20 Micro pyramidal mesh processed by the fly-cut method. The second-order submicron grooves with different spacings are shown as (a) 800 nm, (b) 600 nm, (c) 400 nm and (d) 200 nm. They have high-quality exit and entry sections of a micropyramid. ([2] [Open access].)

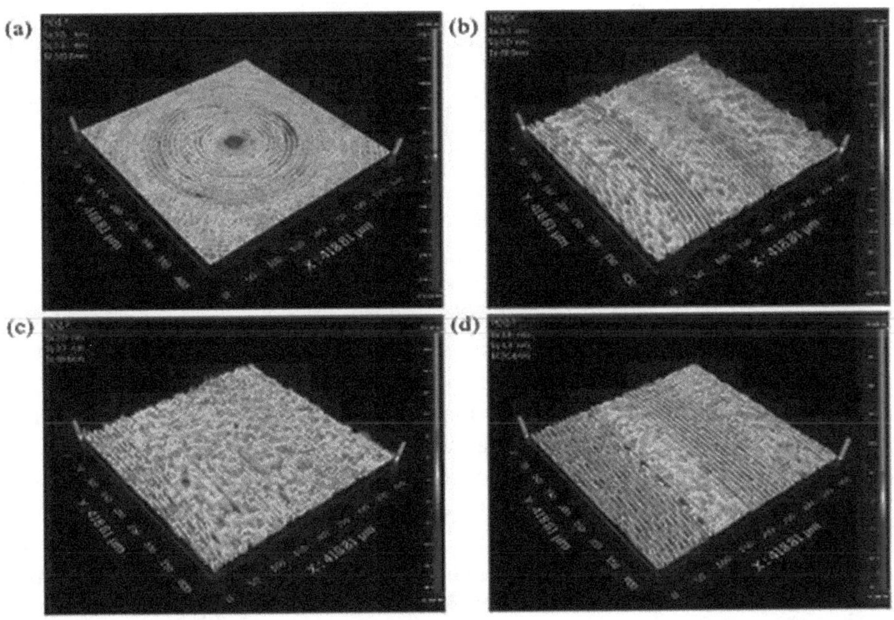

FIGURE 8.21 Surface morphology and measured roughness of convex shape with optimal parameters at several degrees of inclination: 0°, 5°, 10°, and 15°. ([27] [Open access].)

large length-to-diameter ratio are used in machining complex surfaces, and system deformation is one of the main causes of errors in this process. To reduce errors caused by force, the cutting direction angle, cutter stroke, and inclination angles are usually optimized during the process planning stage. As a result, the round-nose end mills can have a poor machining efficiency because material removal rates vary significantly with the cutting angle. Based on the instantaneous cutting force models when the cutting groove passes through the cutting contact point and rigidity of the machining system, this study investigates the mechanism of influence of machining angles on a force-induced error. To evaluate the division of force error and equal volume sphere of removed material, the force error/efficiency index (FEI) is determined for machining angles [4]. The FEI is dimensionless; the lower the FEI, the higher the processing efficiency and the smaller the error caused by force. The critical FEI divides the set of machining angles into two subsets and is determined taking into account the constraints on force error and required material removal rate for the best selection of machining angles. The machining angles in the ideal subset will have a higher machining accuracy and efficiency after the feed rate planning process, while the machining angles in the other subset will have a lower machining accuracy and efficiency. Through five-axis machining and free-form surface machining experiment, round head end mill is used to validate the superiority and efficiency of the proposed FEI approach. This tool can improve the processing efficiency while limiting the inaccuracy caused by force.

8.5.1 Three-Axis Case

Using a non-spherical tool can significantly improve the machining efficiency when cutting complex free-form parts because it allows you to adaptively adjust the position of the tool to maximize its contact area with the part surface. However, to change the position of the tool, it is necessary to change its orientation, which requires the use of a five-axis machine, which is much more expensive than the traditional three-axis machine. In addition, a complex free-form surface with a large change in the curvature requires radical change in the tool axis to match the change in the curvature, resulting in faster machine tool rotary axes and achieving high speeds. In this study, we propose a splitting-based 3+2-axis non-spherical tool machining method of standard complex free-form surface to overcome these problems. 3+2-axis machining requires a limited number of unique tool orientations; therefore, an indexed three-axis machine is sufficient, eliminating the need for an expensive five-axis machine. In addition, the kinematics and dynamics of the machine will be significantly improved, increasing machining accuracy, thanks to the significantly increased rigidity of three linear axes. Both computer simulations and physical machining experiments are conducted and results confirm that the non-spherical cutter along with adaptive adjustment of the feed direction and contact tool position can lead to significant improvements in machining efficiency (more than 50% in our study). The experiments compared to using a conventional spherical cutter, complex surface treatments, lined or embossed, are required for injection molds, forging, stamping dies and aircraft engine parts. There are two factors involved here: complex part geometry and difficult-to-cut materials. Modern finishing technology is high-speed milling in three axes with ball cutters or five axes with ball or flat-end mills. Dies and molds require good roughness and an accurate final shape. The best tool paths and cutting conditions can be determined using the milling operation model. Here are some examples of modern processing methods. Then, we set up a power-controlled polishing system, which includes a control system,

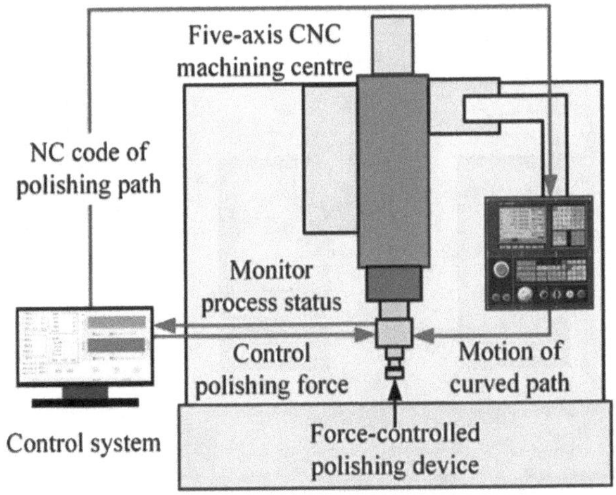

FIGURE 8.22 Force-controlled polishing system diagram. ([39] [Open access].)

a power-controlled polishing device, and a five-axis CNC machining center. The working process of the power-controlled polishing system is shown in Figure 8.22. First, the polishing path model of the complex curved Poly(methyl methacrylate) (PMMA) workpiece is generated by the interprocess communication (IPC) of the control system, which also sends the CNC code of the polishing path to the machining center. Then, the power-controlled polishing device installed on the spindle of the machining center rotates the spindle to complete the power-controlled polishing, following the complex curved polishing path [39].

Optical microscopy is used to investigate the microstructural differences in multi-pass aging (AGE), and the results are shown in Figure 8.23. In addition, the grain size statistics results are obtained using the free and open-source image processing software ImageJ 1.53v. It is obvious that with increasing of AGE passes, the grain size is significantly improved. The initial average grain size of annealed samples is 86 μm. Meanwhile, the cross-sectional shape of grains is dominated by isometric polygons. After two AGE iterations, fibrous grains are dominant and average grain size decreases to 16 μm. The increase in the number of extrusion passes could lead to an average grain size from 0.6 to 1.3 μm obtained in four and six passes, respectively [31].

8.5.2 Five-Axis Case

Five-axis surface finishing methodology automatically selects cutting tools using curvature-based machining techniques. The ideal rounded end mill is selected from a standard cutting tool library to minimize machine errors and maximize material removal rates. Cutter length, cutter radius, and corner radius are the studied tool parameters. To accommodate the change in the surface curvature, the greatest sweep of the inclined tool silhouette is proposed and applied in practice as tool radius selection

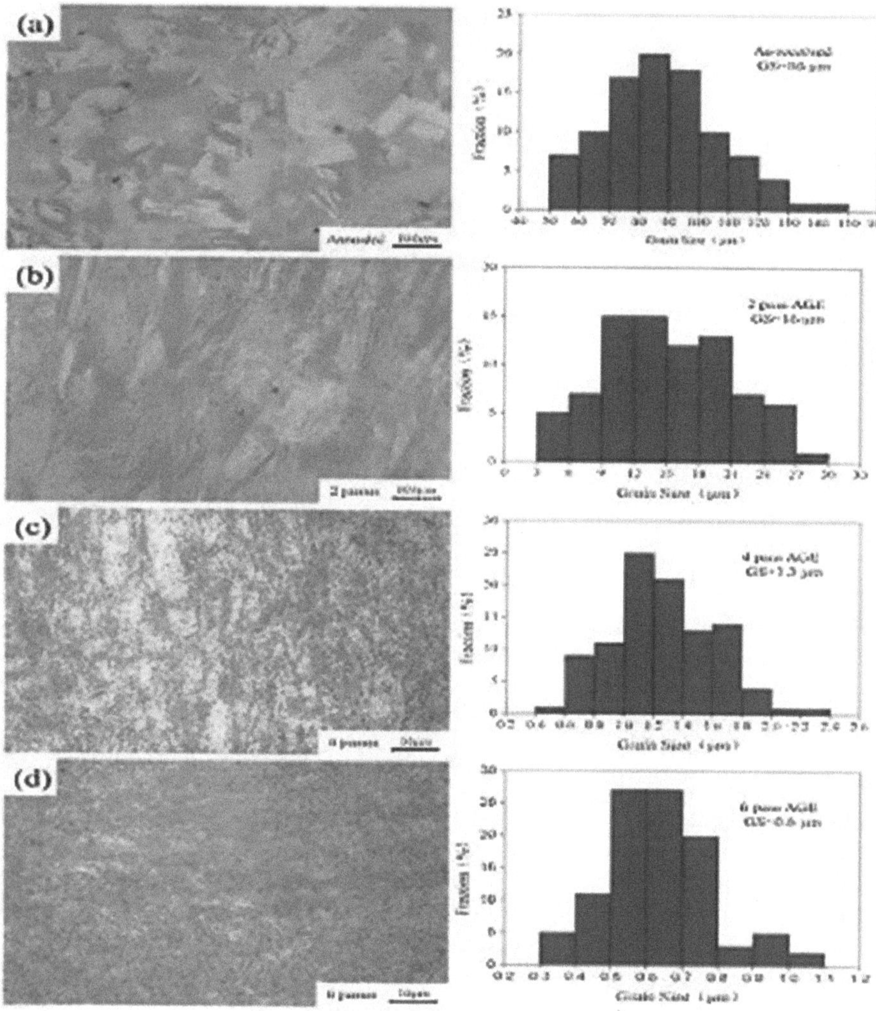

FIGURE 8.23 Microstructure of samples and grain size after AGE. Annealed (a), after two passes (b), after four passes (c), and after six passes (d). ([31] [Open access].)

protocols. We propose algorithms to identify and eliminate global tool interactions and local tool gouges. To identify and eliminate local tool gouging, measure the local distance between the surface and bottom of cutter. Identification and elimination of global tool interaction is solved by calculating the shortest path between the cutter body's axis and part's surface. Distance calculations are accelerated by using the facet technique. The shortest, toughest tool in the library is the result of solving local and global gouging problems. The C programming language is used in ROBLINE to implement these automatic tool selection techniques. This study presents a polishing experiment conducted on a complex curved PMMA part using a five-axis CNC machining center equipped with a controlled-force polishing device to evaluate the effectiveness of the

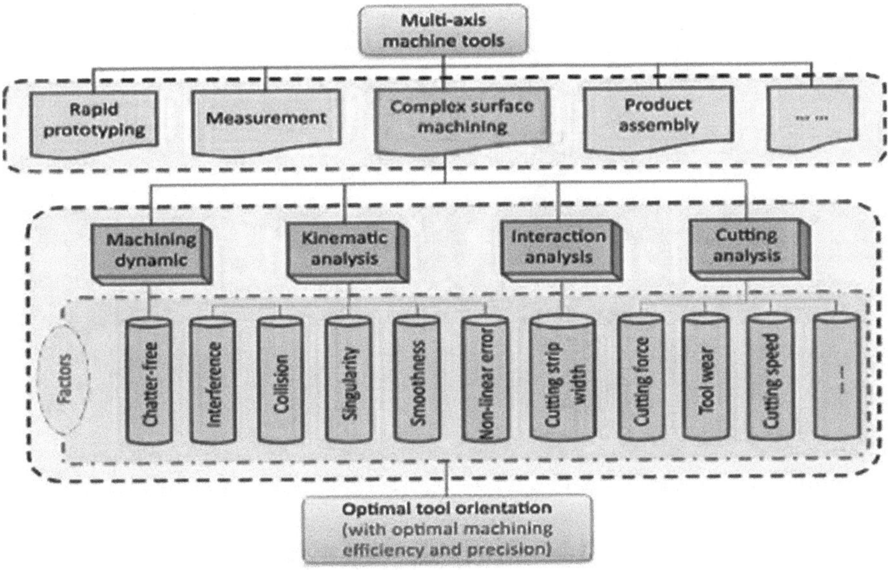

FIGURE 8.24 Polishing process of the PMMA part. ([39] [Open access].)

device [39]. The experiment was conducted using a PMMA part that was double-sided milled by the machining center. Its dimensions were $300 \times 215 \times 50\,mm^3$, and its curvature radius varied from 110 to 250mm. The polishing procedure of the PMMA part is shown in Figure 8.24. The tool orientation in multi-axis machine tool planning is shown in Figure 8.24. An integrated hybrid robot and kinematic circuit make up the five-axis hybrid machining robot (Figure 8.25) [30].

There are two types of hybrid processing robots used in the industrial sector. Top-notch robots such as Exec on and Tricept consist of air-conditioned wrist and 3-DOF parallel mechanism. A ball head milling system with a hybrid robot is shown in Figure 8.26: placement of the instrument within the region is shown in Figure 8.26a, and the location of the workpiece is shown in Figure 8.26b [30].

The machining performance of large and complex parts, including turbine blades and aircraft parts, largely depends on the working space, rigidity and flexibility of machining equipment [1,2]. Hybrid robots have demonstrated significant advantages when processing large and complex products [3]. They do this by combining the advantages of both serial and parallel mechanisms. The second type of the robot, which includes Eco speed, consists of a parallel module with 3 degrees of freedom and two linear guides. Many technologies have been proposed to enhance the computing capabilities of hybrid robots [4,5]. Tool orientation in five-axis end milling can be changed to avoid failures and improve machining efficiency. Although much research has been done on optimizing tool orientation for five-axis machine tools [6,7], work remains to be done on creating smooth tool orientation paths for hybrid robots that take into account their unique properties. Therefore, this work focuses on tool orientation planning in five-axis ball milling by considering the geometric

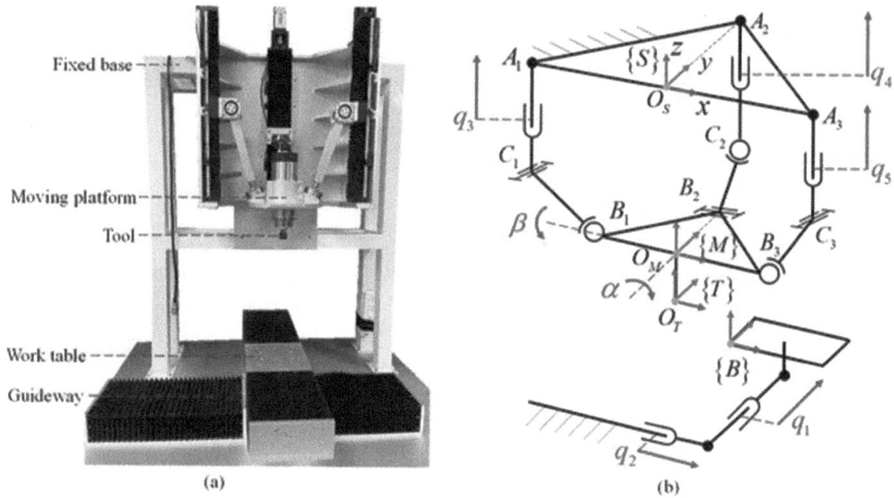

FIGURE 8.25 Five-axis hybrid machining robot is composed of two parts: an integrated hybrid robot and a kinematic circuit. ([30] [Open access].)

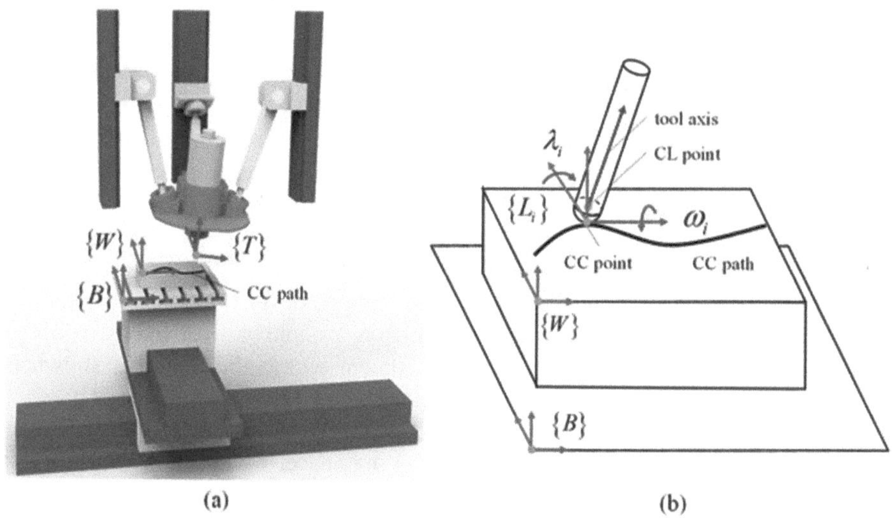

FIGURE 8.26 Ball head milling system with a hybrid robot: (a) workpiece location and (b) placement of the instrument within the region. ([30] [Open access].)

and kinematic aspects of hybrid robots. Creating a safe and efficient tool path is the primary goal of tool orientation planning. Initially, tool orientation smoothing is performed in the workpiece coordinate system (WCS) [8,9]. For example, quaternion representative tool orientation (RTO) interpolation is provided by smooth tool orientation [10]. For each cutter contact point (CC), Jun et al. [11] have created a feasible configuration space (C-space) and ensured a smooth path by iteratively exploring the boundary of feasible C-space. A bisection search was used to alternately change the angles for smoothing and collision avoidance depending on the tool orientation indicated by two angles [12]. However, the resulting tool path cannot ensure smooth movement of joint because tool orientation smoothing methods, as described above, do not take into account the kinematic properties of machining equipment. The tool orientation of five-axis machines is determined by two rotary axes, which are considered to have poor kinematic characteristics. Thus, research [13–15] favors machine coordinate system (MCS) tool orientation planning systems that directly smooth the displacements of two rotating axes. Under certain constraints, two common methods are used to smooth tool orientation paths 4 for machine tools.

8.6 μEDM MILLING

Because micro-erosion drilling is a 2D process, several researchers have looked at 3D milling techniques as a way to overcome the limitations of this type of machining. There are two different ways to create microforms using EDM. Piercing is a typical shape used in EDM; however, it has several disadvantages. One disadvantage is that different tool shapes must be made in advance, which is likely to be expensive and complex. The μEDM milling process is a modern manufacturing method that also uses a spark to remove material. This machining method extracts CNC code to use a round tool to create a specific shape. These programs cover the electrode path for layer-by-layer micro-erosion milling, with the depth limit of each layer typically being around 0.1 μm. Electrode wear must be addressed by a prescribed wear compensation strategy in this machining process as it is important for the accuracy of 3D geometric shapes. The μEDM milling technique for spherical tools ranges from 0.8 to 12 mm in diameter. We developed a numerical model that used the entire three-phase factorial experimental designs to study the effects of voltage, feed rate and capacitance on machining results [14]. Defining the ideal surface roughness (SR) values of 0.04 μm, EWR of 0.044, and material removal rate (MRR) of 0.08 mg/min allowed them to achieve an optimization level of 88% attractiveness. The 3D geometric analysis method [15] for micro-erosion milling has created a "Z" map algorithm to obtain accurate data on differences in tool and workpiece geometry caused by electrode wear during machining. They looked at how processing factors affected the results [14]. A laser scanning micrometer is used in the production [16] to study tool wear in micro-erosion milling. They have calculated the cue coordinates both before and after processing. A comparison of different μEDM milling procedures is carried out [17], taking into account the control parameters (CP) of machined parts and shape of the channel. Their results show that centrifugal force does not affect the tool surface during machining. Since tool wear affects the accuracy of machined surface, numerous methods for assessing and accounting for electrode wear have been modified [18]. To improve the productivity of micro-erosion milling, especially for difficult-to-cut materials. Their technology produced a difficult-to-cut material with an astonishing 30% improvement. A geometric model to simulate the effect of 3D micro-erosion milling on the machining mechanism [19]. The corner radius of flat electrode, electrode wear, and international electrotechnical commission (IEC)

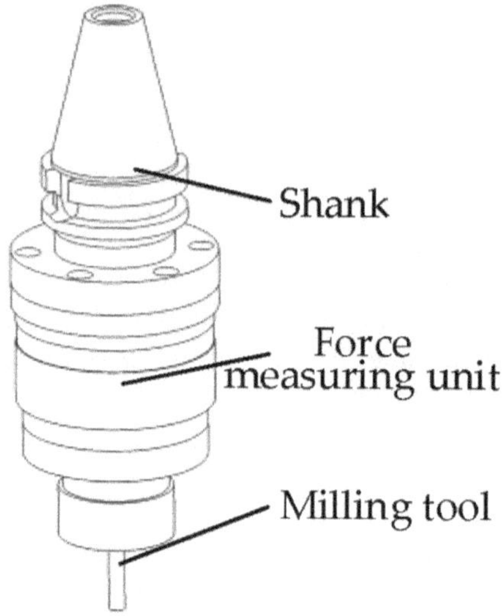

FIGURE 8.27 Milling force-measurement tool system for the cross-beam structure. ([37] [Open access].)

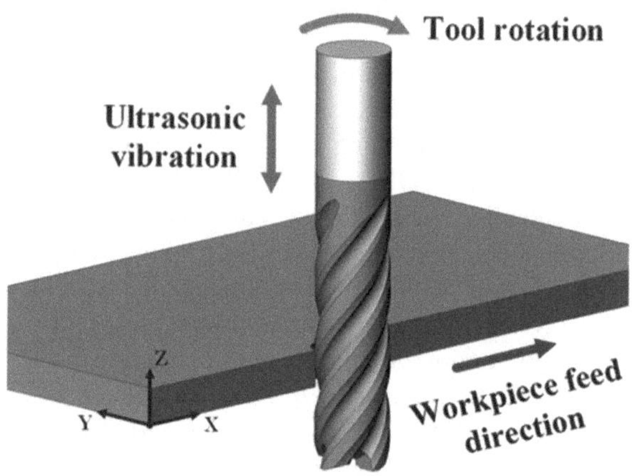

FIGURE 8.28 Tool and workpiece movement in UVAM. ([28] [Open access].)

affected the machining accuracy. A surface milling process based on the constant length compensation method is described [14]. This study examines the limitations of the previously investigated constant tool length compensation approach. As shown in Figure 8.27, the cross-beam milling force-measurement tool system consists of a cutter, a force-measurement unit, and a shank [37].

Figure 8.28 shows ultra-precision vibration-assisted milling (UVAM) is an advanced machining technique specifically tailored for processing CFRP laminates. The ultrasonic vibration direction is parallel to z-axis, the workpiece feed direction is parallel to x-axis and tool rotation direction is always in the XY plane [28].

8.6.1 A Fixed-Length Compensation Method

Increasing the thickness of the Electrical Discharge Machining (EDM) layer is crucial for improving the productivity of Laser Cladding Manufacturing (LCM) and Ultrasonic Welding Machining (UWM). A fixed-length compensation strategy is applied to enhance the probability of obtaining a thicker layer [19]. They have created an incredibly accurate theoretical model of electrode wear for a fixed-length compensation process. High processing accuracy and productivity are achieved through using accurate theoretical models and significant adjustment of layer thickness.

The fixed-length compensation milling technique is used to create a conical end of electrode within a given phase. Therefore, to eliminate this change, more steps must be performed, which greatly complicates the processing process. The electrode and work material dimensions required for the fixed-length compensation procedure have to be appropriate, except layer thickness of less than 85.5 μm. An improved fixed-length compensation method for micro-erosion milling that uses a ball-shaped tool [14]. A tubular electrode is used instead of a cylindrical electrode to improve the fixed-length compensation method [23].

To explain the tiny conical contour of the fixed-length correction factor, an analytical model is created. The investigation has shown that in most cases, the surface deviations remained below 4 m and production error is less than 2%. The workpiece is subjected to sequential linear route micro-EDM milling using these tool electrodes. Figure 8.29 shows the process flow diagram [29].

8.7 PRESENT AND FUTURE

Achieving dimensional perfection of micro parts and reducing processing time are two important conditions for efficient micro-manufacturing in today's highly competitive manufacturing business [1]. Thus, the production of microscopic components

FIGURE 8.29 Illustration of the machining process schematic. ([29] [Open access].)

and parts requires the use of precision machining technologies. The well-known category of micromachining can produce parts of various sizes, from 1 to 999 m, and the amount of material removed is expressed in microns [2]. The need for products made from durable engineering materials has forced scientists to look for alternative production methods. However, there has been an increase in demand for micro components made from innovative materials such as composites, superalloys and ceramics due to their superior mechanical and material properties.

Researchers, as well as life science industries, have identified numerous uses for these materials in mechanical engineering, automotive, aerospace, and other fields. It is difficult to machine complex and unusual materials due to their high strength, fatigue, and hardness. Consequently, one of the most important families of technologies that can assist in fabrication of micro components is non-traditional micromachining procedures. Manufacturing technologies that differ from standard micromachining procedures are called nonconventional micromachining processes. These methods often use new or unusual strategies to achieve microscale precision processing.

Electrical discharge machining (EDM) is one of the non-traditional methods, alongside others such as focused ion beam processing, electrochemical processing, and laser ablation. These methods differ from traditional machining procedures such as milling or turning, and they work particularly well for complex microscale products that require precise measurements and complex geometries. To remove materials, non-traditional processing methods use various types of energy, including chemical, optical, mechanical, electrical, and thermal as opposed to the traditional processing method, which uses sharp-edged cutting tools. There are a number of most important properties, functions, and applications of micromachining processes.

Figure 8.30 illustrates optical micrographs of the scratch region of coatings tested under three different loads. The first, labeled Lc1, shows the initial failure of the coating; the second, labeled Lc2, depicts the delamination of coating from the substrate; and the third, labeled Lc3, is for an intermediate load. It is important to note that the tested coatings have a fairly large critical strength Lc1 and loads Lc1 and Lc2 are differed significantly [24]. This indicates that the coating is quite resistant to failure. In the image of failure that occurred when a load greater than Lc1 is applied, several flakes of coating are visible at the scratch boundaries. Plastic deformation of substrate is associated with cracking of coating and partial delamination [32–36].

The high Cr concentration in the coatings resulted in the formation of Cr2N, which upon heating resulted in sintering of saturated c-(Ti, Cr, Al)N solid solution and formation of metallic Cr. In addition, evidence of precipitation strengthening is obtained. Due to a high Ti concentration, (Cr, Ti)2N solid solution phase is formed. When the Cr concentration in AlN exceeds 41%, a significant decrease in hardness occurs [25].

This is followed by the rapid formation of a dynamically stable wurtzite phase after annealing, which is attributed to the dissolution of Cr–N bonds. The oxidation resistance steadily increased with an increase in the amount of Cr and Al, since these metals provide stable oxide coatings that prevent oxidation. Therefore, when heated to 900°C, $Ti_{0.18}Al_{0.35}Cr_{0.47}N$ coating showed the best oxidation resistance, as shown in Figure 8.31 [25].

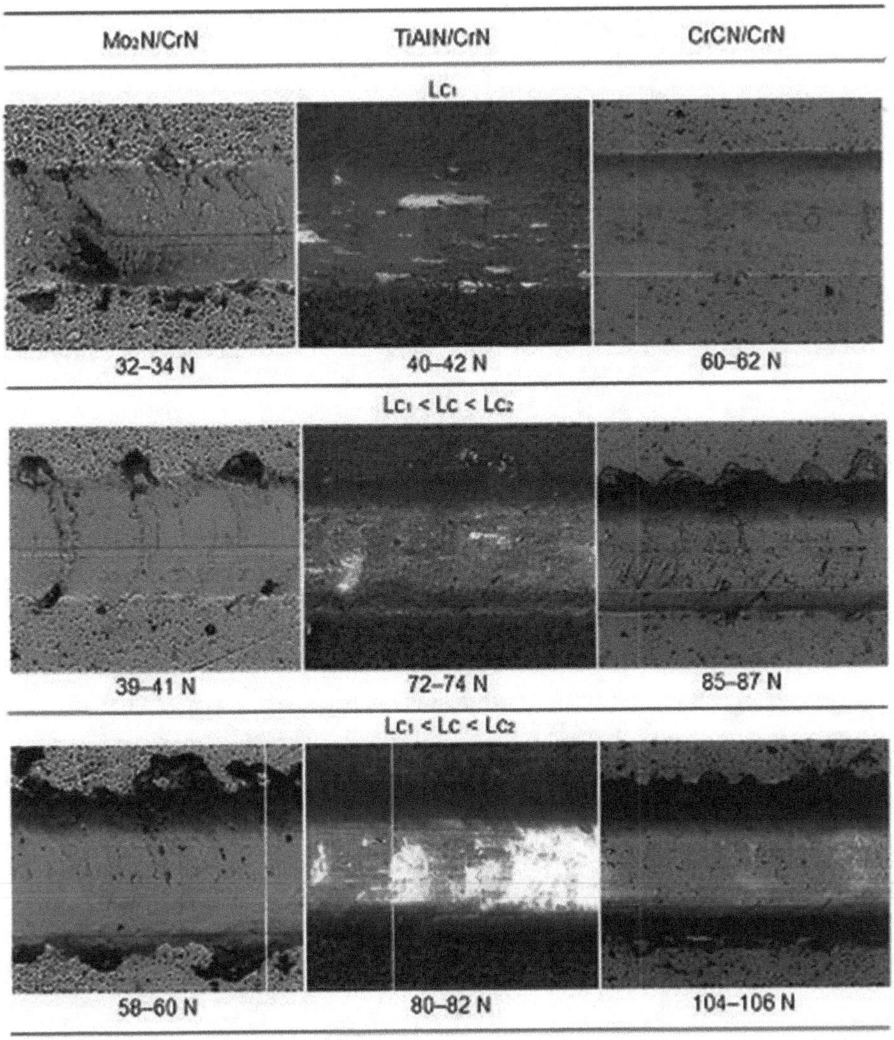

FIGURE 8.30 For the investigated coatings, the results of scratching and optical micrographs of failures at critical loads Lcl, Lcl < Lc < Lc2, and Lc2 are studied. ([24] [Open access].)

On the other hand, the low-chromium $Ti_{0.26}Al_{0.48}Cr_{0.26}N$ coating has proved to be an excellent candidate for cutting tools due to its good high-temperature strength and oxidation resistance. When (Ti,Al)N/CrN coatings are subjected to temperature variations, Figure 8.32 shows that the coating exhibited a spicule interface-oriented decomposition at 700°C with Al atoms attracted to the nearest interface.

As temperature increases, an anisotropic interface-oriented decomposition mechanism occurs along with transboundary interdiffusion (with extensive admixture of Cr and Ti). At this decomposition stage, the columnar grains exhibited distinct

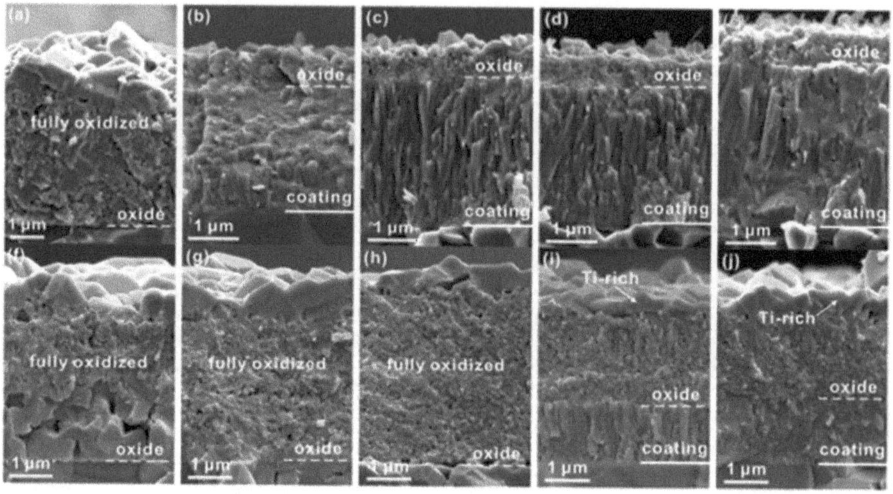

FIGURE 8.31 SEM fracture cross sections of coatings with Al/(Ti+Al) ~0.55 after oxidation at (a, e) 850°C and (f, j) 900°C for 20 hours. Coatings: (a, f) $Ti_{0.46}Al_{0.54}N$, (b, g) $Ti_{0.35}Al_{0.42}Cr_{0.23}N$, (c, h) $Ti_{0.29}Al_{0.36}Cr_{0.35}N$, (d, i) $Ti_{0.26}Al_{0.33}Cr_{0.41}N$, and (e, j) $Ti_{0.24}Al_{0.29}Cr_{0.47}N$. ([25] [Open access].)

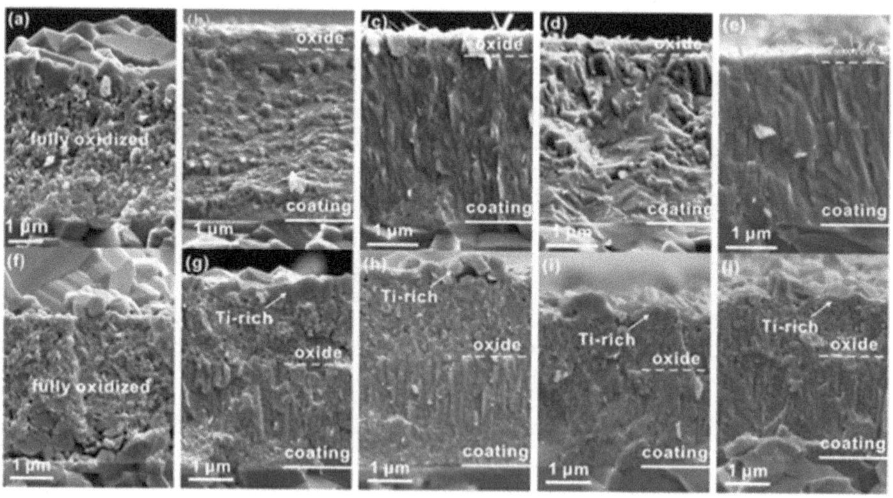

FIGURE 8.32 SEM fracture cross sections of coatings with Al/(Ti+Al) ratio of approximately 0.65 after oxidation at (a, e) 850°C and (f, j) 900°C for 20 hours. Coatings: (a, f) $Ti_{0.34}Al_{0.66}N$, (b, g) $Ti_{0.26}Al_{0.48}Cr_{0.26}N$, (c, h) $Ti_{0.22}Al_{0.40}Cr_{0.38}N$, (d, i) $Ti_{0.19}Al_{0.38}Cr_{0.43}N$, and (e, j) $Ti_{0.18}Al_{0.35}Cr_{0.47}N$. ([25] [Open access].)

compositional gradients, which are oriented opposite to the layer paths. The core layer structure of the coating is completely decomposed at 1000°C, while some areas on the outside did not [25].

PROBLEMS

1. Describe the manufacturing process and application in milling.
2. Explain the five-axis milling applications.
3. Describe the tool-path selection using cutting forces prediction in the manufacture.
4. Describe the fixed-length compensation method.
5. Explain the CAM (Computer-Aided Manufacturing) to control machine tools and automate manufacturing processes.

REFERENCES

1. Ma, Y., Zhang, G., Cao, S., Huo, Z., Han, J., Ma, S., & Huang, Z. (2023). A review of advances in fabrication methods and assistive technologies of micro-structured surfaces. *Processes*, 11(5), 1337.
2. Farouki, R.T., Han, C.Y., & Li, S. (2014). Inverse kinematics for optimal tool orientation control in 5-axis CNC machining. *Comp. Aided Geomet. Design*, 31, 13–26.
3. Mayr, J., Müller, M., & Weikert, S. (2016). Automated thermal main spindle & B-axis error compensation of 5-axis machine tools. *CIRP Annals*, 65, 479–482.
4. Chen, Q., Li, W., Jiang, C., Zhou, Z., & Min, S. (2021). Separation and compensation of geometric errors of rotary axis in 5-axis ultra-precision machine tool by empirical mode decomposition method. *J. Manuf. Process.*, 68, 1509–1523.
5. Kadir, A. A., Xu, X., & Hämmerle, E. (2011). Virtual machine tools and virtual machining—A technological review. *Robot Comp. Integr. Manuf.*, 27, 494–508.
6. Campatelli, G., Scippa, A., Lorenzini, L., & Sato, R. (2015). Optimal workpiece orientation to reduce the energy consumption of a milling process. *Int. J. Precision Eng. Manuf. Green Technol.*, 2, 5–13.
7. Soori, M. (2023). Deformation error compensation in 5-axis milling operations of turbine blades. *J. Brazilian Soc. Mech. Sci. Eng.*, 45, 289.
8. Li, C., Liu, X., Li, R., Wu, S., & Song, H. (2019). Geometric error identification and analysis of rotary axes on five-axis machine tool based on precision balls. *Appl. Sci.*, 10(1), 100.
9. Fu, G., Shi, J., Xie, Y., Gao, H., & Deng, X. (2020). Closed-loop mode geometric error compensation of five-axis machine tools based on the correction of axes movements. *Int. J. Adv. Manuf. Technol.*, 110, 365–382.
10. Liu, Y., Wan, M., Xing, W.-J., Xiao, Q.-B., & Zhang, W.-H. (2018). Generalized actual inverse kinematic model for compensating geometric errors in five-axis machine tools. *Int. J. Mech. Sci.*, 145, 299–317.
11. Xing, K., Achiche, S., & Mayer, J. (2019). Five-axis machine tools accuracy condition monitoring based on volumetric errors and vector similarity measures. *Int. J. Machine Tools Manuf.*, 138, 80–93.
12. Liu, D., Zhang, Y., Luo, M., & Zhang, D. (2019). Investigation of tool wear and chip morphology in dry trochoidal milling of titanium alloy Ti-6Al-4V. *Materials*, 12, 1937.
13. Flynn, J. M., Shokrani, A., Vichare, P., Dhokia, V., & Newman, S. T. (2018). A new methodology for identifying location errors in 5-axis machine tools using a single ballbar set-up. *Int. J. Adv. Manuf. Technol.*, 99, 53–71.
14. Mareš, M., Horejš, O., & Havlík, L. (2020). Thermal error compensation of a 5-axis machine tool using indigenous temperature sensors and CNC integrated Python code validated with a machined test piece. *Precision Eng.*, 66, 21–30.
15. Lian, M., Feng, X., Xu, B., Fu, L., & Jiang, K. (2022). Simulation of temperature field in micro-EDM assisted machining of micro-holes in printed circuit boards. *Micromachines*, 13(5), 776.

16. Gdula, M., Burek, J., Zylka, L., & Plodzien, M. (2018). Five-axis milling of sculptured surfaces of the turbine blade. *Aircraft Eng. Aerospace Technol.*, 90, 146–157.
17. Gdula, M. (2020). Empirical models for surface roughness and topography in 5-axis milling based on analysis of lead angle and curvature radius of sculptured surfaces. *Metals*, 10, 932.
18. Lu, L., Zhang, J., Fuh, J. Y. H., Han, J., & Wang, H. (2020). Time-optimal tool motion planning with tool-tip kinematic constraints for robotic machining of sculptured surfaces. *Robot. Comp. Integr. Manuf.*, 65, 101969.
19. Pelayo, G. U., Olvera-Trejo, D., Luo, M., de Lacalle, L. L., & Elías-Zuñiga, A. (2021). Surface roughness prediction with new barrel-shape mills considering runout: Modelling and validation. *Measurement*, 173, 108670.
20. Yang, J., Aslan, D., & Altintas, Y. (2018). A feedrate scheduling algorithm to constrain tool tip position and tool orientation errors of five-axis CNC machining under cutting load disturbances. *CIRP J. Manuf. Sci. Technol.*, 23, 78–90.
21. Tunc, L. T. (2019). Smart tool path generation for 5-axis ball-end milling of sculptured surfaces using process models. *Robot. Comp. Integr. Manuf.*, 56, 212–221.
22. Wang, D., Ren, J., & Tian, W. (2020). A method for the prediction of cutting force for 5-axis ball-end milling of workpieces with curved surfaces. *Int. J. Adv. Manuf. Technol.*, 107, 2023–2039.
23. Warcholinski, B., Gilewicz, A., & Tarnowska, M. (2023). The surface assessment and the properties of selected multilayer coatings. *Lubricants*, 11(9), 371.
24. Dabees, S., Mirzaei, S., & Sobola, D. (2022). Characterization and evaluation of engineered coating techniques for different cutting tools—Review. *Materials*, 15, 5633.
25. Qu, Z., Wang, X., Shen, X., & Zhou, H. (2022). Study of the Cu(111) surface by scanning tunneling microscopy: The morphology evolution, reconstructions, superstructures and line defects. *Nanomaterials*, 12(23), 4278.
26. Zhang, Y., Ren, J., & Zhou, J. (2023). Effect of duty cycle on cutting force for ultrasonic vibration-assisted milling carbon fiber-reinforced polymer laminates. *Materials*, 16(23), 7457.
27. Jing, Q., Zhang, Y., Kong, L., Xu, M., & Ji, F. (2021). An investigation into accumulative difference mechanism in time and space for material removal in micro-EDM milling. *Micromachines*, 12(6), 711.
28. Fan, J., Li, J., Liu, W., & Wang, C. (2023). The 5-axis hybrid machining robot is composed of two parts: An integrated hybrid robot and a kinematic circuit. *Processes*, 11(8), 2305.
29. Wang, P., Bai, Q., Cheng, K., Zhao, L., & Zhang, Y. (2023). Multi-objective optimization of micro-milling parameters—The trade-offs between machining quality, efficiency, and sustainability in the fabrication of thin-walled microstructures. *Appl. Sci.*, 13(16), 9392.
30. Varga, J., Demko, M., Kaščák, Ľ., Ižol, P., Vrabeľ, M., & Brindza, J. (2024). Influence of tool inclination and effective cutting speed on roughness parameters of machined shaped surfaces. *Machines*, 12(5), 318.
31. Ji, L., He, X., Li, W., Tian, F., Shi, W., Zhou, L., & Agarwal, R. (2024). Research progress of advanced design method, numerical simulation, and experimental technology of pumps in deep-sea resource exploitation. *Water*, 16(13), 1881.
32. Bai, Y., An, Q., Suo, S., Wang, W., & Jia, X. (2023). An analytical model for the normal contact stiffness of mechanical joint surfaces based on parabolic cylindrical asperities. *Materials*, 16(5), 1883.
33. Chan, T. C., Chang, C. C., Ullah, A., & Lin, H. H. (2023). Study on kinematic structure performance and machining characteristics of 3-axis machining center. *Appl. Sci.*, 13(8), 4742.8.
34. Bai, Y., An, Q., Suo, S., Wang, W., & Jia, X. (2023). An analytical model for the normal contact stiffness of mechanical joint surfaces based on parabolic cylindrical asperities. *Materials*, 16(5), 1883.

35. Chan, T. C., Chang, C. C., Ullah, A., & Lin, H. H. (2023). Study on kinematic structure performance and machining characteristics of 3-axis machining center. *Appl. Sci.*, 13(8), 4742.
36. Ji, L., He, X., Li, W., Tian, F., Shi, W., Zhou, L., & Agarwal, R. (2024). Research progress of advanced design method, numerical simulation, and experimental technology of pumps in deep-sea resource exploitation. *Water*, 16(13), 1881.
37. Song, X., Wu, W., Zhao, Y., Cheng, Y., & Liu, L. (2023). Structural design and optimization of the milling force measurement tool system embedded with thin-film strain sensors. *Micromachines*, 14(12), 2133.
38. Nguyen, T. T., Vu, T. T., & Nguyen, T. D. (2024). Surface topography in cutting-speed-direction ultrasonic-assisted turning. *Micromachines*, 15(6), 668.
39. Meng, X., Wang, Y., Yin, X., Fu, H., Sun, S., & Sun, Y. (2024). Investigation of force-controlled polishing of complex curved PMMA parts on a machining center. *Machines*, 12(4), 259.

Index

FSC
www.fsc.org
MIX
Papier | Fördert
gute Waldnutzung
FSC® C083411

Kolibri 360 GmbH
Ferdinand-Jühlke-Straße 7
99095 Erfurt, Deutschland
produktsicherheit@kolibri360.de